Ewing Sarcoma: Basic Biology, Clinical Challenges and Future Perspectives

Guest Editors

Stefan Burdach
Uta Dirksen
Poul H. Sorensen

MDPI

Basel • Beijing • Wuhan • Barcelona • Belgrade • Novi Sad • Cluj • Manchester

Guest Editors
Stefan Burdach
Translational Pediatric Cancer Research Action
Institute of Pathology and Department of Pediatrics
TUM School of Medicine
Technical University of Munich
Munich
Germany

Uta Dirksen
Pediatrics III
West German Cancer Centre
University Hospital Essen
Essen
Germany

Poul H. Sorensen
Department of Pathology and Laboratory Medicine
University of British Columbia
Vancouver
Canada

Editorial Office
MDPI AG
Grosspeteranlage 5
4052 Basel, Switzerland

This is a reprint of the Special Issue, published open access by the journal *Cancers* (ISSN 2072-6694), freely accessible at: https://www.mdpi.com/journal/cancers/special_issues/ewing_sarcoma_clinical_perspectives.

For citation purposes, cite each article independently as indicated on the article page online and as indicated below:

Lastname, A.A.; Lastname, B.B. Article Title. *Journal Name* **Year**, *Volume Number*, Page Range.

ISBN 978-3-7258-3447-1 (Hbk)
ISBN 978-3-7258-3448-8 (PDF)
https://doi.org/10.3390/books978-3-7258-3448-8

Contents

About the Editors

Stefan Burdach

Stefan Burdach is a pediatrician, oncologist, and cancer researcher. His scientific focus is on malignant sarcomas in children and on cell therapy. From 2003 to 2021, he was Chief of Staff at Schwabing Children's Hospital of the Technical University of Munich (TUM), after spending research sojourns from 1984 to 1987 at Harvard University, Stanford University, and with Don Thomas at the Fred Hutchinson Cancer Research Center in Seattle. His tenure track appointment followed in Düsseldorf in 1989. In 1997, he was appointed full Professor of Pediatrics at Martin Luther University Halle-Wittenberg. In 2003, he became full Professor and Chair of the Department of Pediatrics at TUM. In January 2022, he was appointed as the inaugural Academy of Translational Medicine Visiting Professor at the University of British Columbia. Dr. Burdach's scientific merits include four decades of bringing together basic science and clinical application in the field of pediatric oncology and hematology, as well as significant accomplishments in establishing research collaboration across the Atlantic. Dr. Burdach's work has not only elucidated important mechanisms of metastatic spread but has also developed targeted immunotherapies to potentially target the metastatic process. He was acknowledged as a pioneer of translational medicine even before this term was coined. As the author of over 500 publications with over 7,000 citations (Web of Science), he acts as a reviewer for the German Research Foundation, Helmholtz Association, German Cancer Aid, and the German José Carreras Leukemia Foundation, among others. Awards (selection): 1987: Hill Memorial Award, Stanford University; 1991: Reinhard–Heynen and Emmi–Heynen Prize, Heinrich Heine University Düsseldorf; 2009: Kind–Philipp Prize of the Society for Pediatric Oncology and Hematology in the Stifterverband für die deutsche Wissenschaft for his work on stem cells and cancer mechanisms; 2023: Federal Cross of Merit, 1st Class.

Uta Dirksen

Uta Dirksen is a pediatrician, oncologist, clinical cancer researcher, and university professor. She has held an endowed professorship for pediatric oncology with the Faculty of Medicine at the University of Duisburg-Essen since 2017. Dirksen studied medicine at the Universities of Münster and Duisburg from 1984 to 1986 and at the University of Düsseldorf from 1986 to 1991. She then worked at the Center for Pediatric and Adolescent Medicine at Düsseldorf University Hospital until 2005. She received her doctorate there in 1997 (Advisor: Stefan Burdach) and qualified as a specialist in pediatrics in 2001. In 2005, she moved to Münster University Hospital, where she specialized in oncology and pediatric hematology. Her work has repeatedly won awards. In 2015, she took over from Heribert Jürgens at Münster University Hospital as head of the Ewing sarcoma study group of the Society for Pediatric Oncology and Hematology. As part of the Children's Cancer Program, Uta Dirksen has been involved in setting up appropriate structures for the treatment of childhood cancer in Eritrea since 2015. She is also a member of the advisory board of the Sarcoma Conference. Since 2018, she has been the deputy chair of the Society for Pediatric Oncology and Hematology. In 2017, she was awarded an endowed professorship for pediatric oncology by the Foundation for Children with Cancer, which aims to strengthen clinical and translational research in pediatric oncology at Essen University Hospital. She is Deputy Director of the Department of Pediatrics at Essen University Hospital.

Poul H. Sorensen

Poul H. Sorensen is a molecular pathologist specializing in the biology of pediatric cancers. Dr. Sorensen holds the Johal Endowed Chair in Childhood Cancer Research at the University of British Columbia (UBC) and is a Professor of Pathology at UBC. His group has discovered many novel chromosomal translocation-associated alterations in childhood cancer. His current work is focused on how cancer cells respond to acute microenvironmental stress. More recently, the group has been using proteomics to identify surface proteins that are strongly expressed in high-risk childhood cancers, but not in normal tissues, for the development of antibody–drug conjugates (ADCs) and chimeric antigen receptor (CAR) T and NK cells. Positions and Employment: 1999–present: Johal Endowed Chair in Pediatric Oncology Basic and Translational Research, Department of Pediatrics and Pathology, UBC; 2003–present: Professor, Department of Pathology and Laboratory Medicine, UBC; 2014–present: Distinguished Scientist, BC Cancer Research Centre. Selected Scientific Achievements: ETV6-NTRK3 gene fusion in diverse solid tumors. He was the first to discover recurrent NTRK fusions in human tumors, namely the ETV6-NTRK3 chimeric tyrosine kinase in pediatric sarcomas. EWS-ERG gene fusion in Ewing sarcoma. His identification of the t(21;22) gene fusion, EWS-ERG, representing ~10-15% of Ewing sarcoma cases, was the first known example of different but related gene fusions occurring in phenotypically and clinically similar human solid tumors, suggesting the concept of genetic redundancy in gene fusion driver lesions. Selected Honors: 2016 August-Wilhelm Scheer Gastprofessorship, Technical University Munich, Germany; 2019 Distinguished Achievement Award, Faculty of Medicine, University of British Columbia, Canada; 2019 Fellow of the Royal Society of Canada, Life Science Division; 2021 AACR Team Science Award (St. Baldrick's Foundation / Stand Up 2 Cancer /AACR PCDT); 2021 Order of British Columbia, Canada.

Preface

Next to osteosarcoma, Ewing sarcoma is the second most common primary bone tumor in children, adolescents, and young adults (AYAs), with ~1.5 cases per million children and AYAs globally and a male excess of ~1.5. Substantial incidence variation across populations ranges from ~1 per million in predominantly African 10-to-19 year-old boys to ~8 per million in predominantly European boys in the same age range. This highly malignant cancer is characterized by unique ews/ets translocations, which constitute prototypic oncogenic drivers. However, these oncofusions do not determine clinical biology and outcome. Patient fate is mostly determined by metastasis; both the high propensity and the complex spreading process are far from being completely understood. In addition, prognostic parameters for localized disease are limited. Ewing sarcomas are amongst the childhood cancers with a low mutational rate, rendering reliable biomarkers elusive. However, mutation rates increase with relapses and cumulative mutagenic therapy exposure. That said, a silent tumor genome generally limits targeted therapy approaches, which applies particularly to Ewing sarcoma. Nevertheless, precision oncology approaches aim at enhancing the therapeutic index of conventional chemotherapy with novel small molecules targeting epigenetics, metabolism, and stress responses. This all constitutes an urgent medical need in relation to Ewing sarcoma, given the risk of secondary malignancies after treatment with chemotherapy and radiotherapy. While the risk of secondary malignancies after Ewing sarcoma is considerable, the occurrence of Ewing sarcoma as a secondary malignant neoplasm is rare.

Although their genome is generally silent, Ewing sarcomas may reactivate endogenous retro-elements. Their activation is linked to a particular inflammatory response and a prometastatic modulation of the microenvironment. Inflammation is a characteristic feature of Ewing sarcoma and confers inferior prognosis. On the other hand, these tumors are characterized, at least in most cases, by immune inertia, i.e., a scarcity of T cell infiltrates and a predominance of immunosuppressive myeloid signatures (M0/M2), resulting in the antagonistic processes of inflammation and immunosuppression. What seems to be an oxymoron at first glance turns out to be an immunoregulatory escape by turning the tumor's periphery into an immune desert. Mechanistically, chronic inflammation induces suppressive myeloid cells to shield the tumor against adaptive immune attack, hampering the efficacy of chimeric antigen receptor or T cell receptor transgenic T cells. Overcoming these immunosuppressive mechanisms may enhance immunotherapeutic efficacy. Several approaches have been pursued in this regard. The manipulation of inflammatory agents and mechanisms can neutralize the immunosuppressive tumor microenvironment. The utilization of oncolytic viruses, genetically engineered to depend on their lytic cycle, on metastatic drivers can induce immunogenic cell death and, particularly when in combination with cell cycle inhibitors, also have the potential to overcome barriers to immunotherapy in Ewing sarcoma. Several approaches have been pursued to increase the longevity of genetically engineered therapeutic T cells, including the use of orthotopic instead of retroviral TCR replacement by transgenic TCRs or CARs.

This series of 10 articles (7 original Articles, 1 Communication, and 2 Reviews) is presented by international leaders in the field of Ewing sarcoma research.

The article "Correlation of Transcriptomics and FDG-PET SUVmax Indicates Reciprocal Expression of Stemness-Related Transcription Factor and Neuropeptide Signaling Pathways in Glucose Metabolism of Ewing Sarcoma" uses artificial intelligence and sophisticated computer algorithms to integrate functional genomics, which are assessed by gene expression, and functional imaging, which are assessed by FDG-PET, in order to better characterize these highly malignant tumors and to provide additional biomarkers, as well as prognostic parameters, for both localized and metastatic disease. The identified genes and pathways might serve as a starting point for prospective experimental and clinical studies of new therapeutic interventions. Thus, this study provides new opportunities for future research to improve the outcome of patients with poor survival rates, which will hopefully become a reality in the near future.

The article "Exploiting the Stemness and Chemoresistance Transcriptome of Ewing Sarcoma to Identify Candidate Therapeutic Targets and Drug-Repurposing Candidates" revealed ABCG1 as an additional potential cell surface marker of progression. The authors used functional assays and transcriptomic analyses to characterize the cells that are responsible for progression and relapse. They explored a data bank to find known drugs that bind to these targets. This approach may, at least in part, overcome some limits of targeted therapy. In the future, after the preclinical validation of efficacy and specificity in Ewing sarcoma, some of these drugs may be assessed as combination treatments in clinical trials, with the goal of improving outcomes.

The article "Identification of Factors Driving Doxorubicin-Resistant Ewing Tumor Cells to Survival" identified biomarkers of resistance to doxorubicin in primary cultures of Ewing sarcoma cells using single-cell transcriptomic and proteomic analyses. Through this investigation, it was confirmed that MGST1 and tCOL6A2 are both produced by doxorubicin-resistant cells. This constitutes an interesting contribution to precision oncology, aimed at enhancing the therapeutic index of conventional chemotherapy.

The article "Secondary Malignancies after Ewing Sarcoma" shows a cumulative incidence (CI) of 14% at 30 years. While the CI for hematologic secondary malignant neoplasms (SMNs) remained stable during this time, solid SMNs increased over time and were higher for metastatic patients than in localized EwS patients. The use of radiation doses $\geq$ 60 Gy correlated with the occurrence of SMNs.

In contrast, the article "Ewing Sarcoma as Secondary Malignant Neoplasm" shows EwS as an SMN in 1.1% of all patients with EwS. Survival is similar to that of primary EwS.

The review "Ewing Sarcoma Meets Epigenetics, Immunology, and Nanomedicine" addresses the long-term toxicities and failures of conventional cytotoxic treatment with recent advances in nanomedicine to provide novel delivery drug systems. In addition, it bridges the topics of epigenetic and immunologic therapeutic strategies.

The article "C-Reactive Protein Pretreatment-Level Evaluation for Ewing's Sarcoma Prognosis" shows that a CRP pretreatment value >0.5 mg/dL represents a sensitive prognostic risk factor for distant metastasis and poor prognosis, as well as confirming inflammation as a characteristic feature of Ewing sarcoma, conferring inferior prognosis.

The communication "Routine EWS Fusion Analysis in the Oncology Clinic to Identify Cancer-Specific Peptide Sequence Patterns That Span Breakpoints in Ewing Sarcoma and DSRCT" reports amino acid fusion sequences from the EWS gene and the fusion partner gene (FLI1, ERG, FEV, and WT1) to obtain fusion neoantigen data used in cancer vaccine trials. These findings might help to overcome immune inertia.

The article "T Cells Directed against the Metastatic Driver Chondromodulin-1 in Ewing Sarcoma: Comparative Engineering with CRISPR/Cas9 vs. Retroviral Gene Transfer for Adoptive Transfer" assessed the feasibility of endogenous TCR orthotopic replacement with a TCR containing a CHM1 targeting sequence via CRISPR/Cas9, evaluating the tumor recognition and cytotoxicity function of CRISPR/Cas9 engineered T cells and comparing the prevention of endogenous TCR expression in CRISPR/Cas9 vs. retrovirally engineered T cells. It shows that both engineered T cell products specifically recognize tumor cells and elicit cytotoxicity in vitro, with CRISPR/Cas9-engineered T cells providing a more prolonged cytotoxic activity. Thus, orthotopic TCR replacement may increase the longevity of genetically engineered therapeutic T cells.

Finally, the review "Current State of Immunotherapy and Mechanisms of Immune Evasion in Ewing Sarcoma" assesses the commonalities that EwS shares with other immunologically cold solid malignancies. It is of interest that EwS and osteosarcoma (OS) were among the first tumors treated with immunotherapy. While OS exhibits recurrent somatic copy-number alterations, EwS possesses one of the lowest mutation rates among cancers, as it is driven by a single oncogenic fusion protein. In spite of these differences, both EwS and OS are allied with low immunogenicity. The article discusses the mechanisms of immune escape in these tumors, including the low presentation of tumor-specific antigens, low expression levels of MHC-I antigen-presenting molecules, the accumulation of immunosuppressive M2 macrophages and myeloid proinflammatory cells, and the release of extracellular vesicles (EVs) that are capable of reprogramming the tumor microenvironment.

The editors are most grateful to the authors of this series of unique articles, which represent a collaborative, international effort that reflects the scope and spirit of the Ewing Sarcoma Research Community by merging basic and clinical biology with diagnostic and therapeutic approaches in a truly translational approach.

Stefan Burdach, Uta Dirksen, and Poul H. Sorensen
Guest Editors

cancers

MDPI

Review

Ewing Sarcoma Meets Epigenetics, Immunology and Nanomedicine: Moving Forward into Novel Therapeutic Strategies

Sara Sánchez-Molina [1,2,†], Elisabet Figuerola-Bou [1,2,†], Víctor Sánchez-Margalet [3], Luis de la Cruz-Merino [4], Jaume Mora [1,2], Enrique de Álava Casado [5,6,7,*], Daniel José García-Domínguez [3,4,‡] and Lourdes Hontecillas-Prieto [3,4,*,‡]

1 Developmental Tumor Biology Laboratory, Institut de Recerca Sant Joan de Déu, Hospital Sant Joan de Déu, Esplugues de Llobregat, 08950 Barcelona, Spain
2 Pediatric Cancer Center Barcelona, Hospital Sant Joan de Déu, Esplugues de Llobregat, 08950 Barcelona, Spain
3 Clinical Laboratory, Department of Medical Biochemistry and Molecular Biology, School of Medicine, Virgen Macarena University Hospital, University of Seville, 41009 Seville, Spain
4 Oncology Service, Department of Medicines, School of Medicine, Virgen Macarena University Hospital, University of Seville, 41009 Seville, Spain
5 Institute of Biomedicine of Seville (IBiS), Hospital Universitario Virgen del Rocío/CSIC/University of Seville/CIBERONC, 41013 Seville, Spain
6 Pathology Unit, Hospital Universitario Virgen del Rocío/CSIC/University of Seville/CIBERONC, 41013 Seville, Spain
7 Department of Normal and Pathological Cytology and Histology, School of Medicine, University of Seville, 41009 Seville, Spain
* Correspondence: enrique.alava.sspa@juntadeandalucia.es (E.d.Á.C.); lhontecillas-ibis@us.es (L.H.-P.)
† These authors contributed equally to this work as the first authors.
‡ These authors contributed equally to this work as the last authors.

Citation: Sánchez-Molina, S.; Figuerola-Bou, E.; Sánchez-Margalet, V.; de la Cruz-Merino, L.; Mora, J.; de Álava Casado, E.; García-Domínguez, D.J.; Hontecillas-Prieto, L. Ewing Sarcoma Meets Epigenetics, Immunology and Nanomedicine: Moving Forward into Novel Therapeutic Strategies. *Cancers* **2022**, *14*, 5473. https://doi.org/10.3390/cancers14215473

Academic Editors: Stefan Burdach, Uta Dirksen, Poul H. Sorensen and Verrecchia Franck

Received: 26 September 2022
Accepted: 3 November 2022
Published: 7 November 2022

Simple Summary: Ewing Sarcoma treatment is traditionally based on chemotherapy, surgery, and radiotherapy. Although these standard of care regimens are efficient at early disease stages, many patients fail to respond appropriately, which has prompted the search for more efficacious and specific treatments. A deeper understanding of the basic molecular mechanisms underlying the biology of both tumor cells and the tumor microenvironment, as well as advances in drug delivery, has led to the development of different approaches to improve the treatment in Ewing Sarcoma patients. Thus, epigenetic, and immunotherapy-based drugs, along with nanotechnology delivery strategies, represent novel preclinical and clinical studies in the treatment of Ewing Sarcoma. In this review, we provide a comprehensive overview of these emerging therapeutic strategies and summarize the potential of the latest preclinical and clinical trials in Ewing Sarcoma research. Finally, we underline the value and future directions of these new treatments.

Abstract: Ewing Sarcoma (EWS) is an aggressive bone and soft tissue tumor that mainly affects children, adolescents, and young adults. The standard therapy, including chemotherapy, surgery, and radiotherapy, has substantially improved the survival of EWS patients with localized disease. Unfortunately, this multimodal treatment remains elusive in clinics for those patients with recurrent or metastatic disease who have an unfavorable prognosis. Consistently, there is an urgent need to find new strategies for patients that fail to respond to standard therapies. In this regard, in the last decade, treatments targeting epigenetic dependencies in tumor cells and the immune system have emerged into the clinical scenario. Additionally, recent advances in nanomedicine provide novel delivery drug systems, which may address challenges such as side effects and toxicity. Therefore, therapeutic strategies stemming from epigenetics, immunology, and nanomedicine yield promising alternatives for treating these patients. In this review, we highlight the most relevant EWS preclinical and clinical studies in epigenetics, immunotherapy, and nanotherapy conducted in the last five years.

Keywords: Ewing Sarcoma; epigenetic; immunotherapy; nanotherapy

1. Introduction

Ewing Sarcoma (EWS) is a rare and highly aggressive bone and soft tissue tumor that affects children, adolescents, and young adults with a peak of incidence in the second decade of life. The prognosis of EWS has improved considerably, with current multimodal therapy including chemotherapy, surgery, and radiation, with a 65–70% cure rate for localized disease. However, older patients (>18 years), metastatic patients at diagnosis and patients with early relapsing tumors still have a poor prognosis, with a 5-year survival rate of less than 30% [1,2]. Therefore, the higher therapeutic challenge remains on how to control the systemic disease and improve the survival rates, especially in those patients with worse prognosis.

EWS tumor cells are characterized by a fusion gene involving one member of the *FET* family of genes and one of the *ETS* family of transcription factors, *EWSR1-FLI1* being the most common [1,3]. Fusion genes have been demonstrated to be essential for tumorigenesis and, therefore, are attractive therapeutic targets that can be addressed through direct and indirect molecular targeted approaches [4]. Nevertheless, the lack of specific enzymatic activity of EWSR1-FLI1 challenges a direct targeted pharmacological inhibition. Moreover, indirect inhibition of oncogene activity by the perturbation of downstream targets, although it has presented successful integration in preclinical models, remains elusive in clinics [5].

Advances in the molecular mechanisms underlying the epigenetic remodeling of chromatin mediated by the fusion oncogene and the immune system have led to the development of novel therapeutic approaches. Epigenetic changes driven by EWSR1-FLI1 have been reported in the tumorigenesis of EWS. Indeed, EWSR1-FLI1 rewires chromatin and reprograms gene expression causing both induction and repression of selected gene pathways [6–8]. Therefore, epigenetic-based treatments provide a prominent option for treating this aggressive tumor by reversing the effect in the epigenome induced by the fusion gene. Moreover, based on the experience gained from adult cancer, immunotherapy studies have been translated to pediatric tumors including EWS.

There is a pressing requirement to develop targeted therapies or drug carriers that can deliver therapeutic agents with higher efficiency to lower the dosage needed and minimize side effects. On this basis, nanotechnology plays a prominent role in modern medicine, by potentially overcoming the deficiencies of conventional methods of administering chemotherapy and ultimately improving clinical outcomes [9].

In this article, we will revise the ongoing preclinical and clinical studies of the last five years focusing on epigenetics, immunotherapy and nanotherapy in EWS.

2. Epigenetic and Immunotherapy-Based Treatments in EWS: Moving Forward in Targeted Therapies

The ultimate knowledge of the basic aspects of the epigenetics and immunotherapy of cancer has made significant strides, leading to the development of a wide variety of new therapeutic agents. Here, we summarize the newest epigenetic and immune-based treatments in EWS.

2.1. Epigenetic Therapy

Epigenetics encompasses the reversible molecular processes affecting chromatin that define cellular identity by maintaining on and off states of transcription without alterations in the DNA sequence. Upon sequencing studies, different groups reported EWS as a tumor with paucity in the mutational rate, implicating epigenetics behind EWSR1-FLI1 as a tumorigenic factor [10]. As a result, many publications have shed light on the role of the EWS epigenome both in the understanding of the molecular mechanisms involved in tumor development and in the identification of novel targets for new and combinational therapies [11].

Epigenetics is critical to induce the proper environment for EWSR1-FLI1 establishment, as cells with higher plasticity will provide more significant opportunities for reprogramming by the oncogene [12,13]. Besides, the oncogene that interacts directly with DNA presents scaffolding properties that mediate protein–protein interactions with important epigenetic regulators of chromatin structure, rewiring the complete epigenome and, ultimately, their expression programs [14]. EWSR1-FLI1 behaves as a pioneer factor by directly recruiting chromatin remodelers to GGAA microsatellites, where it induces the formation of de novo active super-enhancers in regions that were previously repressed [11]. Finally, the repressive role of the oncogene is described by its capability to displace endogenous transcription factors [7]. Understanding the epigenetic mechanisms that permit cancer cells to quickly adapt, and their reversibility, therefore, constitutes a great opportunity for the development of new strategies to treat cancer [15]. The following sections will focus on those epigenetic drugs that can be translated into the clinics, which include targeting DNA methylation, nucleosome remodelers, histone post-translational modifications and their modifiers (Table 1).

Table 1. Summarizing the open clinical trials (last 5 years) targeting epigenetic factors. Source: ClinicalTrials.gov (accessed on 1 September 2022).

Molecular Mechanism	Molecular Target	Drug	Clinical Trial Identifier	Patients	Phase	Status/Ref
DNA methylation	IDH	Ivodesinib	NCT04195555	Advanced Solid Tumors, Lymphoma, or Histiocytic disorders with IDH1 mutations	II	Recruiting
Nucleosome remodeling	LSD1/NURD	Seclidemstat + topotecan and cyclophos-phamide	NCT03600649	Ewing Sarcoma (EWS); Myxoid Liposarcoma; Sarcomas with FET-family translocation	I	Recruiting
		Seclidemstat	NCT05266196	EWS; Myxoid Liposarcoma; Desmoplastic Small Round Cell Tumor; Extraskeletal Myxoid Chondrosarcoma; Angiomatoid Fibrous Histiocytoma; Clear Cell Sarcoma; Myoepithelial Tumor; Low Grade Fibromyxoid Sarcoma; Sclerosing Epithelioid Fibrosarcoma	I/II	Enrolling
		INCB059872	NCT03514407	Refractory or relapsed EWS	Ib	Terminated
		INCB059872	NCT02712905	Solid Tumors and Hematologic Malignancy	I/II	Terminated

Table 1. *Cont.*

Molecular Mechanism	Molecular Target	Drug	Clinical Trial Identifier	Patients	Phase	Status/Ref
	SWI/SNF	Trabectedin + radiation	NCT05131386	Osteosarcoma; Chondrosarcoma; EWS; Rhabdomyosarcoma; Desmoplastic Small Round Cell Tumor	II	Recruiting
		Trabectedin + irinotecan	NCT04067115	EWS	I	Recruiting
		Lurbinectedin with or without irinotecan	NCT05042934	Metastatic and recurrent EWS	I/II	Withdrawn
		Lurbinectedin + irinotecan	NCT02611024	Advanced Solid Tumors; Glioblastoma; Soft Tissue Sarcoma (Excluding GIST) Endometrial Carcinoma; Epithelial Ovarian; Carcinoma; Mesothelioma; Gastroenteropancreatic Neuroendocrine Tumor; SCLC; Gastric Carcinoma; Pancreatic Adenocarcinoma; Colorectal Carcinoma; Neuroendocrine Tumors	I/II	Recruiting
Histone writer	EZH2	Tazemetostat	NCT03213665	Relapsed or refractory: Brain tumors; Solid Tumors; non-Hodgkin Lymphoma; histiocytic disorders with EZH2, SMARCB1, or SMARCA4 gene mutations	II	Active, not recruiting
Histone eraser	HDAC	Vorinostat + chemotherapy	NCT04308330	EWS; Rhabdomyosarcoma; Wilms Tumor; Neuroblastoma; Hepatoblastoma; Germ Cell Tumor	I	Recruiting
Histone reader	BET	BMS-986158 and BMS-986378	NCT03936465	Pediatric Cancer	I	Recruiting

2.1.1. DNA Methylation

DNA methylation at cytosine (5-methylcytosine, 5mC) is an essential process in embryonic development and cell differentiation [16]. Disruption of the DNA methylation pattern is a common trait of different cancers, including EWS, where hypermethylation of key genes correlates with more aggressive behavior and hypomethylation was reported in active enhancers [17,18]. DNA methyltransferases (DNMT) and ten-eleven translocation

(TET) methylcytosine dioxygenases, responsible for DNA demethylation, have been major targets for epigenetic drug development. Despite their high efficiency, DNMT inhibitors (DNMTi), such as azacitidine and decitabine, presented toxicity in phase I clinical trials and low doses in combination with other agents were further tested [19]. Recently, the novel non-nucleoside DNMTi MC3343 has been described to induce a specific depletion of DNMT1 that induces DNA damage without alterations in DNA methylation [20].

Besides, non-epigenetic drugs were reported to affect TET enzymes and histone demethylases. Mutations that disrupt isocitrate dehydrogenase IDH1/2 enzymatic function produce a reduction in α-ketoglutarate (αKG) and an increase in the oncometabolite 2-hydroxyglutarate (2HG). In particular, 2HG inhibits TET enzymes resulting in DNA hypermethylation; thus, drugs inhibiting mutant IDH1/2 reactivate αKG and restore methylation levels [6]. On this basis, ivodesinib, an inhibitor of mutated IDH1, is actually in phase II clinical trial for refractory and recurrent pediatric solid tumors including EWS (NCT04195555).

2.1.2. Nucleosome Remodeling

Nucleosome remodeling refers to the ATP-dependent multiprotein complexes that affect nucleosome positioning and structure, influencing transcription regulation. Among these complexes, EWSR1-FLI1 recruits the nucleosome remodeling and deacetylase (NuRD) complex. This complex contains histone deacetylases (HDAC), lysine specific demethylase 1 (LSD1) and chromodomain-helicase-DNA-binding protein 3/4 (CHD3/4) and directly binds to EWSR1-FLI1 promoting transcriptional repression in EWS [21]. The inhibition of LSD1 with the non-competitive reversible LSD1 inhibitors HCI-2509 and HCI-2528 was effective in targeting EWS cell lines, while their efficiency was dependent on *EWSR1-FLI1* expression [21]. HCI-2509 delayed tumor growth in monotherapy [22] and its efficiency was not altered by the previous inhibition of EWS cell lines with the irreversible inhibitor GSK-LSD1, suggesting that HCI-2509 disrupts the LSD1 interaction with EWSR1-FLI1 [23]. Nevertheless, the latest studies have reported LSD1 colocalization at EWSR1-FLI1 active super-enhancers, correlating with HCI-2509 disruption not only of repression but also gene activation [24]. SP-2577 (seclidemstat), another LSD1 inhibitor, inhibited the growth of three out of eight EWS xenograft models [25]. At present, there are four clinical trials: (i) a phase I evaluating the safety-dose escalation and expansion of seclidemstat with topotecan and cyclophosphamide in patients with relapsed or refractory EWS (NCT03600649); (ii) a phase I/II as a continuation of a previous one, which allows the patient continued access to the drug (NCT05266196); (iii) a phase I study evaluating the safety and preliminary antitumor activity of INCB059872, another selective and oral LDS1 inhibitor, in refractory or relapsed EWS patients (NCT03514407); and (iv) a dose-escalation and dose expansion study of INCB059872 in advanced solid malignancies including EWS (NCT02712905).

Among a panel of pediatric sarcoma cell lines, EWS cells were the most sensitive to trabectedin, an antitumor drug derived from the sea squirt that binds to the minor groove of DNA, reversing the gene signature of EWSR1-FLI1 by interference with its transcription factor activity [26]. EWSR1-FLI1 can also recruit the mammalian switch/sucrose non-fermenting (SWI/SNF) nucleosome remodeler to enhancers containing GGAA microsatellites facilitating chromatin opening and activation of EWSR1-FLI1-targets [27]. Later studies demonstrated that trabectedin evicted the SWI/SNF complex from chromatin and redistributed EWSR1-FLI1 within the nucleus, disrupting its function as a pioneer factor [28]. Although the phase I clinical trial in children with refractory solid tumors concluded that trabectedin was safe, a phase II study was unsuccessful [29,30]. A new phase II clinical trial combines trabectedin with radiation in advanced and metastatic EWS (NCT05131386), and another three evaluate the combination of trabectedin or its derivative lurbinectedin with irinotecan based on their synergy (NCT04067115, NCT05042934, NCT02611024 and [31]).

2.1.3. Histone Modifications and Modifiers

Histone tails undergo a variety of post-translational covalent modifications that affect their interaction with DNA. The different histone modifications constitute a code where synergistic or antagonistic interactions determine chromatin accessibility to transcription factors and ultimately transcription activation or repression [32]. The enzymatic activities behind this histone code involve writers that settle these modifications (including histone acetyltransferases (HAT) or histone methyltransferases (HMT)), erasers, which eliminate them (including histone demethylases (HDM) or HDAC), and finally, readers that recognize and mediate an epigenetic signal.

Histone Writers: Polycomb Group and G9a Methyltransferase

The polycomb group (PcG) proteins segregate in two transcriptional repressive complexes, PRC1 and PRC2. PRC1 contains the E3 ubiquitin ligase enzyme RING1A or RING1B, while PRC2 consists of HMT activity from EZH1 or EZH2. Despite the repressive role of PRC1, RING1B has been described to be a transcriptional activator in various cancer entities [33,34]. In EWS, RING1B is highly expressed and is necessary for the expression of critical EWSR1-FLI1 targets by facilitating oncogene recruitment to active enhancers. Inhibition of aurora kinase (AURK) B by AZD1152 has been proposed as an excellent strategy to impair RING1B activity at active enhancers [35]. Moreover, EWS cells were highly sensitive to both AURKA and B inhibitors and their combination with focal adhesion kinase (FAK) inhibitors reduced the tumor growth in EWS mouse models [36].

The PRC2 subunit EZH2 is overexpressed in EWS and its knockdown inhibited tumor growth and metastasis in vivo [37,38]. Consequently, different EZH2 inhibitors have been evaluated in EWS in order to target PCR2 activity, such as the non-specific inhibitor 3-deazaneplanocin A (DZNep) and the specific inhibitor tazemetostat. DZNep treatment produced a cell cycle arrest in vitro and growth suppression in EWS mice [39]. The tolerability of tazemetostat is being evaluated in a phase II clinical trial in pediatric patients with gain of function mutations of EZH2 including EWS (NCT03213665). Nevertheless, tazemetostat showed no activity in four xenograft models of EWS [40]. Besides, EZH2 inhibitors combined with immunotherapy might offer a new therapeutic opportunity. It has been observed that GSK126, another selective EZH2 inhibitor, as well as tazemetostat, enhance the surface expression of disialoganglioside (GD2) in EWS cell lines, which sensitizes EWS cells to cytolysis by GD2-specific chimeric antigen receptor (CAR) T-cell immune therapy (see next chapter) [41].

Finally, G9a, an HMT that dimethylates H3K9, has been found to be overexpressed in different cancer types. Specifically, its overexpression in EWS correlated with poor prognosis and metastasis [42]. Indeed, the G9a inhibition with BIX01294 was proved effective in disrupting migration, invasion, adhesion, colony formation, and vasculogenic mimicry via the upregulation of *NEU1*. Decrease in metastasis and tumor growth with BIX01294 was proven in two in vivo models of EWS metastasis [42].

Histone Erasers: Deacetylases and Demethylases

HDAC antagonize the enzymatic activity of HAT by removing histone acetylation. EWSR1-FLI1 was shown to globally repress HAT activity while stimulating HDAC [43]. Consistently, several HDAC inhibitors (HDACi) were screened in EWS, including FK228 (romidepsin) and MS-275 (entinostat), which presented antitumor activity in vitro and in vivo in EWS, as well as vorinostat (SAHA) and sodium butyrate (NaB) [43–45]. Lessnick et al., showed that both vorinostat and depletion of HDAC2/3 reversed expression patterns of EWSR1-FLI1-repressed targets, indicating that the oncogene relies on HDAC for its repressive role in transcription [21]. Nevertheless, the first initial preclinical testing of vorinostat retrieved no objective responses for any of the solid tumors tested, including EWS [46]. Besides, entinostat, a selective HDAC1 and HDAC3 inhibitor, significantly reduced tumor burden and increased survival in preclinical xenograft models inducing cell cycle arrest and apoptosis. However, only the knockdown of HDAC3 was critical for EWS survival [47].

Further studies have revealed the potential of HDACi in combination with other drugs. A screening of 43 epigenetic drugs revealed that the most sensitive agents in EWS cell lines were related to HDAC inhibition, being BML-281, a specific inhibitor of HDAC6, the drug with the lower IC50. BML-281 increased acetylation levels of specificity protein 1 (SP1), reducing its binding to the *EWSR1-FLI1* promoter and causing repression of the oncogene and its associated targets [48]. Furthermore, the combination of the HDAC6 inhibitor ACY-1215 with doxorubicin reduced tumor growth in EWS xenografts [48]. On the other hand, *HDAC1* and *HDAC2* knockouts demonstrated a reduction in invasiveness and tumor growth in xenografts [49]. Since the effect in tumor growth resembled EZH2 inhibition [37], the HDACi romidepsin was combined with the embryonic ectoderm development (EED) inhibitor (A-395), which inactivates the PRC2 complex. This combination treatment was superior to monotherapy blocking the proliferation and tumor growth of SK-N-MC or EW7 xenograft models [49]. In addition, the combination of SAHA with HCI-2509 decreased cell proliferation, triggering cell cycle arrest and apoptosis, reducing *EWSR1-FLI1* expression by regulation of the *EWSR1* promoter and altering tumor growth [50]. Along the same line, the combination of romidepsin with HCI-2509 has also proved to be synergistic [51]. Currently, a phase I clinical trial combining vorinostat with chemotherapy in refractory or relapsed solid tumors is open (NCT04308330). Interestingly, HDACi could be chemically modified to have a second pharmacophore, like fimepinostat, which is a hybrid inhibitor of phosphatidylinositol 3-kinase (PI3K) and HDACs. This drug not only reduced EWSR1-FLI1 protein by affecting its stability but also cell viability and tumor growth in sarcoma xenograft models [52].

Regarding histone demethylation, the Jumonji-domain HDM KDM3B demethylates H3K9me2, and has been described as a novel oncogene downstream of EWSR1-FLI1 [53]. KDM3B and its direct target, the cell adhesion molecule MCAM, were positively implicated in cell migration and invasion, and their knockout reduced metastasis in vivo [54]. Indeed, EWS cell lines were sensitive to the pan-selective Jumonji HDM inhibitor JIB-04, which increased methylation levels of H3K4me3, H3K9me2, and H3K27me3 and affected the whole EWSR1-FLI1 transcriptome. JIB-04 induced DNA damage via CDKN1A and decreased tumor growth in xenograft models [55]. Besides, a drug screening revealed that EWS cell lines were sensitive to the H3K27me3 demethylase inhibitor GSK J4. This drug sensitized EWS cell lines to chemotherapy and synergized in vivo with the cyclin-dependent kinase (CDK) 7/12/13 inhibitor THZ1 [56]. Nevertheless, these new epigenetic drugs have not yet reached into the clinics.

Histone Readers: Bromodomains

The bromodomain and extra-terminal (BET) family consists of four conserved mammalian members (BRD containing 2 (BRD2), BRD3, BRD4, and BRDT) that interact through bromodomains with acetylated lysine residues [57,58]. The first BET inhibitor described was JQ1, a molecule that competitively binds to bromodomains, preventing the interaction between BET proteins and acetylated histones. In EWS, both JQ1 and depleted BRD proteins suppressed the *EWSR1-FLI1* gene signature. Besides, JQ1 compromised cell proliferation, angiogenesis, and tumor growth in EWS xenograft models [59,60]. BMS-986158 and BMS-986378, another two BET inhibitors, have now entered clinical trials as investigational drugs for evaluating their efficacy for pediatric brain and solid tumors (NCT03936465).

2.2. Immunotherapy

Immunotherapy is a treatment that boosts the immune system response against cancer or blocks any mechanism that prevents antitumor immunity. The local tumor microenvironment (TME) and the host immune system define the tumor immunophenotype, which is generally divided into hot and cold tumors. Whereas hot tumors resemble an immune-inflamed phenotype characterized by infiltration of T lymphocytes, cold tumors present an immune-desert or immune-excluded phenotype with the absence or exclusion of T-cells [61]. EWS exemplifies an immune cold tumor with very poor infiltration of immune cells or inflammatory infiltrates due to immune escape, immune privilege, or immune inhibition by the TME. Tumor cells resemble a deficient expression of human leukocyte antigens (HLA) that prevents recognition of tumor-associated antigens by effector T-cells and antigen presenting cells. Consistently, self or tumor-reactive T-cells extracted from EWS patients show an exhausted phenotype that failed to activate despite the presence of high doses of antigen [62]. In the same lines, immune-inhibitory ligands, such as HLA-G were found locally expressed on tumor cells and on infiltrating lymphocytes, which promote direct inhibition of the immune response by natural killer (NK) cells as well as the induction and expansion of myeloid-derived suppressor cells (MDSCs) [63,64]. Besides, large populations of MDSCs were shown to inhibit EWS immune responses to therapy [65]. While a better understanding of the interplay between EWS and TME is being developed, novel immunotherapy strategies are focused on increasing the number of T-cells driving them into the tumor and reversing the immunosuppressive TME [66]. These therapies include immune checkpoint inhibitors, adoptive cell therapy, antibody-based immunotherapy, and cancer vaccines, which are addressed below (Figure 1).

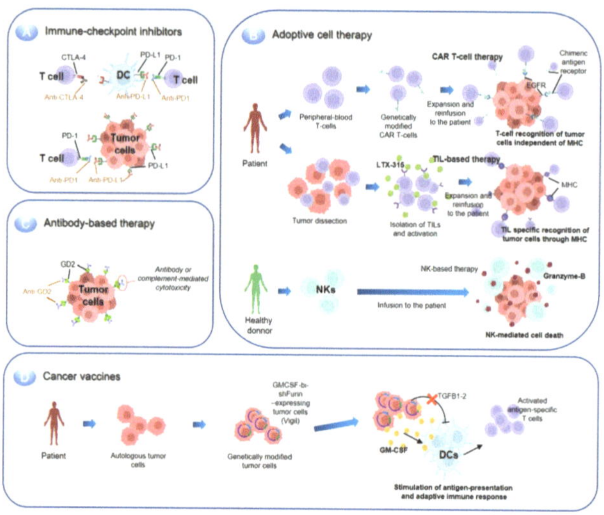

Figure 1. Immune therapies currently explored in EWS. (**A**) Immune checkpoint inhibitors block the interaction of immune checkpoint molecules (e.g., PD-1 or CTLA-4) with its inhibitory ligands to stimulate the immune response. (**B**) Adoptive cell therapy involves the infusion of modified autologous T-cells or allogenic NK cells. T-cells can be genetically modified to express a chimeric antigen receptor (CAR) specific of a tumor-associated antigen (e.g., EGFR) that can be recognized by major histocompatibility complex (MHC)-independent mechanisms. In contrast, T-cells isolated from tumors can be stimulated with an oncolytic peptide (e.g., LTX-315) and reinfusioned back to mediate an antitumoral MHC-dependent response. Transfer of NK cells from healthy donors is based on the innate ability of NK cells to kill tumor cells through various mechanisms such as granzyme B release. (**C**) Antibody-based therapies involve the use of specific antibodies targeting tumor-associated antigens (e.g., GD2). (**D**) Cancer vaccines stimulate the immune system response of the host through various mechanisms. The VIGIL vaccine in EWS is based on the tumor cells engineered to express GM-CSF and a bifunctional shRNA that prevents immunosuppression by TGFβ1-2 release. Reinfusion of these tumor cells, thus, promotes antigen-presentation and the adaptive immune response.

2.2.1. Immune Checkpoint Inhibitors

Immune checkpoint molecules are inhibitory and stimulatory ligand–receptor pairs that exert an inhibitory or stimulatory effect on immune responses. They are usually expressed in T-cells to maintain self-tolerance and regulate the magnitude of the immune response. Additionally, these molecules have been described as participating in immune evasion in cancer [67]. Blocking the interaction of checkpoint molecules by immune checkpoint inhibitors (ICI) is currently under research to increase T-cell activation and proliferation, causing T-cell cytotoxicity towards tumor cells. ICI treatment typically targets PD1 or CTLA4 immune checkpoint molecules, which have shown promising clinical efficacy in various solid tumors, including melanoma [68,69]. Three trials have studied the efficacy of ICI in pediatric sarcomas showing no benefit for EWS patients. In a phase I trial, ipilimumab (anti-CTLA4) was evaluated in children and adolescents with sarcoma, however, it showed no remarkable benefit considering the small sample size [70]. Next, a multicentric study evaluated pembrolizumab (anti-PD1) in advanced sarcomas, reporting an objective response in only 18 and 5% of soft tissue and bone sarcoma, respectively, although no response in the 13 EWS patients was observed [71]. The last trial studied the combination of both anti-PD1 and anti-CTLA4 and confirmed the limited efficacy of anti-PD1 in monotherapy, while reporting modest benefits of the combination in some sarcoma subtypes beyond EWS (5% and 16% overall response rate, respectively) [72]. The tumor mutation burden contributes to the immune recognition of cancer cells and, together with the expression of both PD-1 and PD-L1, seem to predict the response to ICI treatment [73,74]. On this basis, the low mutation rates of EWS and the fact that these tumors have a low expression of PD1 or its ligands (25.7% and 19.2%, respectively) might explain the poor response of these tumors to ICI. Moreover, another study reported PD-L1 expression in 33% of EWS, which significantly anticorrelated with survival [75,76].

New therapeutic strategies beyond ICI focus on combining these agents. VEGF promotes an immunosuppressive microenvironment and contributes to ICI resistance in cancer [77]. Consistently, clinical trials are combining pembrolizumab with VEGFR inhibitors (NCT02636725, NCT05182164). The combination of pembrolizumab with the VEGFR inhibitor axitinib has shown low toxicity and preliminary activity in a phase II trial, although no remarkable response was reported for EWS patients [78] (NCT02636725). Another phase II study is assessing the efficacy of combining pembrolizumab with cabozantinib, a receptor tyrosine kinase inhibitor, in patients with advanced sarcomas (NCT05182164). Additionally, a phase I/II trial with sequential administration of nivolumab (anti-PD1) and escalating doses of the mTOR-inhibitor ABI-009 has been conducted with EWS patients in which the efficacy and safety of the treatment will be evaluated (NCT03190174). The last results of this study showed no dose-limiting toxicities [79]. NKTR-214 is an engineered version of the interleukin 2 receptor (IL-2R) with a polyethylene glycol chain (bempegaldesleukin or BEMPEG) that reduces IL-2 binding to CD25 over CD122. Consequently, a sustained activation of antitumor CD8^{+} T-cells and NK cells, which is associated with tumor regression, is promoted [80]. Novel studies indicate the benefit of combining this therapy with ICI [81]. On this basis, a non-randomized two part open-label trial is evaluating the safety, tolerability, and dose level of the combinatory treatment of nivolumab with BEMPEG, as well as the efficacy of the combination in children and young adults with recurrent or refractory tumors including EWS (NCT04730349). However, trials with this combination have been discontinued recently.

Finally, B7 homolog 3 (B7-H3) is a checkpoint inhibitory protein of the B7-CD28 family that is overexpressed in multiple cancer types including osteosarcoma, whose expression is associated with poor survival [82]. Enoblituzumab (MGA271) is a humanized IgG1 monoclonal antibody targeting B7-H3 that is being trialed in children with relapsed or refractory malignant solid tumors with high expression of B7-H3, including osteosarcoma, EWS, neuroblastoma, rhabdomyosarcoma, Wilms Tumor and desmoplastic small round cell tumors (NCT02982941). This phase I trial will determine its safety, tolerability, immunogenicity, and preliminary antitumor activity in these tumor entities.

2.2.2. Adoptive Cell Therapy

In contrast to ICI therapy, which is intended to reinvigorate a suppressed or poor immune response against tumor, adoptive cell therapy (ACT) or cellular immunotherapy evades T-cell activation steps. On this basis, ACT involves the infusion of tumor-resident or peripheral blood-modified immune cells to promote an antitumor response, which includes the transfer of modified T-cells and NK cells.

Transfer of T-Cells

Tumor-infiltrating lymphocytes (TIL) are T-cells found in malignant tissues whose function and localization are critical to eventual tumor control or progression [83]. Consistent with the immune cold phenotype of EWS, a poor number of TILs are closely associated with deficient HLA expression in tumor cells that protects against immune recognition. Moreover, low expression of HLA-I is associated with poor survival in EWS patients [75,84]. Consequently, ACT therapies are seeking reinvigorating strategies, such as the infiltration of pre-stimulated TILs or genetically modified T-cells, for the patient. TILs' collection and expansion from tumors is feasible, and reinfusion has shown cytotoxic responses against tumor [85]. Nevertheless, the pre-treatment conditioning of T-cells is important to enhance engraftment and persistence of transferred cell populations. This strategy is currently being explored in phase I/II clinical trials with advanced and metastatic sarcomas, in which TILs' reinfusion to the patient is co-administered with or without a high dose of IL-2 (NCT04052334, NCT03449108). A phase II trial investigated the treatment of TILs with an oncolytic peptide (LTX-315), resulting in a feasible and tolerable combination with manageable toxicity in various metastatic sarcomas [86]. Other strategies focus on the infusion of T cells with a genetically modified T cell receptor (TCR) recognizing HLA-I restricted antigens uniquely expressed by tumor cells, which permits to identify intracellular antigens. The first clinical use of TCR transgenic T cells recognizing EWS-derived peptides (allorestricted) in EWS patients was directed against chondromodulin-1 (CHM1), a transmembrane glycoprotein directly activated by EWSR1-FLI1 that promotes metastatic spread [87]. Transfer of the HLA-A*02:01/CHM1^{319} TCR transgenic CD8^{+} T cells to three refractory EWS patients was well tolerated and was associated with disease regression, although this has not gone into clinical trials yet [88]. Furthermore, transferred T cells home into the affected bone marrow and persist, which gives hope to those patients with bone marrow metastasis that do not survive irrespective of therapy [89]. Other TCR-based therapies targeting the tumor-restricted expression of cancer testis antigens like NY-ESO-1 has been extensively studied in the context of sarcomas, with promising clinical results in synovial sarcoma [90,91]. Two phase I clinical trials with NY-ESO-1-based TCR therapies are currently ongoing in bone and soft tissue sarcomas (NCT03462316 and NCT03240861).

On the other hand, CAR therapies are based on the engineering of T-cells expressing a novel receptor designed to combine the effector properties of T-cells and the ability of antibodies to recognize pre-defined surface antigens of cancer cells with a high degree of specificity [92]. CAR-based therapies have been highly efficient for hematologic malignancies and around 470 clinical trials are now running [93,94]. However, multiple facts constraint its success in solid tumors, which includes T-cell limited survival and expansion, activation-induced cell death, T-cell exhaustion, trogocytosis, antigen loss, and unintended gene transduction of tumor cells [95,96]. Furthermore, designing CAR therapies is challenging in heterogeneous tumors such as EWS, where minimal "universally" membrane-expressed targets exist. GD2, aforementioned, has a 40–90% expression in primary EWS and thus has been used as a CAR-based target [97,98]. GD2-specific CAR T-cells were highly effective in patients with high-risk neuroblastoma [99], although no antitumor effect against GD2-positive EWS xenograft models was reported. However, investigators found MDSCs inhibited human CAR T cell responses in sarcomas and treatment with retinoic acid reduced the immunosuppressive capacity of MDSCs. These results suggested that retinoids enhanced the clinical efficacy of CAR therapies in sarcomas [65]. Novel therapies in tumors expressing high GD2, including EWS, explore the clinical effect of a GD2-CAR therapy in combination with chemotherapy with or without a previous lymphodepletion regimen (NCT03373097 and NCT03635632, respectively). Moreover, the combinatory effect of CAR-T-cells (targeting multiple markers like GD2), with low dose chemotherapy followed by maintenance with sarcoma vaccines is in a phase I/II trial (NCT04433221). Further approaches have designed CARs against the ICI molecule B7-H3, which has shown potent antitumor activity in EWS xenograft models [100]. Consistently, B7-H3-based CARs are now in phase I clinical trials in pediatric solid tumors including EWS (NCT04897321, NCT04483778). Finally, the epithelial growth factor receptor (EGFR) is another target for CAR therapy in EWS and its inhibition has an antitumor activity in vitro [101]. A phase I trial using EGFR-CAR (EGFR806) is recruiting relapsed patients with preliminary data indicating acceptable toxicity and antitumor activity in children and young adults (NCT03618381) [102].

Challenging clinical aspects of CAR therapies is the high toxicity reported, partially explained by the expression of CAR-targeting antigens in healthy tissues. The design of new generation CARs might overcome this issue. In this regard, larger phase I/II clinical trials are being conducted to study the safety and efficacy of 4th generation CAR T-cell therapies in various tumors, including EWS (NCT03356782).

Transfer of Natural Killer Cells

NK cells were named for their ability to kill cancer cells autonomously without antigen presentation. These cells express numerous inhibitory, activating, adhesion, and cytokine receptors that permit the direct recognition of cell-stress signals or foreign antigens to self-activate or suppress its cytolytic activity [103]. Considering the lack of neoepitopes in pediatric tumors, the innate ability of NK cells to recognize activating ligands on tumors is beneficial. A preclinical study showed that chemoresistant sarcoma cell lines, including EWS, were sensitive to NK cell killing in vitro and in vivo [104]. Moreover, investigators showed that EWS cells and primary tumors were susceptible to NK cytotoxicity through the expression of ligands for the activating NK cell receptors NKG2D and DNAM-1 and the use of cytokines increased the effectivity [105]. Additionally, transduced NK cells with a GD2-specific CAR has shown to enhance their ability to lyse cells in EWS in vitro [106]. On this basis, a phase I clinical trial explores this antitumor strategy by transplanting allogenic (donated) and previously stimulated NK cells in pediatric patients with solid tumors or leukemia (NCT01287104). NK cells usually are infused from a histocompatible donor. A phase I trial including EWS patients proposes using NK cells from unmatched healthy donors stimulated with the interleukin 15 agonist ALT-803, an experimental procedure that has not yet been approved by FDA (NCT02890758). Finally, results from a phase II clinical trial with the infusion of autologous NK cells in combination with sirolimus (mTOR inhibitor) maintenance strategy in relapsed patients have shown good tolerance with 45% 2-year overall survival (OS) and 25% of progression-free survival (PFS) in EWS patients [107]. The technical improvements of the last years in the expansion of NK cells ex vivo as well as the development of new platforms (like CARs or bispecific NK-cell engagers) that increase target specificity of NK cells makes this a promising immunotherapy strategy not only in sarcomas but in other pediatric tumors [108,109].

2.2.3. Antibody-Based Immunotherapy

Treatment based on the usage of monoclonal antibodies (mAbs) emerged at least 30 years ago and are standard-of-care treatment nowadays for malignancies like breast cancer [110]. These therapies are based on the specific binding of mAbs targeting tumor-specific antigens, including the TME, which produces the killing of tumor cells through various mechanisms, as reviewed by Weiner [111]. Many studies aimed to use mAbs-based therapies in EWS clinical trials, as summarized in Table 2. For instance, mAbs targeting the insulin growth factor 1 (IGF-1) pathway have been explored extensively. The IGF-1 pathway is pivotal in EWS pathogenesis with studies showing that inhibition of IGF-1R reduced cell migration and tumor growth in vitro and in vivo [112–115]. However, clinical trials with anti-IGF-1 have shown an overall response rate of only 10–14% and a median PFS of less than 2 years [116–118]. Moreover, a randomized phase III clinical trial (NCT02306161) evaluated the use of ganitumab (targeting IGF-1R) with interval-compressed conventional chemotherapy in metastatic EWS patients, but this study was closed due to increased toxicities and lack of clinical benefit [119]. Further trials with ganitumab include its combination with palbociclib (NCT04129151), although lack of clinical benefit was reported [120]. Other mAbs targeting the IGF pathway have been analyzed in preclinical and clinical studies with relatively low response rates, as reviewed by Casey et al. [121]. Apart from IGF-1, mAbs targeting the VEGF pathway alone or in combination with chemotherapy have also been explored in sarcomas like EWS. Consistently, a randomized phase II clinical trial evaluated whether the addition of bevacizumab (targeting VEGF-R) to vincristine, cyclophosphamide and topototecan chemotherapy regimens could improve survival. However, the benefit to add bevacizumab was unclear [122]. Recent studies with the anti-murine DC101 targeting VEGF-R2 further support the rationale to target this pathway in EWS. They showed the administration of DC101 caused a delay in tumor growth of sarcoma PDX like EWS and its addition to chemotherapy resulted in an improvement of the anti-tumoral response [123]. A phase I clinical trial with the humanized version of DC101, ramucirumab, has been conducted in a range of pediatric patients with recurrent or refractory solid tumors, whose results are still missing (NCT02564198). Olaratumab (IMC-3G3) exemplifies another mAb-based therapy clinically explored in EWS, which targets the plateled-derived growth factor receptor (PDGFR). A phase I and randomized phase II study in patients with unresectable or metastatic soft tissue sarcoma reported to improve OS nearly 12 months when received olaratumab with doxorubicin compared to doxorubicin alone (NCT01185964) [124]. However, this was not confirmed in the following phase III trial (NCT02451943, ANNOUNCE) [125]. Subsequent trials evaluating the second-line addition of olaratumab to gemcitabine and docetaxel in advanced soft tissue sarcomas indicated no statistical significant improvement in the OS between the two arms (NCT02659020). However, the combination resulted favorable in the PFS and an objective response in both cohorts [126]. Additional studies have reported olaratumab combined with pembrolizumab is safe and well tolerated in patients with advanced soft tissue sarcomas, although further studies with an increased sample size are needed to evaluate the efficacy of these regimens (NCT03126591) [127].

Dinutuximab is a humanized GD2-mAb that was proved to benefit the survival of high-risk neuroblastoma patients and is now used for maintenance therapy. Hu14.18K322A is a derivative of dinutuximab developed to reduce allogeneic reactions. NCT02159443 trial will evaluate the presence of pretreatment anti-therapeutic antibodies that might influence Hu14.18K322A response in EWS and other malignancies.

Table 2. Summarizing the novel mAb-based therapies in EWS that are in clinical trials. Source: ClinicalTrials.gov (accessed on 1 Septembre 2022).

Molecular Target	Molecular Mechanism	Drug	Clinical Trial Identifier	Patients	Phase	Status/Ref
IGF1R	mAb + targeted therapy	Ganitumab + Palbociclib (targets CDK4 and CDK6)	NCT04129151	EWS; Relapsed EWS	II	Active, not recruiting [120]
	mAb + chemotherapy	Ganitumab + variouschemotherapy regimens (vincristine, vincristine sulfate, ifosfamide, etoposide, etoposide sulphate, doxorubicin, doxorubicin hydrochloride, cyclophosphamide)	NCT02306161	Metastatic EWS; Metastatic Bone Malignant neoplasm; Metastatic malignant lung neoplasm; Metastatic and peripheral PNET	III	Active, not recruiting [119]
GD2	mAb	Hu14.18K322A	NCT02159443	EWS; Melanoma; Neuroblastoma; Osteosarcoma	I	Completed
	ADC	^{131}I-3F8	NCT00445965	Brain and CNS tumors; Intraocular melanoma and melanoma; Lung cancer; Metastatic Cancer; Neuroblastoma; Ovarian Cancer; Sarcoma; Small intestine cancer; Retinoblastoma	II	Active not recruiting
AXL	ADC with or without ICI	BA3011 (CAB-AXL-ADC) with or without PD-1 inhibitor	NCT03425279	Sarcomas and refractory sarcomas; EWS; Non small cell lung cancer; Melanoma; Solid Tumor	I/II	Active, recruiting
B7-H3	ADC	^{131}I-8H9	NCT00089245	Brain and CNS tumors; Sarcoma; Neuroblastoma	I	Active, recruiting [128]
Endosialin	mAb + chemotherapy	Ontuxizumab (MORAb-004) + gemcitabine and docetaxel	NCT01574716	Metastasic soft tissue sarcomas	II	Completed [129]
PDGFR	mAb + ICI	Olaratumab (LY3012207) + Pembrolizumab (MK3475)	NCT03126591	Soft Tissue Sarcoma	I	Active, not recruiting [127]
	mAb + chemotherapy	Olaratumab (LY3012207) + gemcitabine and docetaxel (ANNOUNCE 2)	NCT02659020	Soft Tissue Sarcoma	I/II	Completed [126]

In the last decades, mAb technology has improved by conjugating antibodies with various antitumor effector molecules, including cytotoxic drugs (named antibody-drug conjugates or ADC) and radioconjugates (RIC). ADCs are comprised of a mAb bound to a cytotoxic drug that selectively binds to target cells and directly delivers the toxic payload [130]. Endoglin (CD105) is a coreceptor of the TGFβ family associated with poor prognosis in EWS [131]. Targeting endoglin with a first-in-class ADC conjugated to a tubulin polymerization inhibitor showed potent preclinical activity in EWS, although this has not been explored clinically [132]. AXL has a high expression in EWS and was associated with worse OS. Moreover, its inhibition chemosensitized EWS cell lines to vincristine or doxorubicin [133]. Now a phase II clinical trial with EWS patients is exploring usage of AXL-mAb (BA3011) conjugated with monomethyl auristatin E (MMAE), a chemotherapy agent that blocks tubulin polymerization (NCT03425279). On the other hand, RIC is a combination of mAbs to radioisotopes that can be used for both radioimaging of tumor cells and pharmacologically targeting tumor cells [134]. Endosialin (TEM-1) is a cell surface glycoprotein expressed in advanced sarcomas that promotes tumor growth through the platelet-derived growth factor (PDGF) signaling [135]. Ontuxizumab (MORAb-004), a mAb that targets endosialin and blocks PDGF signaling, has been studied in phase I/II trials in EWS; however, no objective response was observed [129,136]. Recently, the endosialin-RIC [^{111}In] CHX-DTPA-scFv78-Fc (with indium-111) has been evaluated preclinically in EWS cell lines showing potential to translate into clinics [137]. Finally, two phase I/II clinical studies with GD2 or B7-H3- targeting RIC are explored in sarcoma patients with dissemination in the central nervous system or leptomeningeal space (NCT00445965, NCT00089245). The B7-H3 study, including 9 pediatric sarcoma patients, has shown that intraventricular administration of the RIC mAb-therapy was safe and had a favorable dosimetry in the central nervous system, suggesting this might have clinical utility in patients with this type of dissemination [128].

2.2.4. Cancer Vaccines

Cancer vaccines stimulate the immune system typically by recognizing tumor-associated antigens and include peptide or dendritic cell vaccines loaded with tumor lysate or pulsed with antigenic peptides. On this basis, a vaccine made of patient-derived dendritic immune cells loaded with autologous tumor lysate or tumor antigens ex vivo has been shown to activate an antitumor response, although poor response in phase I/II studies in EWS and other soft tissue sarcomas [138]. New efforts are focused on combining these vaccines with chemotherapeutic regimens. Vigil (formally known as FANG) is an autologous cancer cell vaccine that is engineered to express granulocyte-macrophage colony-stimulating factor (GM-CSF), which stimulates antigen-presentation in combination with a bifunctional shRNA-furin that prevents cleavage of TGF-β and reduces its local immunosuppressive effect. A prospective non-randomized study of advanced EWS patients reported 1-year survival of 73% for Vigil-treated patients compared to 23% in the control group historically treated with conventional therapy [139]. Given the very low toxicity reported, a randomized phase III clinical trial combining temozolomide and irinotecan with or without Vigil in relapsed EWS patients is currently under investigation, representing one of the few phase III trials for these patients (NCT03495921).

3. Nanotherapy: A Refined Target-Specific Drug Delivery System

Nanomedicine is a novel therapeutic strategy based on the application of nanotechnology to medicine through the development and use of nanoparticles (NPS). NPS have nanoscale dimensions (ranging from 1 to 100 nm in diameter) [140] with specific nanomaterial properties, which include surface charge, size, morphology, and area that compromise their activities and effects [141]. Considering all these variables, NPS can be classified based on (i) structure (flat, spherical, crystalline, etc.); (ii) dimensionality (one-, two- and three-dimensional NPS) [142]; (iii) porosity (porous and non-porous materials) [143]; and, (iv) chemical composition (organic, inorganic, carbon-based nanomaterials and hybrid nanostructures [142].

Among the fundamental advantages of nanomedicine usage is the improvement of diagnostic sensitivity, imaging, and radiation therapy, as well as a more precise and efficacy delivery of pharmaceutical agents to the targeted tissue [144,145]. Therefore, its main application in cancer is the delivery of chemotherapeutic drugs reducing side effects to the minimum and getting the maximum clinical benefit. For this reason, the number of preclinical and clinical studies has considerably increased in recent years.

3.1. Ewing Sarcoma Nano-Systems

The side effects associated with the administration of chemotherapy drugs and the innate and/or acquired chemotherapy resistance in EWS cells remain challenging. On this basis, drug delivery systems involving NPS refine some of these aspects, although the number of studies is still limited. The following sections describe novel preclinical and clinical studies in EWS based on the chemical composition of NPS (Figure 2 and Table 3).

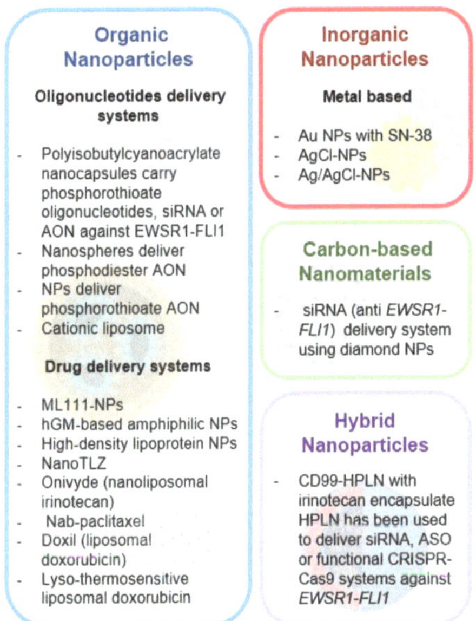

Figure 2. Summarizing the Nanoparticles and nanomaterials used in EWS studies.

3.1.1. Organic NPS

Organic NPS, also named polymers, are the most widely used NPS in biomedicine, including cancer, due to their biodegradable and non-toxic properties. They include micelles, dendrimers, liposomes, hydrogels, among others [146]. EWS studies carried out to date with organic NPS can be divided into two groups: oligonucleotide and drug delivery systems.

Oligonucleotide Delivery Systems

On this basis, polyisobutylcyanoacrylate nanocapsules have demonstrated their ability to inhibit *EWSR1-FLI1*. These NPS allow to carry phosphorothioate oligonucleotides [147] or small interfering RNAs (siRNAs) against *EWSR1-FLI1* [148], which inhibited tumor growth on EWS mice xenografts, and *EWSR1-FLI1* expression [147,148]. Furthermore, phosphorotionate NPS and phosphodiester nanospheres carrying antisense oligonucleotides (AON) against *EWSR1-FLI1* were also used, showing both a reduction of EWS tumor growth in vivo when injected intratumorally [149]. Rao et al., have developed a bifunctional shRNA (bi-shRNA) target sequence against the *EWSR1-FLI1* fusion gene that was complexed with a cationic liposome (PBI-shRNA *EWSR1-FLI1* LPX), resulting in 85–92% of *EWSR1-FLI1* knockdown (protein and RNA) in vitro. PBI-shRNA *EWSR1-FLI1* LPX was used in EWS xenograft mice, confirming its efficacy and safety. However, side effects were observed including temperature elevation on the first day, transient liver enzyme elevation at high doses and occasional limited hypertension [150]. Considering the results of the bi-shRNA LPX system in EWS and a previous clinical trial in lung cancer (NCT00059605) [151], a phase I active clinical trial was developed that involves pediatric patients (over 8 years old) with advanced EWS (NCT02736565).

Table 3. Summary of NPS used in clinical trials (last 5 years) in EWS. Source: ClinicalTrials.gov (accessed on 1 September 2022).

Molecular Mechanism	Interventions	Clinical Trial Identifier	Patients	Phase	Status
Oncogene driver inhibition	Biological: pbi-shRNA™ EWS/FLI1 Type 1 LPX	NCT02736565	EWS	I	Active, not recruiting
DNA damage by topoisomerase inhibition	Onivyde + Talazoparib (Arm A) or Temozolomide (Arm B)	NCT04901702	Recurrent Solid Tumors: EWS; Hepatoblastoma; Neuroblastoma; Osteosarcoma; Rhabdomyosarcoma; Wilms Tumor. Refractory Solid Tumors: EWS; Hepatoblastoma; Malignant Germ Cell Tumor; Malignant Solid Neoplasm; Neuroblastoma; Osteosarcoma; Peripheral Primitive Neuroectodermal Tumor; Rhabdoid Tumor; Rhabdomyosarcoma	I/II	Active, not recruiting
	MM-398 (Irinotecan Sucrosofate Liposome) + cyclophosphamide	NCT02013336	Recurrent or Refractory Solid Tumors: EWS; Rhabdomyosarcoma; Neuroblastoma; Osteosarcoma	I	Recruiting
Depolymerization of microtubules (paclitaxel)	Nab-paclitaxel	NCT03275818	Desmoplastic Small Round Cell, Adult; Desmoplastic Small Round Cell, childhood; EWS; Desmoid	II	Completed
	Nab-paclitaxel	NCT01962103	Neuroblastoma; Rhabdomyosarcoma; EWS; Epitheliod Sarcoma, Soft Tissue Sarcoma, Spindle Cell Melanoma; Melanoma; Osteosarcoma; Histiocytoma; Fibrosarcoma; Dermatofibrosarcoma	I/II	Completed [152]
	Nab-paclitaxel + Gemcitabine	NCT03507491	Cancer	I	Recruiting
	Nab-Paclitaxel + Gemcitabine	NCT02945800	Osteosarcoma; EWS; Rhabdomyosarcoma; Soft Tissue Sarcoma	II	Recruiting
DNA damage by intercalation, disruption of topoisomerase-II and generation of free radicals (doxorubicin)	Disulfiram + Copper Gluconate and Liposomal Doxorubicin	NCT05210374	Relapsed Sarcomas (including EWS)	I	Not yet recruiting
	Liposomal Doxorubicin + MR-HIFU Hyperthermia	NCT02557854	EWS; Rhabdomyosarcoma; Wilms Tumor; Neuroblastoma; Hepatoblastoma; Germ Cell Tumor	I	Withdrawn
	Temsirolimus + liposomal doxorubicin	NCT00949325	Sarcoma (including EWS)	I/II	Completed [153]
	Lyso-thermosensitive liposomal doxorubicin (LTLD) + MR-HIFU Hyperthermia	NCT04791228	EWS; Malignant Epithelial Neoplasm; Rhabdomyosarcoma; Wilms Tumor; Hepatic Tumor; Germ Cell Tumor	II	Not yet recruiting
	Lyso-thermosensitive liposomal doxorubicin + Magnetic resonance high intensity focused ultrasound	NCT02536183	Rhabdomyosarcoma; EWS; Osteosarcoma; Neuroblastoma; Wilms Tumor; Hepatic Tumor; Germ Cell Tumors	I	Recruiting

Drug Delivery Systems

Several studies have used NPS to carry anticancer drugs in order to improve drug kinetics and achieve better therapeutic results. On this basis, a small molecule uncharacterized compound ML111 was found to inhibit in vitro the proliferation of six established EWS cell lines with nanomolar potency [154]. Sabei et al., have reported that ML111 encapsulated into a hydrophobic core of PEG-PCL-based polymeric NPS (ML111-NPS) was able to internalize into EWS cell lines and specifically inhibit their viability without altering nonmalignant human cell lines [155]. Moreover, a synergistic effect on the viability of EWS cells resulted from combining ML111-NPS with vincristine in vitro, compared to nonmalignant cells. Moreover, a regression of EWS tumors was observed when using ML111-NPS in vivo, both in monotherapy and in combination with vincristine. No toxicity effects were identified in mice organs with ML111-NPS alone, and with the combination there was a reduction of side effects associated with vincristine [155]. Besides, the use of a hydrolyzed galactomannan (hGM)-based amphiphilic NPS for selective intratumoral accumulation in pediatric sarcoma was also investigated [156]. Coupling of these NPS with the tyrosine kinase inhibitor imatinib could target glucose transporter-1 (GLUT-1), both in rhabdomyosarcoma cells and in EWS PDX with different GLUT-1 expression levels with a 7.5% of efficiency [157], which make them a potential tool against GLUT-1-expressing tumors. Furthermore, Bell et al., employed biomimetic high-density lipoprotein (HDL) NPS. These HDL NPS were able to bind both HDL receptors and scavenger receptor type B-1 (SCARB1), depriving tumor cells HDL and cholesterol, and blocking proliferation in hedgehog-driven EWS cells and medulloblastoma [158].

PARP inhibitors such as talazoparib (TLZ) or olaparib did not show activity in EWS [159,160], although they potentiate the treatment with the DNA alkylating agent temozolomide (TMZ). A nanoformulation of TLZ (NanoTLZ) was reported to be more effective and well tolerated in vivo, while its combination with TMZ elicited an increase in the maximum tolerated dose of TMZ for EWS treatment [159]. Nevertheless, another study showed that the TC71 TLZ-resistant EWS cell line was not affected by frequently administered oral TLZ nor affected by the long-acting PEGylated TLZ [161].

Onivyde (MM-398 or PEP02) is a nanoliposomal formulation of the DNA topoisomerase I inhibitor irinotecan, which is used to treat several solid tumors, although it has a complex pharmacokinetics [162]. Onivyde showed an improvement on the antitumor activity, biodistribution, and a reduction of toxicity in EWS xenografts compared to the current clinical formulation of irinotecan [162]. Currently, a recruiting phase I clinical trial studies the highest dose of MM-398 that can be given safely when combined with cyclophosphamide in patients with recurrent or refractory pediatric solid tumors (NCT02013336). Indeed, an active phase I trial is being conducted with combinations of onivyde with TLZ or TMZ (NCT04901702) to determine the highest tolerable doses of the two combinations (NCT04901702).

A recent work evaluated the albumin-bound (nab)-paclitaxel NPS in PDXs of EWS, rhabdomyosarcoma, and osteosarcoma [163]. These NPS bind to tumor cells that express SPARC, a secreted acidic protein and rich in cysteine that shows a high affinity to bind albumin. Nab-paclitaxel was less bound in SPARC-knocked down (SPARC-KD) compared to SPARC-WT cells [163]. EWS PDX with high expression of SPARC was associated to accumulation of nab-paclitaxel showing better drug responses compared to tumors with lower SPARC levels. Consistently, pediatric tumors that express SPARC were able to accumulate nab-paclitaxel for more extended periods of time [163]. Nab-paclitaxel is being evaluated in several clinical trials including an active phase II clinical trial in monotherapy for patients with EWS and other tumors (NCT03275818). A completed phase I/II multicenter trial (NCT01962103, [152]) showed in EWS patients that the overall response rate was 0%, the disease control rate was 30.8% (4 stable disease), the median PFS was 13.0; and the 1-year OS rate was 48%. The safety of nab-paclitaxel in pediatric patients was confirmed; however, limited activity was observed [164]. Finally, two clinical trials are recruiting patients with pediatric relapsed and refractory solid tumors (NCT03507491);

and patients with recurrent/refractory sarcoma (NCT02945800), in which nab-paclitaxel and gemcitabine will be given. However, the results for the EWS arm of NCT02945800 have been published. This clinical trial of nab-paclitaxel and gemcitabine displayed limited activity in a small cohort of EWS patients confirming only one partial response. Moreover, two partial responses after two cycles was observed, but due to the side effects or the progression of the disease, these two patients were withdrawn [165]. The response rate of 9% was similar to other study in EWS patients treated with gemcitabine and docetaxel [165].

Liposomal doxorubicin was designed to increase its therapeutic efficacy while decreasing toxicity. A phase I clinical trial (not yet recruiting) purpose to evaluate disulfiram with copper gluconate and liposomal doxorubicin in treatment-refractory sarcomas (NCT05210374). Another phase I trial is currently running to determine whether delivery of a liposomal doxorubicin called doxil prior to MR-HIFU (magnetic resonance-guided high intensity focused ultrasound) hyperthermia will be safe for the treatment of pediatric and young adult patients with recurrent and refractory solid tumors. Unfortunately, this trial is withdrawn because of the lack of enrollment (NCT02557854). Also-thermosensitive liposomal doxorubicin (LTLD) is the first heat-activated formulation of a liposomal drug carrier to be utilized in human clinical trials. There are two clinical trials with EWS patients. A phase I, recruiting trial that combined LTLD and MR-HIFU in pediatric refractory solid tumors (NCT02536183); and a phase II trial, in which LTLD with MR-HIFU hyperthermia followed by ablation will be studied in subjects with refractory/relapsed solid tumors (NCT04791228).

Finally, complete phase I and II clinical trials showed that combinations of liposomal doxorubicin and temsirolimus were safe and showed efficacy for patients with recurrent sarcoma (NCT00949325 and [153,166]).

3.1.2. Inorganic NPS

Inorganic NPS are metal-based (gold, iron, lead, silver) and metal oxide-based (aluminum oxide, zinc oxide, etc.) particles [146]. Metal-based NPS of gold and silver (Au and Ag NPS, respectively) have been reported to have antitumor effects [156,167,168]; however, Ag NPS can induce general toxicity in non-target organs [169]. These NPS have been also evaluated in the context of EWS at preclinical level. Naumann et al., have developed Au-NPS where selective SN-38 activation in cancer cells is mediated by the EWS specific mRNAs BIRC5 (survivin) and EWSR1-FLI1. In this system, the gold particle is conjugated to the specific mRNA where the complementary SN38-conjugated oligonucleotide anneals. SN38 release will be dependent on the presence of the EWS specific mRNA. The viability of EWS cells treated with SN38-survivin Au-NPS and SN38-EWS/FLI1 Au-NPS was significantly reduced in four EWS cell lines and in murine xenografts [170]. The antitumor activity of silver chloride and silver/silver chloride NPS (AgCl and Ag/AgCl NPS, respectively) has been also investigated in EWS. Treatment of EWS cell lines and a non-tumor cell line with Ag NPS caused a reduction in cell viability specific for tumor cells. Both AgCl and Ag/Ag-NPS increased the percentage of apoptotic cells and ROS production, accompanied with a loss of mitochondrial membrane potential, and lysosomal damage. These effects were specific for tumor cells with minimal effects shown on healthy cells [171].

3.1.3. Carbon-Based Nanomaterials

Carbon-based NPS include fullerenes, carbon nanofibers, diamonds, carbon nanotubes, and graphene. These NPS display multiple properties that make them suitable for drug delivery systems and cancer therapy, as well as imaging, biosensing, or diagnosis [156]. Alhaddad et al., investigated the ability of a siRNA delivery system using diamond NPS [172]. These diamond NPS were coated with a cationic polymer and encapsulated siRNA to inhibit *EWSR1-FLI1*. Because diamond NPS have intrinsic fluorescent properties its internalization into EWS cell lines was efficacious and could be observed directly. Following the internalization, *EWSR1-FLI1* inhibition was observed at mRNA and protein levels in vitro. Finally, cell toxicity was low after treatment with diamond NPS [172].

3.1.4. Hybrid NPS

Hybrid NPS are formed by polymer and organic- or inorganic-based NPS systems that combine the properties of single systems. Consistently, they have lower circulation time and bioavailability, more stability and therapeutic efficacy, being a viable alternative when compared to single systems [173,174].

Hybrid polymerization liposomal NPS (HPLNs) has been developed antibody encapsulated with irinotecan (CD99-HPLN/Ir). Low doses of this hybrid system have shown reduced EWS tumors in xenograft mice and complete tumor ablation, which was more efficacious compared to onivyde and doxil NPS systems. Drug bioavailability was improved six-fold with HPLN and encapsulated irinotecan without CD99 and twelve-fold with CD99-HPLN/Ir in respect to onivyde. Consistently, irinotecan toxic side effects were minimized [173]. Along the same lines, HPLN has been used to deliver siRNA, ASO, or functional CRISPR-Cas9 systems against the fusion oncogene *EWSR1-FLI1*. In vitro experiments resulted in an efficient *EWSR1-FLI1* reduction being the most effective HPLN/CRISPR-Cas9 system. Moreover, CD99-HPLN with CRISPR-Cas9 against encapsulated *EWSR1-FLI1*, reduced EWS tumor growth in vivo [175]. Although promising preclinical results, the potential of these hybrid NPS systems remains to be further evaluated in clinical studies.

4. Conclusions and Future Perspectives

The standard therapy for EWS patients based on cytotoxic chemotherapy and radiotherapy has reached a plateau, especially for that subgroup of patients with the worst prognosis. Moreover, patients who survive face debilitating and often life-threatening health consequences as a result of the high toxicity of these therapies. Therefore, and especially considering the young age and potential lifespan of the patients, there is urgent need for finding new therapies to improve the outcome of these patients [176]. Epigenetic-based therapies have changed the targeting focus from extracellular and intracellular signaling to chromatin, where these pathways integrate regulating gene expression in a reversible manner that offers the opportunity for phenotype conversion. The inhibition of epigenetic complexes that regulate the expression and protein stability of the oncogene itself, as well as those cofactors that participate in the modulation of its activity, shows significant achievements in the control of the disease in preclinical studies. Nevertheless, the epigenetic drugs used in clinics have reported modest antitumor efficacy in monotherapy leading to the development of new epigenetic approaches based on the usage of second generation drugs and combinatorial strategies that promote synergistic effects. Further mechanistic approaches should explore differences in drug efficiency between targeting specific enzymatic domains and the effects of depleting the whole protein or inducing its degradation. Immunotherapy, unlike other approaches, induces a therapeutic response not only limited to disrupt a single oncogenic event. Despite promising results of immunotherapy in adult tumors, their application in EWS and other sarcomas has demonstrated poor therapeutic activity due to the immune-cold nature of these tumors. However, different immune strategies are being developed searching for efficient combinations with standard or new targeted therapies, including epigenetic drugs. Besides, other research strategies are focused on the development of more targeted approaches and the reversion of the cold immune landscape of EWS into a hot phenotype. Indeed, our understanding of the crosstalk between the tumor and the tissue microenvironment as well as the basic aspects of the vasculature and hypoxia of EWS would help future direction in immunomodulation therapies [177]. Both for epigenetics and immunotherapy, the introduction of CRISPR screenings to define novel targetable tumor dependencies will postulate promising combinatorial strategies to explore in clinical trials. On the other hand, nanomedicine has evolved to face the lack of specificity, drug resistance and high toxicity rates of both standard regimens and these new therapeutic alternatives. Consistently, NPS as drug delivery systems have reduced both the toxicity associated with cytotoxic drugs and the tolerated dose. This fact raises the possibility to rescue the usage of effective drugs that were discarded in clinics for their side effects by coupling them to novel NPS systems.

Finally, the highly heterogeneity in EWS tumors promote more limitations that also affects the treatment. Firstly, the possibility of developing an appropriate in vivo model, which could contribute to the discrepancy between preclinical and clinical results. However, latest research in patient derived organoid, which recapitulate genetic and phenotypic characteristics of their tissue of origin, support the inclusion of this models in preclinical validation as predictors of response [178]. Secondly, the necessity on finding specific and universally expressed membrane biomarkers that might improve treatment specificity. The discovery of new biomarkers with prognostic value and response to treatment with a more accurate classification of patients, would benefit the creation of a specific treatment plan, also referred to as personalized medicine, that might benefit survival of EWS patients. Taken together, these new therapeutic alternatives and the more effective delivery of drugs by NPS represent a new horizon in treating EWS patients, which is expected to benefit patient survival.

Author Contributions: S.S.-M., E.F.-B., D.J.G.-D. and L.H.-P. were involved in the conception and drafting of the manuscript. All authors have critically reviewed the manuscript and have approved the final version for publication. All authors have read and agreed to the published version of the manuscript.

Funding: This work was supported by FIS-FEDER PI2000003 to E.d.Á.C.; L.H.-P. is supported by the Consejería de Salud y Familias, Junta de Andalucía (RH-0047-2021).

Conflicts of Interest: The authors declare no conflict of interest. The authors have designed and conceived both figures. The figures require no copyright permission from third parties.

References

1. De Alava, E.L.S.L.; Stamenkovic, I. Ewing sarcoma. In *WHO Classification of Tumours of Soft Tissue and Bone*, 5th ed.; IARC Press: Lyon, France, 2020.
2. Mora, J.; Castaneda, A.; Perez-Jaume, S.; Lopez-Pousa, A.; Maradiegue, E.; Valverde, C.; Martin-Broto, J.; Garcia Del Muro, X.; Cruz, O.; Cruz, J.; et al. GEIS-21: A multicentric phase II study of intensive chemotherapy including gemcitabine and docetaxel for the treatment of Ewing sarcoma of children and adults: A report from the Spanish sarcoma group (GEIS). *Br. J. Cancer* **2017**, *117*, 767–774. [CrossRef] [PubMed]
3. Delattre, O.; Zucman, J.; Plougastel, B.; Desmaze, C.; Melot, T.; Peter, M.; Kovar, H.; Joubert, I.; de Jong, P.; Rouleau, G.; et al. Gene fusion with an ETS DNA-binding domain caused by chromosome translocation in human tumours. *Nature* **1992**, *359*, 162–165. [CrossRef] [PubMed]
4. Mitelman, F.; Johansson, B.; Mertens, F. The impact of translocations and gene fusions on cancer causation. *Nat. Rev. Cancer* **2007**, *7*, 233–245. [CrossRef] [PubMed]
5. Grunewald, T.G.P.; Cidre-Aranaz, F.; Surdez, D.; Tomazou, E.M.; de Alava, E.; Kovar, H.; Sorensen, P.H.; Delattre, O.; Dirksen, U. Ewing sarcoma. *Nat. Rev. Dis. Primers* **2018**, *4*, 5. [CrossRef] [PubMed]
6. Nacev, B.A.; Jones, K.B.; Intlekofer, A.M.; Yu, J.S.E.; Allis, C.D.; Tap, W.D.; Ladanyi, M.; Nielsen, T.O. The epigenomics of sarcoma. *Nat. Rev. Cancer* **2020**, *20*, 608–623. [CrossRef]
7. Riggi, N.; Knoechel, B.; Gillespie, S.M.; Rheinbay, E.; Boulay, G.; Suva, M.L.; Rossetti, N.E.; Boonseng, W.E.; Oksuz, O.; Cook, E.B.; et al. EWS-FLI1 Utilizes Divergent Chromatin Remodeling Mechanisms to Directly Activate or Repress Enhancer Elements in Ewing Sarcoma. *Cancer Cell* **2014**, *26*, 668–681. [CrossRef]
8. Tomazou, E.M.; Sheffield, N.C.; Schmidl, C.; Schuster, M.; Schonegger, A.; Datlinger, P.; Kubicek, S.; Bock, C.; Kovar, H. Epigenome mapping reveals distinct modes of gene regulation and widespread enhancer reprogramming by the oncogenic fusion protein EWS-FLI1. *Cell Rep.* **2015**, *10*, 1082–1095. [CrossRef]
9. Yang, S.; Wallach, M.; Krishna, A.; Kurmasheva, R.; Sridhar, S. Recent Developments in Nanomedicine for Pediatric Cancer. *J. Clin. Med.* **2021**, *10*, 1437. [CrossRef]
10. Crompton, B.D.; Stewart, C.; Taylor-Weiner, A.; Alexe, G.; Kurek, K.C.; Calicchio, M.L.; Kiezun, A.; Carter, S.L.; Shukla, S.A.; Mehta, S.S.; et al. The genomic landscape of pediatric Ewing sarcoma. *Cancer Discov.* **2014**, *4*, 1326–1341. [CrossRef]
11. Riggi, N.; Suva, M.L.; Stamenkovic, I. Ewing's Sarcoma. *N. Engl. J. Med.* **2021**, *384*, 154–164. [CrossRef]
12. Chen, T.; Dent, S.Y. Chromatin modifiers and remodellers: Regulators of cellular differentiation. *Nat. Rev. Genet.* **2014**, *15*, 93–106. [CrossRef]
13. Suva, M.L.; Riggi, N.; Bernstein, B.E. Epigenetic reprogramming in cancer. *Science* **2013**, *339*, 1567–1570. [CrossRef]
14. Perry, J.A.; Seong, B.K.A.; Stegmaier, K. Biology and Therapy of Dominant Fusion Oncoproteins Involving Transcription Factor and Chromatin Regulators in Sarcomas. *Annu. Rev. Cancer Biol.* **2019**, *3*, 299–321. [CrossRef]
15. Dawson, M.A. The cancer epigenome: Concepts, challenges, and therapeutic opportunities. *Science* **2017**, *355*, 1147–1152. [CrossRef]

16. Chen, Z.; Zhang, Y. Role of Mammalian DNA Methyltransferases in Development. *Annu. Rev. Biochem.* **2020**, *89*, 135–158. [CrossRef]
17. Sheffield, N.C.; Pierron, G.; Klughammer, J.; Datlinger, P.; Schonegger, A.; Schuster, M.; Hadler, J.; Surdez, D.; Guillemot, D.; Lapouble, E.; et al. DNA methylation heterogeneity defines a disease spectrum in Ewing sarcoma. *Nat. Med.* **2017**, *23*, 386–395. [CrossRef]
18. Park, H.R.; Jung, W.W.; Kim, H.S.; Park, Y.K. Microarray-based DNA methylation study of Ewing's sarcoma of the bone. *Oncol. Lett.* **2014**, *8*, 1613–1617. [CrossRef]
19. Hu, C.; Liu, X.; Zeng, Y.; Liu, J.; Wu, F. DNA methyltransferase inhibitors combination therapy for the treatment of solid tumor: Mechanism and clinical application. *Clin. Epigenetics* **2021**, *13*, 166. [CrossRef]
20. Cristalli, C.; Manara, M.C.; Valente, S.; Pellegrini, E.; Bavelloni, A.; De Feo, A.; Blalock, W.; Di Bello, E.; Pineyro, D.; Merkel, A.; et al. Novel Targeting of DNA Methyltransferase Activity Inhibits Ewing Sarcoma Cell Proliferation and Enhances Tumor Cell Sensitivity to DNA Damaging Drugs by Activating the DNA Damage Response. *Front. Endocrinol.* **2022**, *13*, 876602. [CrossRef]
21. Sankar, S.; Bell, R.; Stephens, B.; Zhuo, R.; Sharma, S.; Bearss, D.J.; Lessnick, S.L. Mechanism and relevance of EWS/FLI-mediated transcriptional repression in Ewing sarcoma. *Oncogene* **2013**, *32*, 5089–5100. [CrossRef]
22. Sankar, S.; Theisen, E.R.; Bearss, J.; Mulvihill, T.; Hoffman, L.M.; Sorna, V.; Beckerle, M.C.; Sharma, S.; Lessnick, S.L. Reversible LSD1 inhibition interferes with global EWS/ETS transcriptional activity and impedes Ewing sarcoma tumor growth. *Clin. Cancer Res.* **2014**, *20*, 4584–4597. [CrossRef] [PubMed]
23. Pishas, K.I.; Drenberg, C.D.; Taslim, C.; Theisen, E.R.; Johnson, K.M.; Saund, R.S.; Pop, I.L.; Crompton, B.D.; Lawlor, E.R.; Tirode, F.; et al. Therapeutic Targeting of KDM1A/LSD1 in Ewing Sarcoma with SP-2509 Engages the Endoplasmic Reticulum Stress Response. *Mol. Cancer* **2018**, *17*, 1902–1916. [CrossRef] [PubMed]
24. Theisen, E.R.; Selich-Anderson, J.; Miller, K.R.; Tanner, J.M.; Taslim, C.; Pishas, K.I.; Sharma, S.; Lessnick, S.L. Chromatin profiling reveals relocalization of lysine-specific demethylase 1 by an oncogenic fusion protein. *Epigenetics* **2021**, *16*, 405–424. [CrossRef] [PubMed]
25. Kurmasheva, R.T.; Erickson, S.W.; Han, R.; Teicher, B.A.; Smith, M.A.; Roth, M.; Gorlick, R.; Houghton, P.J. In vivo evaluation of the lysine-specific demethylase (KDM1A/LSD1) inhibitor SP-2577 (Seclidemstat) against pediatric sarcoma preclinical models: A report from the Pediatric Preclinical Testing Consortium (PPTC). *Pediatr. Blood Cancer* **2021**, *68*, e29304. [CrossRef] [PubMed]
26. Grohar, P.J.; Griffin, L.B.; Yeung, C.; Chen, Q.R.; Pommier, Y.; Khanna, C.; Khan, J.; Helman, L.J. Ecteinascidin 743 interferes with the activity of EWS-FLI1 in Ewing sarcoma cells. *Neoplasia* **2011**, *13*, 145–153. [CrossRef] [PubMed]
27. Boulay, G.; Sandoval, G.J.; Riggi, N.; Iyer, S.; Buisson, R.; Naigles, B.; Awad, M.E.; Rengarajan, S.; Volorio, A.; McBride, M.J.; et al. Cancer-Specific Retargeting of BAF Complexes by a Prion-like Domain. *Cell* **2017**, *171*, 163–178.e19. [CrossRef]
28. Harlow, M.L.; Chasse, M.H.; Boguslawski, E.A.; Sorensen, K.M.; Gedminas, J.M.; Kitchen-Goosen, S.M.; Rothbart, S.B.; Taslim, C.; Lessnick, S.L.; Peck, A.S.; et al. Trabectedin Inhibits EWS-FLI1 and Evicts SWI/SNF from Chromatin in a Schedule dependent Manner. *Clin. Cancer Res.* **2019**, *25*, 3417–3429. [CrossRef]
29. Baruchel, S.; Pappo, A.; Krailo, M.; Baker, K.S.; Wu, B.; Villaluna, D.; Lee-Scott, M.; Adamson, P.C.; Blaney, S.M. A phase 2 trial of trabectedin in children with recurrent rhabdomyosarcoma, Ewing sarcoma and non-rhabdomyosarcoma soft tissue sarcomas: A report from the Children's Oncology Group. *Eur. J. Cancer* **2012**, *48*, 579–585. [CrossRef]
30. Lau, L.; Supko, J.G., Blaney, S.; Hershon, L.; Seibel, N.; Krailo, M.; Qu, W.; Malkin, D.; Jimeno, J.; Bernstein, M.; et al. A phase I and pharmacokinetic study of ecteinascidin-743 (Yondelis) in children with refractory solid tumors. A Children's Oncology Group study. *Clin. Cancer Res.* **2005**, *11*, 672–677. [CrossRef]
31. Grohar, P.J.; Segars, L.E.; Yeung, C.; Pommier, Y.; D'Incalci, M.; Mendoza, A.; Helman, L.J. Dual targeting of EWS-FLI1 activity and the associated DNA damage response with trabectedin and SN38 synergistically inhibits Ewing sarcoma cell growth. *Clin. Cancer Res.* **2014**, *20*, 1190–1203. [CrossRef]
32. Jenuwein, T.; Allis, C.D. Translating the histone code. *Science* **2001**, *293*, 1074–1080. [CrossRef]
33. Chan, H.L.; Beckedorff, F.; Zhang, Y.; Garcia-Huidobro, J.; Jiang, H.; Colaprico, A.; Bilbao, D.; Figueroa, M.E.; LaCava, J.; Shiekhattar, R.; et al. Polycomb complexes associate with enhancers and promote oncogenic transcriptional programs in cancer through multiple mechanisms. *Nat. Commun.* **2018**, *9*, 3377. [CrossRef]
34. Rai, K.; Akdemir, K.C.; Kwong, L.N.; Fiziev, P.; Wu, C.J.; Keung, E.Z.; Sharma, S.; Samant, N.S.; Williams, M.; Axelrad, J.B.; et al. Dual Roles of RNF2 in Melanoma Progression. *Cancer Discov.* **2015**, *5*, 1314–1327. [CrossRef]
35. Sanchez-Molina, S.; Figuerola-Bou, E.; Blanco, E.; Sanchez-Jimenez, M.; Taboas, P.; Gomez, S.; Ballare, C.; Garcia-Dominguez, D.J.; Prada, E.; Hontecillas-Prieto, L.; et al. RING1B recruits EWSR1-FLI1 and cooperates in the remodeling of chromatin necessary for Ewing sarcoma tumorigenesis. *Sci. Adv.* **2020**, *6*, eaba3058. [CrossRef]
36. Wang, S.; Hwang, E.E.; Guha, R.; O'Neill, A.F.; Melong, N.; Veinotte, C.J.; Conway Saur, A.; Wuerthele, K.; Shen, M.; McKnight, C.; et al. High-throughput Chemical Screening Identifies Focal Adhesion Kinase and Aurora Kinase B Inhibition as a Synergistic Treatment Combination in Ewing Sarcoma. *Clin. Cancer Res.* **2019**, *25*, 4552–4566. [CrossRef]
37. Richter, G.H.; Plehm, S.; Fasan, A.; Rossler, S.; Unland, R.; Bennani-Baiti, I.M.; Hotfilder, M.; Lowel, D.; von Luettichau, I.; Mossbrugger, I.; et al. EZH2 is a mediator of EWS/FLI1 driven tumor growth and metastasis blocking endothelial and neuro-ectodermal differentiation. *Proc. Natl. Acad. Sci. USA* **2009**, *106*, 5324–5329. [CrossRef]

38. Riggi, N.; Suva, M.L.; Suva, D.; Cironi, L.; Provero, P.; Tercier, S.; Joseph, J.M.; Stehle, J.C.; Baumer, K.; Kindler, V.; et al. EWS-FLI-1 expression triggers a Ewing's sarcoma initiation program in primary human mesenchymal stem cells. *Cancer Res.* **2008**, *68*, 2176–2185. [CrossRef]
39. Tanaka, M.; Yamazaki, Y.; Kanno, Y.; Igarashi, K.; Aisaki, K.; Kanno, J.; Nakamura, T. Ewing's sarcoma precursors are highly enriched in embryonic osteochondrogenic progenitors. *J. Clin. Investig.* **2014**, *124*, 3061–3074. [CrossRef]
40. Kurmasheva, R.T.; Sammons, M.; Favours, E.; Wu, J.; Kurmashev, D.; Cosmopoulos, K.; Keilhack, H.; Klaus, C.R.; Houghton, P.J.; Smith, M.A. Initial testing (stage 1) of tazemetostat (EPZ-6438), a novel EZH2 inhibitor, by the Pediatric Preclinical Testing Program. *Pediatr. Blood Cancer* **2017**, *64*, e26218. [CrossRef]
41. Kailayangiri, S.; Altvater, B.; Lesch, S.; Balbach, S.; Gottlich, C.; Kuhnemundt, J.; Mikesch, J.H.; Schelhaas, S.; Jamitzky, S.; Meltzer, J.; et al. EZH2 Inhibition in Ewing Sarcoma Upregulates GD2 Expression for Targeting with Gene-Modified T Cells. *Mol. Ther.* **2019**, *27*, 933–946. [CrossRef]
42. Garcia-Dominguez, D.J.; Hajji, N.; Lopez-Alemany, R.; Sanchez-Molina, S.; Figuerola-Bou, E.; Moron Civanto, F.J.; Rello-Varona, S.; Andres-Leon, E.; Benito, A.; Keun, H.C.; et al. Selective histone methyltransferase G9a inhibition reduces metastatic development of Ewing sarcoma through the epigenetic regulation of NEU1. *Oncogene* **2022**, *41*, 2638–2650. [CrossRef] [PubMed]
43. Sakimura, R.; Tanaka, K.; Nakatani, F.; Matsunobu, T.; Li, X.; Hanada, M.; Okada, T.; Nakamura, T.; Matsumoto, Y.; Iwamoto, Y. Antitumor effects of histone deacetylase inhibitor on Ewing's family tumors. *Int. J. Cancer* **2005**, *116*, 784–792. [CrossRef] [PubMed]
44. Jaboin, J.; Wild, J.; Hamidi, H.; Khanna, C.; Kim, C.J.; Robey, R.; Bates, S.E.; Thiele, C.J. MS-27-275, an inhibitor of histone deacetylase, has marked in vitro and in vivo antitumor activity against pediatric solid tumors. *Cancer Res.* **2002**, *62*, 6108–6115.
45. Sonnemann, J.; Dreyer, L.; Hartwig, M.; Palani, C.D.; Hongle, T.T.; Klier, U.; Broker, B.; Volker, U.; Beck, J.F. Histone deacetylase inhibitors induce cell death and enhance the apoptosis-inducing activity of TRAIL in Ewing's sarcoma cells. *J. Cancer Res. Clin. Oncol.* **2007**, *133*, 847–858. [CrossRef] [PubMed]
46. Keshelava, N.; Houghton, P.J.; Morton, C.L.; Lock, R.B.; Carol, H.; Keir, S.T.; Maris, J.M.; Reynolds, C.P.; Gorlick, R.; Kolb, E.A.; et al. Initial testing (stage 1) of vorinostat (SAHA) by the pediatric preclinical testing program. *Pediatr. Blood Cancer* **2009**, *53*, 505–508. [CrossRef]
47. Ma, Y.; Baltezor, M.; Rajewski, L.; Crow, J.; Samuel, G.; Staggs, V.S.; Chastain, K.M.; Toretsky, J.A.; Weir, S.J.; Godwin, A.K. Targeted inhibition of histone deacetylase leads to suppression of Ewing sarcoma tumor growth through an unappreciated EWS-FLI1/HDAC3/HSP90 signaling axis. *J. Mol. Med.* **2019**, *97*, 957–972. [CrossRef]
48. Garcia-Dominguez, D.J.; Hajji, N.; Sanchez-Molina, S.; Figuerola-Bou, E.; de Pablos, R.M.; Espinosa-Oliva, A.M.; Andres-Leon, E.; Terron-Camero, L.C.; Flores-Campos, R.; Pascual-Pasto, G.; et al. Selective inhibition of HDAC6 regulates expression of the oncogenic driver EWSR1-FLI1 through the EWSR1 promoter in Ewing sarcoma. *Oncogene* **2021**, *40*, 5843–5853. [CrossRef]
49. Schmidt, O.; Nehls, N.; Prexler, C.; von Heyking, K.; Groll, T.; Pardon, K.; Garcia, H.D.; Hensel, T.; Gurgen, D.; Henssen, A.G.; et al. Class I histone deacetylases (HDAC) critically contribute to Ewing sarcoma pathogenesis. *J. Exp. Clin. Cancer Res.* **2021**, *40*, 322. [CrossRef]
50. Garcia-Dominguez, D.J.; Hontecillas-Prieto, L.; Rodriguez-Nunez, P.; Pascual-Pasto, G.; Vila-Ubach, M.; Garcia-Mejias, R.; Robles, M.J.; Tirado, O.M.; Mora, J.; Carcaboso, A.M.; et al. The combination of epigenetic drugs SAHA and HCI-2509 synergistically inhibits EWS-FLI1 and tumor growth in Ewing sarcoma. *Oncotarget* **2018**, *9*, 31397–31410. [CrossRef]
51. Welch, D.; Kahen, E.; Fridley, B.; Brohl, A.S.; Cubitt, C.L.; Reed, D.R. Small molecule inhibition of lysine-specific demethylase 1 (LSD1) and histone deacetylase (HDAC) alone and in combination in Ewing sarcoma cell lines. *PLoS ONE* **2019**, *14*, e0222228. [CrossRef]
52. Pedot, G.; Marques, J.G.; Ambuhl, P.P.; Wachtel, M.; Kasper, S.; Ngo, Q.A.; Niggli, F.K.; Schafer, B.W. Inhibition of HDACs reduces Ewing sarcoma tumor growth through EWS-FLI1 protein destabilization. *Neoplasia* **2022**, *27*, 100784. [CrossRef]
53. Parrish, J.K.; Sechler, M.; Winn, R.A.; Jedlicka, P. The histone demethylase KDM3A is a microRNA-22-regulated tumor promoter in Ewing Sarcoma. *Oncogene* **2015**, *34*, 257–262. [CrossRef]
54. Sechler, M.; Parrish, J.K.; Birks, D.K.; Jedlicka, P. The histone demethylase KDM3A, and its downstream target MCAM, promote Ewing Sarcoma cell migration and metastasis. *Oncogene* **2017**, *36*, 4150–4160. [CrossRef]
55. Parrish, J.K.; McCann, T.S.; Sechler, M.; Sobral, L.M.; Ren, W.; Jones, K.L.; Tan, A.C.; Jedlicka, P. The Jumonji-domain histone demethylase inhibitor JIB-04 deregulates oncogenic programs and increases DNA damage in Ewing Sarcoma, resulting in impaired cell proliferation and survival, and reduced tumor growth. *Oncotarget* **2018**, *9*, 33110–33123. [CrossRef]
56. Heisey, D.A.R.; Jacob, S.; Lochmann, T.L.; Kurupi, R.; Ghotra, M.S.; Calbert, M.L.; Shende, M.; Maves, Y.K.; Koblinski, J.E.; Dozmorov, M.G.; et al. Pharmaceutical Interference of the EWS-FLI1-driven Transcriptome By Cotargeting H3K27ac and RNA Polymerase Activity in Ewing Sarcoma. *Mol. Cancer* **2021**, *20*, 1868–1879. [CrossRef]
57. Alqahtani, A.; Choucair, K.; Ashraf, M.; Hammouda, D.M.; Alloghbi, A.; Khan, T.; Senzer, N.; Nemunaitis, J. Bromodomain and extra-terminal motif inhibitors: A review of preclinical and clinical advances in cancer therapy. *Future Sci. OA* **2019**, *5*, FSO372. [CrossRef]
58. Shorstova, T.; Foulkes, W.D.; Witcher, M. Achieving clinical success with BET inhibitors as anti-cancer agents. *Br. J. Cancer* **2021**, *124*, 1478–1490. [CrossRef]
59. Bid, H.K.; Phelps, D.A.; Xaio, L.; Guttridge, D.C.; Lin, J.; London, C.; Baker, L.H.; Mo, X.; Houghton, P.J. The Bromodomain BET Inhibitor JQ1 Suppresses Tumor Angiogenesis in Models of Childhood Sarcoma. *Mol. Cancer* **2016**, *15*, 1018–1028. [CrossRef]

60. Gollavilli, P.N.; Pawar, A.; Wilder-Romans, K.; Natesan, R.; Engelke, C.G.; Dommeti, V.L.; Krishnamurthy, P.M.; Nallasivam, A.; Apel, I.J.; Xu, T.; et al. EWS/ETS-Driven Ewing Sarcoma Requires BET Bromodomain Proteins. *Cancer Res.* **2018**, *78*, 4760–4773. [CrossRef]
61. Zhang, J.; Huang, D.; Saw, P.E.; Song, E. Turning cold tumors hot: From molecular mechanisms to clinical applications. *Trends. Immunol.* **2022**, *43*, 523–545. [CrossRef]
62. Altvater, B.; Kailayangiri, S.; Theimann, N.; Ahlmann, M.; Farwick, N.; Chen, C.; Pscherer, S.; Neumann, I.; Mrachatz, G.; Hansmeier, A.; et al. Common Ewing sarcoma-associated antigens fail to induce natural T cell responses in both patients and healthy individuals. *Cancer Immunol. Immunother. CII* **2014**, *63*, 1047–1060. [CrossRef] [PubMed]
63. Berghuis, D.; de Hooge, A.S.; Santos, S.J.; Horst, D.; Wiertz, E.J.; van Eggermond, M.C.; van den Elsen, P.J.; Taminiau, A.H.; Ottaviano, L.; Schaefer, K.L.; et al. Reduced human leukocyte antigen expression in advanced-stage Ewing sarcoma: Implications for immune recognition. *J. Pathol.* **2009**, *218*, 222–231. [CrossRef] [PubMed]
64. Spurny, C.; Kailayangiri, S.; Altvater, B.; Jamitzky, S.; Hartmann, W.; Wardelmann, E.; Ranft, A.; Dirksen, U.; Amler, S.; Hardes, J.; et al. T cell infiltration into Ewing sarcomas is associated with local expression of immune-inhibitory HLA-G. *Oncotarget* **2018**, *9*, 6536–6549. [CrossRef] [PubMed]
65. Long, A.H.; Highfill, S.L.; Cui, Y.; Smith, J.P.; Walker, A.J.; Ramakrishna, S.; El-Etriby, R.; Galli, S.; Tsokos, M.G.; Orentas, R.J.; et al. Reduction of MDSCs with All-trans Retinoic Acid Improves CAR Therapy Efficacy for Sarcomas. *Cancer Immunol. Res.* **2016**, *4*, 869–880. [CrossRef] [PubMed]
66. Binnewies, M.; Roberts, E.W.; Kersten, K.; Chan, V.; Fearon, D.F.; Merad, M.; Coussens, L.M.; Gabrilovich, D.I.; Ostrand-Rosenberg, S.; Hedrick, C.C.; et al. Understanding the tumor immune microenvironment (TIME) for effective therapy. *Nat. Med.* **2018**, *24*, 541–550. [CrossRef] [PubMed]
67. Zhang, Y.; Zheng, J. Functions of Immune Checkpoint Molecules Beyond Immune Evasion. *Adv. Exp. Med. Biol.* **2020**, *1248*, 201–226. [CrossRef]
68. Robert, C.; Schachter, J.; Long, G.V.; Arance, A.; Grob, J.J.; Mortier, L.; Daud, A.; Carlino, M.S.; McNeil, C.; Lotem, M.; et al. Pembrolizumab versus Ipilimumab in Advanced Melanoma. *N. Engl. J. Med.* **2015**, *372*, 2521–2532. [CrossRef]
69. Weber, J.S.; D'Angelo, S.P.; Minor, D.; Hodi, F.S.; Gutzmer, R.; Neyns, B.; Hoeller, C.; Khushalani, N.I.; Miller, W.H., Jr.; Lao, C.D.; et al. Nivolumab versus chemotherapy in patients with advanced melanoma who progressed after anti-CTLA-4 treatment (CheckMate 037): A randomised, controlled, open-label, phase 3 trial. *Lancet Oncol.* **2015**, *16*, 375–384. [CrossRef]
70. Merchant, M.S.; Wright, M.; Baird, K.; Wexler, L.H.; Rodriguez-Galindo, C.; Bernstein, D.; Delbrook, C.; Lodish, M.; Bishop, R.; Wolchok, J.D.; et al. Phase I Clinical Trial of Ipilimumab in Pediatric Patients with Advanced Solid Tumors. *Clin. Cancer Res.* **2016**, *22*, 1364–1370. [CrossRef]
71. Tawbi, H.A.; Burgess, M.; Bolejack, V.; Van Tine, B.A.; Schuetze, S.M.; Hu, J.; D'Angelo, S.; Attia, S.; Riedel, R.F.; Priebat, D.A.; et al. Pembrolizumab in advanced soft-tissue sarcoma and bone sarcoma (SARC028): A multicentre, two-cohort, single-arm, open-label, phase 2 trial. *Lancet Oncol.* **2017**, *18*, 1493–1501. [CrossRef]
72. D'Angelo, S.P.; Mahoney, M.R.; Van Tine, B.A.; Atkins, J.; Milhem, M.M.; Jahagirdar, B.N.; Antonescu, C.R.; Horvath, E.; Tap, W.D.; Schwartz, G.K.; et al. Nivolumab with or without ipilimumab treatment for metastatic sarcoma (Alliance A091401): Two open-label, non-comparative, randomised, phase 2 trials. *Lancet Oncol.* **2018**, *19*, 416–426. [CrossRef]
73. Rizvi, N.A.; Hellmann, M.D.; Snyder, A.; Kvistborg, P.; Makarov, V.; Havel, J.J.; Lee, W.; Yuan, J.; Wong, P.; Ho, T.S.; et al. Cancer immunology. Mutational landscape determines sensitivity to PD-1 blockade in non-small cell lung cancer. *Science* **2015**, *348*, 124–128. [CrossRef]
74. McGranahan, N.; Furness, A.J.; Rosenthal, R.; Ramskov, S.; Lyngaa, R.; Saini, S.K.; Jamal-Hanjani, M.; Wilson, G.A.; Birkbak, N.J.; Hiley, C.T.; et al. Clonal neoantigens elicit T cell immunoreactivity and sensitivity to immune checkpoint blockade. *Science* **2016**, *351*, 1463–1469. [CrossRef]
75. Machado, I.; Lopez-Guerrero, J.A.; Scotlandi, K.; Picci, P.; Llombart-Bosch, A. Immunohistochemical analysis and prognostic significance of PD-L1, PD-1, and CD8+ tumor-infiltrating lymphocytes in Ewing's sarcoma family of tumors (ESFT). *Virchows Arch.* **2018**, *472*, 815–824. [CrossRef]
76. Kim, C.; Kim, E.K.; Jung, H.; Chon, H.J.; Han, J.W.; Shin, K.H.; Hu, H.; Kim, K.S.; Choi, Y.D.; Kim, S.; et al. Prognostic implications of PD-L1 expression in patients with soft tissue sarcoma. *BMC Cancer* **2016**, *16*, 434. [CrossRef]
77. Lapeyre-Prost, A.; Terme, M.; Pernot, S.; Pointet, A.L.; Voron, T.; Tartour, E.; Taieb, J. Immunomodulatory Activity of VEGF in Cancer. *Int. Rev. Cell Mol. Biol.* **2017**, *330*, 295–342. [CrossRef]
78. Wilky, B.A.; Trucco, M.M.; Subhawong, T.K.; Florou, V.; Park, W.; Kwon, D.; Wieder, E.D.; Kolonias, D.; Rosenberg, A.E.; Kerr, D.A.; et al. Axitinib plus pembrolizumab in patients with advanced sarcomas including alveolar soft-part sarcoma: A single-centre, single-arm, phase 2 trial. *Lancet Oncol.* **2019**, *20*, 837–848. [CrossRef]
79. Chua-Alcala, V.S.; Kim, K.; Assudani, N.; Al-Shihabi, A.; Quon, D.; Wong, S.; Chawla, S.P.; Gordon, E.M. Initial results of a phase III investigation of safety/efficacy of nivolumab and ABI-009 (nab-sirolimus) in advanced undifferentiated pleomorphic sarcoma (UPS), liposarcoma (LPS), chondrosarcoma (CS), osteosarcoma (OS), and Ewing sarcoma. *J. Clin. Oncol.* **2019**, *37*, 21. [CrossRef]
80. Charych, D.H.; Hoch, U.; Langowski, J.L.; Lee, S.R.; Addepalli, M.K.; Kirk, P.B.; Sheng, D.; Liu, X.; Sims, P.W.; VanderVeen, L.A.; et al. NKTR-214, an Engineered Cytokine with Biased IL2 Receptor Binding, Increased Tumor Exposure, and Marked Efficacy in Mouse Tumor Models. *Clin. Cancer Res.* **2016**, *22*, 680–690. [CrossRef]

81. Sharma, M.; Khong, H.; Fa'ak, F.; Bentebibel, S.E.; Janssen, L.M.E.; Chesson, B.C.; Creasy, C.A.; Forget, M.A.; Kahn, L.M.S.; Pazdrak, B.; et al. Bempegaldesleukin selectively depletes intratumoral Tregs and potentiates T cell-mediated cancer therapy. *Nat. Commun.* **2020**, *11*, 661. [CrossRef]
82. Wang, L.; Zhang, Q.; Chen, W.; Shan, B.; Ding, Y.; Zhang, G.; Cao, N.; Liu, L.; Zhang, Y. B7-H3 is overexpressed in patients suffering osteosarcoma and associated with tumor aggressiveness and metastasis. *PLoS ONE* **2013**, *8*, e70689. [CrossRef] [PubMed]
83. Paijens, S.T.; Vledder, A.; de Bruyn, M.; Nijman, H.W. Tumor-infiltrating lymphocytes in the immunotherapy era. *Cell. Mol. Immunol.* **2021**, *18*, 842–859. [CrossRef] [PubMed]
84. Yabe, H.; Tsukahara, T.; Kawaguchi, S.; Wada, T.; Torigoe, T.; Sato, N.; Terai, C.; Aoki, M.; Hirose, S.; Morioka, H.; et al. Prognostic significance of HLA class I expression in Ewing's sarcoma family of tumors. *J. Surg. Oncol.* **2011**, *103*, 380–385. [CrossRef] [PubMed]
85. Mullinax, J.E.; Hall, M.; Beatty, M.; Weber, A.M.; Sannasardo, Z.; Svrdlin, T.; Hensel, J.; Bui, M.; Richards, A.; Gonzalez, R.J.; et al. Expanded Tumor-infiltrating Lymphocytes From Soft Tissue Sarcoma Have Tumor-specific Function. *J. Immunother.* **2021**, *44*, 63–70. [CrossRef] [PubMed]
86. Nielsen, M.; Monberg, T.; Albieri, B.; Sundvold, V.; Rekdal, O.; Junker, N.; Svane, I.M. LTX-315 and adoptive cell therapy using tumor-infiltrating lymphocytes in patients with metastatic soft tissue sarcoma. *J. Clin. Oncol.* **2022**, *40*, 11567. [CrossRef]
87. Von Heyking, K.; Calzada-Wack, J.; Gollner, S.; Neff, F.; Schmidt, O.; Hensel, T.; Schirmer, D.; Fasan, A.; Esposito, I.; Muller-Tidow, C.; et al. The endochondral bone protein CHM1 sustains an undifferentiated, invasive phenotype, promoting lung metastasis in Ewing sarcoma. *Mol. Oncol.* **2017**, *11*, 1288–1301. [CrossRef]
88. Thiel, U.; Schober, S.J.; Einspieler, I.; Kirschner, A.; Thiede, M.; Schirmer, D.; Gall, K.; Blaeschke, F.; Schmidt, O.; Jabar, S.; et al. Ewing sarcoma partial regression without GvHD by chondromodulin-I/HLA-A*02:01-specific allorestricted T cell receptor transgenic T cells. *Oncoimmunology* **2017**, *6*, e1312239. [CrossRef]
89. Thiel, U.; Wawer, A.; von Luettichau, I.; Bender, H.U.; Blaeschke, F.; Grunewald, T.G.; Steinborn, M.; Roper, B.; Bonig, H.; Klingebiel, T.; et al. Bone marrow involvement identifies a subgroup of advanced Ewing sarcoma patients with fatal outcome irrespective of therapy in contrast to curable patients with multiple bone metastases but unaffected marrow. *Oncotarget* **2016**, *7*, 70959–70968. [CrossRef]
90. D'Angelo, S.P.; Melchiori, L.; Merchant, M.S.; Bernstein, D.; Glod, J.; Kaplan, R.; Grupp, S.; Tap, W.D.; Chagin, K.; Binder, G.K.; et al. Antitumor Activity Associated with Prolonged Persistence of Adoptively Transferred NY-ESO-1 (c259)T Cells in Synovial Sarcoma. *Cancer Discov.* **2018**, *8*, 944–957. [CrossRef]
91. Robbins, P.F.; Morgan, R.A.; Feldman, S.A.; Yang, J.C.; Sherry, R.M.; Dudley, M.E.; Wunderlich, J.R.; Nahvi, A.V.; Helman, L.J.; Mackall, C.L.; et al. Tumor regression in patients with metastatic synovial cell sarcoma and melanoma using genetically engineered lymphocytes reactive with NY-ESO-1. *J. Clin. Oncol.* **2011**, *29*, 917–924. [CrossRef]
92. Feins, S.; Kong, W.; Williams, E.F.; Milone, M.C.; Fraietta, J.A. An introduction to chimeric antigen receptor (CAR) T-cell immunotherapy for human cancer. *Am. J. Hematol.* **2019**, *94*, S3–S9. [CrossRef]
93. Brentjens, R.J.; Davila, M.L.; Riviere, I.; Park, J.; Wang, X.; Cowell, L.G.; Bartido, S.; Stefanski, J.; Taylor, C.; Olszewska, M.; et al. CD19-targeted T cells rapidly induce molecular remissions in adults with chemotherapy-refractory acute lymphoblastic leukemia. *Sci. Transl. Med.* **2013**, *5*, 177ra138. [CrossRef]
94. Maude, S.L.; Frey, N.; Shaw, P.A.; Aplenc, R.; Barrett, D.M.; Bunin, N.J.; Chew, A.; Gonzalez, V.E.; Zheng, Z.; Lacey, S.F.; et al. Chimeric antigen receptor T cells for sustained remissions in leukemia. *N. Engl. J. Med.* **2014**, *371*, 1507–1517. [CrossRef]
95. Brown, C.E.; Mackall, C.L. CAR T cell therapy: Inroads to response and resistance. *Nat. Rev. Immunol.* **2019**, *19*, 73–74. [CrossRef]
96. Shah, N.N.; Fry, T.J. Mechanisms of resistance to CAR T cell therapy. *Nat. Rev. Clin. Oncol.* **2019**, *16*, 372–385. [CrossRef]
97. Dobrenkov, K.; Ostrovnaya, I.; Gu, J.; Cheung, I.Y.; Cheung, N.K. Oncotargets GD2 and GD3 are highly expressed in sarcomas of children, adolescents, and young adults. *Pediatr. Blood Cancer* **2016**, *63*, 1780–1785. [CrossRef]
98. Kailayangiri, S.; Altvater, B.; Meltzer, J.; Pscherer, S.; Luecke, A.; Dierkes, C.; Titze, U.; Leuchte, K.; Landmeier, S.; Hotfilder, M.; et al. The ganglioside antigen G(D2) is surface-expressed in Ewing sarcoma and allows for MHC-independent immune targeting. *Br. J. Cancer* **2012**, *106*, 1123–1133. [CrossRef]
99. Yu, A.L.; Gilman, A.L.; Ozkaynak, M.F.; London, W.B.; Kreissman, S.G.; Chen, H.X.; Smith, M.; Anderson, B.; Villablanca, J.G.; Matthay, K.K.; et al. Anti-GD2 antibody with GM-CSF, interleukin-2, and isotretinoin for neuroblastoma. *N. Engl. J. Med.* **2010**, *363*, 1324–1334. [CrossRef]
100. Majzner, R.G.; Theruvath, J.L.; Nellan, A.; Heitzeneder, S.; Cui, Y.; Mount, C.W.; Rietberg, S.P.; Linde, M.H.; Xu, P.; Rota, C.; et al. CAR T Cells Targeting B7-H3, a Pan-Cancer Antigen, Demonstrate Potent Preclinical Activity Against Pediatric Solid Tumors and Brain Tumors. *Clin. Cancer Res.* **2019**, *25*, 2560–2574. [CrossRef]
101. Kersting, N.; Kunzler Souza, B.; Araujo Vieira, I.; Pereira Dos Santos, R.; Brufatto Olguins, D.; Jose Gregianin, L.; Tesainer Brunetto, A.; Lunardi Brunetto, A.; Roesler, R.; Brunetto de Farias, C.; et al. Epidermal Growth Factor Receptor Regulation of Ewing Sarcoma Cell Function. *Oncology* **2018**, *94*, 383–393. [CrossRef]
102. Albert, C.M.; Pinto, N.R.; Taylor, M.; Wilson, A.; Rawlings-Rhea, S.; Mgebroff, S.; Brown, C.; Lindgren, C.; Huang, W.; Seidel, K.; et al. STRIvE-01: Phase I study of EGFR806 CAR T-cell immunotherapy for recurrent/refractory solid tumors in children and young adults. *J. Clin. Oncol.* **2022**, *40*, 2541. [CrossRef]
103. Vivier, E.; Ugolini, S.; Blaise, D.; Chabannon, C.; Brossay, L. Targeting natural killer cells and natural killer T cells in cancer. *Nat. Reviews. Immunol.* **2012**, *12*, 239–252. [CrossRef] [PubMed]

104. Cho, D.; Shook, D.R.; Shimasaki, N.; Chang, Y.H.; Fujisaki, H.; Campana, D. Cytotoxicity of activated natural killer cells against pediatric solid tumors. *Clin. Cancer Res.* **2010**, *16*, 3901–3909. [CrossRef] [PubMed]
105. Verhoeven, D.H.; de Hooge, A.S.; Mooiman, E.C.; Santos, S.J.; ten Dam, M.M.; Gelderblom, H.; Melief, C.J.; Hogendoorn, P.C.; Egeler, R.M.; van Tol, M.J.; et al. NK cells recognize and lyse Ewing sarcoma cells through NKG2D and DNAM-1 receptor dependent pathways. *Mol. Immunol.* **2008**, *45*, 3917–3925. [CrossRef] [PubMed]
106. Kailayangiri, S.; Altvater, B.; Spurny, C.; Jamitzky, S.; Schelhaas, S.; Jacobs, A.H.; Wiek, C.; Roellecke, K.; Hanenberg, H.; Hartmann, W.; et al. Targeting Ewing sarcoma with activated and GD2-specific chimeric antigen receptor-engineered human NK cells induces upregulation of immune-inhibitory HLA-G. *Oncoimmunology* **2017**, *6*, e1250050. [CrossRef]
107. Thakar, M.S.; Browning, M.; Hari, P.; Charlson, J.A.; Margolis, D.A.; Logan, B.; Schloemer, N.; Kelly, M.E.; Newman, A.; Johnson, B.; et al. Phase II trial using haploidentical hematopoietic cell transplantation (HCT) followed by donor natural killer (NK) cell infusion and sirolimus maintenance for patients with high-risk solid tumors. *J. Clin. Oncol.* **2020**, *38*, e23551. [CrossRef]
108. Denman, C.J.; Senyukov, V.V.; Somanchi, S.S.; Phatarpekar, P.V.; Kopp, L.M.; Johnson, J.L.; Singh, H.; Hurton, L.; Maiti, S.N.; Huls, M.H.; et al. Membrane-bound IL-21 promotes sustained ex vivo proliferation of human natural killer cells. *PLoS ONE* **2012**, *7*, e30264. [CrossRef]
109. Nayyar, G.; Chu, Y.; Cairo, M.S. Overcoming Resistance to Natural Killer Cell Based Immunotherapies for Solid Tumors. *Front. Oncol.* **2019**, *9*, 51. [CrossRef]
110. Early Breast Cancer Trialists' Collaborative Group. Trastuzumab for early-stage, HER2-positive breast cancer: A meta-analysis of 13 864 women in seven randomised trials. *Lancet Oncol.* **2021**, *22*, 1139–1150. [CrossRef]
111. Weiner, G.J. Monoclonal antibody mechanisms of action in cancer. *Immunol. Res.* **2007**, *39*, 271–278. [CrossRef]
112. O'Neill, A.; Shah, N.; Zitomersky, N.; Ladanyi, M.; Shukla, N.; Uren, A.; Loeb, D.; Toretsky, J. Insulin-like growth factor 1 receptor as a therapeutic target in ewing sarcoma: Lack of consistent upregulation or recurrent mutation and a review of the clinical trial literature. *Sarcoma* **2013**, *2013*, 450478. [CrossRef]
113. Toretsky, J.A.; Kalebic, T.; Blakesley, V.; LeRoith, D.; Helman, L.J. The insulin-like growth factor-I receptor is required for EWS/FLI-1 transformation of fibroblasts. *J. Biol. Chem.* **1997**, *272*, 30822–30827. [CrossRef]
114. Hamilton, G.; Mallinger, R.; Hofbauer, S.; Havel, M. The monoclonal HBA-71 antibody modulates proliferation of thymocytes and Ewing's sarcoma cells by interfering with the action of insulin-like growth factor I. *Thymus* **1991**, *18*, 33–41.
115. Scotlandi, K.; Benini, S.; Nanni, P.; Lollini, P.L.; Nicoletti, G.; Landuzzi, L.; Serra, M.; Manara, M.C.; Picci, P.; Baldini, N. Blockage of insulin like growth factor-I receptor inhibits the growth of Ewing's sarcoma in athymic mice. *Cancer Res.* **1998**, *58*, 4127–4131.
116. Juergens, H.; Daw, N.C.; Geoerger, B.; Ferrari, S.; Villarroel, M.; Aerts, I.; Whelan, J.; Dirksen, U.; Hixon, M.L.; Yin, D.; et al. Preliminary efficacy of the anti-insulin-like growth factor type 1 receptor antibody figitumumab in patients with refractory Ewing sarcoma. *J. Clin. Oncol.* **2011**, *29*, 4534–4540. [CrossRef]
117. Malempati, S.; Weigel, B.; Ingle, A.M.; Ahern, C.H., Carroll, J.M.; Roberts, C.T.; Reid, J.M.; Schmechel, S.; Voss, S.D.; Cho, S.Y.; et al. Phase I/II trial and pharmacokinetic study of cixutumumab in pediatric patients with refractory solid tumors and Ewing sarcoma: A report from the Children's Oncology Group. *J. Clin. Oncol.* **2012**, *30*, 256–262. [CrossRef]
118. Pappo, A.S.; Patel, S.R.; Crowley, J.; Reinke, D.K.; Kuenkele, K.P.; Chawla, S.P.; Toner, G.C.; Maki, R.C.; Meyers, P.A.; Chugh, R.; et al. R1507, a monoclonal antibody to the insulin-like growth factor 1 receptor, in patients with recurrent or refractory Ewing sarcoma family of tumors: Results of a phase II Sarcoma Alliance for Research through Collaboration study. *J. Clin. Oncol.* **2011**, *29*, 4541–4547. [CrossRef]
119. Zollner, S.K.; Amatruda, J.F.; Bauer, S.; Collaud, S.; de Alava, E.; DuBois, S.G.; Hardes, J.; Hartmann, W.; Kovar, H.; Metzler, M.; et al. Ewing Sarcoma-Diagnosis, Treatment, Clinical Challenges and Future Perspectives. *J. Clin. Med.* **2021**, *10*, 1685. [CrossRef]
120. Shulman, D.S.; Merriam, P.; Choy, E.; Guenther, L.M.; Cavanaugh, K.; Kao, P.-C.; Posner, A.; Fairchild, G.; Barker, E.; Stegmaier, K.; et al. Phase 2 trial of palbociclib and ganitumab in patients with relapsed Ewing sarcoma. *J. Clin. Oncol.* **2022**, *40*, e23507. [CrossRef]
121. Casey, D.L.; Lin, T.Y.; Cheung, N.V. Exploiting Signaling Pathways and Immune Targets Beyond the Standard of Care for Ewing Sarcoma. *Front. Oncol.* **2019**, *9*, 537. [CrossRef] [PubMed]
122. Wagner, L.; Turpin, B.; Nagarajan, R.; Weiss, B.; Cripe, T.; Geller, J. Pilot study of vincristine, oral irinotecan, and temozolomide (VOIT regimen) combined with bevacizumab in pediatric patients with recurrent solid tumors or brain tumors. *Pediatr. Blood Cancer* **2013**, *60*, 1447–1451. [CrossRef] [PubMed]
123. Lowery, C.D.; Blosser, W.; Dowless, M.; Renschler, M.; Perez, L.V.; Stephens, J.; Pytowski, B.; Wasserstrom, H.; Stancato, L.F.; Falcon, B. Anti-VEGFR2 therapy delays growth of preclinical pediatric tumor models and enhances anti-tumor activity of chemotherapy. *Oncotarget* **2019**, *10*, 5523–5533. [CrossRef] [PubMed]
124. Tap, W.D.; Jones, R.L.; Van Tine, B.A.; Chmielowski, B.; Elias, A.D.; Adkins, D.; Agulnik, M.; Cooney, M.M.; Livingston, M.B.; Pennock, G.; et al. Olaratumab and doxorubicin versus doxorubicin alone for treatment of soft-tissue sarcoma: An open-label phase 1b and randomised phase 2 trial. *Lancet* **2016**, *388*, 488–497. [CrossRef]
125. Tap, W.D.; Wagner, A.J.; Schoffski, P.; Martin-Broto, J.; Krarup-Hansen, A.; Ganjoo, K.N.; Yen, C.C.; Abdul Razak, A.R.; Spira, A.; Kawai, A.; et al. Effect of Doxorubicin Plus Olaratumab vs Doxorubicin Plus Placebo on Survival in Patients With Advanced Soft Tissue Sarcomas: The ANNOUNCE Randomized Clinical Trial. *JAMA* **2020**, *323*, 1266–1276. [CrossRef] [PubMed]

126. Attia, S.; Villalobos, V.M.; Hindi, N.; Tine, B.A.V.; Wagner, A.J.; Chmielowski, B.; Levy, D.E.; Ceccarelli, M.; Jones, R.L.; Dickson, M.A. Phase (Ph) 1b/2 evaluation of olaratumab in combination with gemcitabine and docetaxel in advanced soft tissue sarcoma (STS). *J. Clin. Oncol.* **2021**, *39*, 11517. [CrossRef]
127. Schöffski, P.; Bahleda, R.; Wagner, A.J.; Burgess, M.; Junker, N.C.; More, M.; Peterson, P.; Ceccarelli, M.; William, T. Results of an open-label, phase 1a/1b study of olaratumab plus pembrolizumab in patients with unresectable, locally advanced or metastatic soft tissue sarcoma. *Ann. Oncol.* **2021**, *32* (Suppl. 7), S1428–S1457. [CrossRef]
128. Kramer, K.; Pandit-Taskar, N.; Donzelli, M.; Wolden, S.L.; Zanzonico, P.; Humm, J.; Haque, S.; Souweidane, M.M.; Lewis, J.; Lyashchenko, S.K.; et al. Intraventricular radioimmunotherapy targeting B7H3 for CNS malignancies. *J. Clin. Oncol* **2019**, *37*, e13592. [CrossRef]
129. Jones, R.L.; Chawla, S.P.; Attia, S.; Schoffski, P.; Gelderblom, H.; Chmielowski, B.; Le Cesne, A.; Van Tine, B.A.; Trent, J.C.; Patel, S.; et al. A phase 1 and randomized controlled phase 2 trial of the safety and efficacy of the combination of gemcitabine and docetaxel with ontuxizumab (MORAb-004) in metastatic soft-tissue sarcomas. *Cancer* **2019**, *125*, 2445–2454. [CrossRef]
130. Chau, C.H.; Steeg, P.S.; Figg, W.D. Antibody-drug conjugates for cancer. *Lancet* **2019**, *394*, 793–804. [CrossRef]
131. Pardali, E.; van der Schaft, D.W.; Wiercinska, E.; Gorter, A.; Hogendoorn, P.C.; Griffioen, A.W.; ten Dijke, P. Critical role of endoglin in tumor cell plasticity of Ewing sarcoma and melanoma. *Oncogene* **2011**, *30*, 334–345. [CrossRef]
132. Puerto-Camacho, P.; Amaral, A.T.; Lamhamedi-Cherradi, S.E.; Menegaz, B.A.; Castillo-Ecija, H.; Ordonez, J.L.; Dominguez, S.; Jordan-Perez, C.; Diaz-Martin, J.; Romero-Perez, L.; et al. Preclinical Efficacy of Endoglin-Targeting Antibody-Drug Conjugates for the Treatment of Ewing Sarcoma. *Clin. Cancer Res.* **2019**, *25*, 2228–2240. [CrossRef]
133. Fleuren, E.D.; Hillebrandt-Roeffen, M.H.; Flucke, U.E.; Te Loo, D.M.; Boerman, O.C.; van der Graaf, W.T.; Versleijen-Jonkers, Y.M. The role of AXL and the in vitro activity of the receptor tyrosine kinase inhibitor BGB324 in Ewing sarcoma. *Oncotarget* **2014**, *5*, 12753–12768. [CrossRef]
134. Sgouros, G.; Bodei, L.; McDevitt, M.R.; Nedrow, J.R. Radiopharmaceutical therapy in cancer: Clinical advances and challenges. *Nat. Rev. Drug Discov.* **2020**, *19*, 589–608. [CrossRef]
135. Rouleau, C.; Smale, R.; Fu, Y.S.; Hui, G.; Wang, F.; Hutto, E.; Fogle, R.; Jones, C.M.; Krumbholz, R.; Roth, S.; et al. Endosialin is expressed in high grade and advanced sarcomas: Evidence from clinical specimens and preclinical modeling. *Int. J. Oncol.* **2011**, *39*, 73–89. [CrossRef]
136. Norris, R.E.; Fox, E.; Reid, J.M.; Ralya, A.; Liu, X.W.; Minard, C.; Weigel, B.J. Phase 1 trial of ontuxizumab (MORAb-004) in children with relapsed or refractory solid tumors: A report from the Children's Oncology Group Phase 1 Pilot Consortium (ADVL1213). *Pediatr. Blood Cancer* **2018**, *65*, e26944. [CrossRef]
137. Cicone, F.; Denoel, T.; Gnesin, S.; Riggi, N.; Irving, M.; Jakka, G.; Schaefer, N.; Viertl, D.; Coukos, G.; Prior, J.O. Preclinical Evaluation and Dosimetry of [(111)In]CHX-DTPA-scFv78-Fc Targeting Endosialin/Tumor Endothelial Marker 1 (TEM1). *Mol. Imaging Biol.* **2020**, *22*, 979–991. [CrossRef]
138. Fedorova, L.; Mudry, P.; Pilatova, K.; Selingerova, I.; Merhautova, J.; Rehak, Z.; Valik, D.; Hlavackova, E.; Cerna, D.; Faberova, L.; et al. Assessment of Immune Response Following Dendritic Cell-Based Immunotherapy in Pediatric Patients With Relapsing Sarcoma. *Front. Oncol.* **2019**, *9*, 1169. [CrossRef]
139. Ghisoli, M.; Barve, M.; Mennel, R.; Lenarsky, C.; Horvath, S.; Wallraven, G.; Pappen, B.O.; Whiting, S.; Rao, D.; Senzer, N.; et al. Three-year Follow up of GMCSF/bi-shRNA(furin) DNA-transfected Autologous Tumor Immunotherapy (Vigil) in Metastatic Advanced Ewing's Sarcoma. *Mol. Ther.* **2016**, *24*, 1478–1483. [CrossRef] [PubMed]
140. Navya, P.N.; Kaphle, A.; Srinivas, S.P.; Bhargava, S.K.; Rotello, V.M.; Daima, H.K. Current trends and challenges in cancer management and therapy using designer nanomaterials. *Nano Converg.* **2019**, *6*, 23. [CrossRef]
141. Tan, K.X.; Barhoum, A.; Pan, S.; Danquah, M.K. Chapter 5—Risks and toxicity of nanoparticles and nanostructured materials. In *Emerging Applications of Nanoparticles and Architecture Nanostructures*; Barhoum, A., Makhlouf, A.S.H., Eds.; Elsevier: Amsterdam, The Netherlands, 2018; pp. 121–139. [CrossRef]
142. Harish, V.; Tewari, D.; Gaur, M.; Yadav, A.B.; Swaroop, S.; Bechelany, M.; Barhoum, A. Review on Nanoparticles and Nanostructured Materials: Bioimaging, Biosensing, Drug Delivery, Tissue Engineering, Antimicrobial, and Agro-Food Applications. *Nanomaterials* **2022**, *12*, 457. [CrossRef] [PubMed]
143. Jeevanandam, J.; Barhoum, A.; Chan, Y.S.; Dufresne, A.; Danquah, M.K. Review on nanoparticles and nanostructured materials: History, sources, toxicity and regulations. *Beilstein J. Nanotechnol.* **2018**, *9*, 1050–1074. [CrossRef] [PubMed]
144. Kemp, J.A.; Kwon, Y.J. Cancer nanotechnology: Current status and perspectives. *Nano Converg.* **2021**, *8*, 34. [CrossRef] [PubMed]
145. Shi, J.; Kantoff, P.W.; Wooster, R.; Farokhzad, O.C. Cancer nanomedicine: Progress, challenges and opportunities. *Nat. Rev. Cancer* **2017**, *17*, 20–37. [CrossRef]
146. Anu Mary Ealia, S.; Saravanakumar, M.P. A review on the classification, characterisation, synthesis of nanoparticles and their application. *IOP Conf. Ser. Mater. Sci. Eng.* **2017**, *263*, 032019. [CrossRef]
147. Lambert, G.; Bertrand, J.R.; Fattal, E.; Subra, F.; Pinto-Alphandary, H.; Malvy, C.; Auclair, C.; Couvreur, P. EWS fli-1 antisense nanocapsules inhibits ewing sarcoma-related tumor in mice. *Biochem. Biophys. Res. Commun.* **2000**, *279*, 401–406. [CrossRef]
148. Toub, N.; Bertrand, J.R.; Tamaddon, A.; Elhamess, H.; Hillaireau, H.; Maksimenko, A.; Maccario, J.; Malvy, C.; Fattal, E.; Couvreur, P. Efficacy of siRNA nanocapsules targeted against the EWS-Fli1 oncogene in Ewing sarcoma. *Pharm. Res.* **2006**, *23*, 892–900. [CrossRef]

149. Maksimenko, A.; Malvy, C.; Lambert, G.; Bertrand, J.R.; Fattal, E.; Maccario, J.; Couvreur, P. Oligonucleotides targeted against a junction oncogene are made efficient by nanotechnologies. *Pharm. Res.* **2003**, *20*, 1565–1567. [CrossRef]
150. Rao, D.D.; Jay, C.; Wang, Z.; Luo, X.; Kumar, P.; Eysenbach, H.; Ghisoli, M.; Senzer, N.; Nemunaitis, J. Preclinical Justification of pbi-shRNA EWS/FLI1 Lipoplex (LPX) Treatment for Ewing's Sarcoma. *Mol. Ther.* **2016**, *24*, 1412–1422. [CrossRef]
151. Lu, C.; Stewart, D.J.; Lee, J.J.; Ji, L.; Ramesh, R.; Jayachandran, G.; Nunez, M.I.; Wistuba, I.I.; Erasmus, J.J.; Hicks, M.E.; et al. Phase I clinical trial of systemically administered TUSC2(FUS1)-nanoparticles mediating functional gene transfer in humans. *PLoS ONE* **2012**, *7*, e34833. [CrossRef]
152. Moreno, L.; Casanova, M.; Chisholm, J.C.; Berlanga, P.; Chastagner, P.B.; Baruchel, S.; Amoroso, L.; Gallego Melcon, S.; Gerber, N.U.; Bisogno, G.; et al. Phase I results of a phase I/II study of weekly nab-paclitaxel in paediatric patients with recurrent/refractory solid tumours: A collaboration with innovative therapies for children with cancer. *Eur. J. Cancer* **2018**, *100*, 27–34. [CrossRef]
153. Thornton, K.A.; Chen, A.R.; Trucco, M.M.; Shah, P.; Wilky, B.A.; Gul, N.; Carrera-Haro, M.A.; Ferreira, M.F.; Shafique, U.; Powell, J.D.; et al. A dose-finding study of temsirolimus and liposomal doxorubicin for patients with recurrent and refractory bone and soft tissue sarcoma. *Int. J. Cancer* **2013**, *133*, 997–1005. [CrossRef]
154. Esfandiari Nazzaro, E.; Sabei, F.Y.; Vogel, W.K.; Nazari, M.; Nicholson, K.S.; Gafken, P.R.; Taratula, O.; Taratula, O.; Davare, M.A.; Leid, M. Discovery and Validation of a Compound to Target Ewing's Sarcoma. *Pharmaceutics* **2021**, *13*, 1553. [CrossRef]
155. Sabei, F.Y.; Taratula, O.; Albarqi, H.A.; Al-Fatease, A.M.; Moses, A.S.; Demessie, A.A.; Park, Y.; Vogel, W.K.; Esfandiari Nazzaro, E.; Davare, M.A.; et al. A targeted combinatorial therapy for Ewing's sarcoma. *Nanomedicine* **2021**, *37*, 102446. [CrossRef]
156. Wojcik, B.; Sawosz, E.; Szczepaniak, J.; Strojny, B.; Sosnowska, M.; Daniluk, K.; Zielinska-Gorska, M.; Balaban, J.; Chwalibog, A.; Wierzbicki, M. Effects of Metallic and Carbon-Based Nanomaterials on Human Pancreatic Cancer Cell Lines AsPC-1 and BxPC-3. *Int. J. Mol. Sci.* **2021**, *22*, 12100. [CrossRef]
157. Zaritski, A.; Castillo-Ecija, H.; Kumarasamy, M.; Peled, E.; Sverdlov Arzi, R.; Carcaboso, A.M.; Sosnik, A. Selective Accumulation of Galactomannan Amphiphilic Nanomaterials in Pediatric Solid Tumor Xenografts Correlates with GLUT1 Gene Expression. *ACS Appl. Mater. Interfaces* **2019**, *11*, 38483–38496. [CrossRef]
158. Bell, J.B.; Rink, J.S.; Eckerdt, F.; Clymer, J.; Goldman, S.; Thaxton, C.S.; Platanias, L.C. HDL nanoparticles targeting sonic hedgehog subtype medulloblastoma. *Sci. Rep.* **2018**, *8*, 1211. [CrossRef]
159. Baldwin, P.; Likhotvorik, R.; Baig, N.; Cropper, J.; Carlson, R.; Kurmasheva, R.; Sridhar, S. Nanoformulation of Talazoparib Increases Maximum Tolerated Doses in Combination With Temozolomide for Treatment of Ewing Sarcoma. *Front. Oncol.* **2019**, *9*, 1416. [CrossRef]
160. Choy, E.; Butrynski, J.E.; Harmon, D.C.; Morgan, J.A.; George, S.; Wagner, A.J.; D'Adamo, D.; Cote, G.M.; Flamand, Y.; Benes, C.H.; et al. Phase II study of olaparib in patients with refractory Ewing sarcoma following failure of standard chemotherapy. *BMC Cancer* **2014**, *14*, 813. [CrossRef]
161. Fontaine, S.D.; Ashley, G.W.; Houghton, P.J.; Kurmasheva, R.T.; Diolaiti, M.; Ashworth, A.; Peer, C.J.; Nguyen, R.; Figg, W.D., Sr.; Beckford-Vera, D.R.; et al. A Very Long-Acting PARP Inhibitor Suppresses Cancer Cell Growth in DNA Repair-Deficient Tumor Models. *Cancer Res.* **2021**, *81*, 1076–1086. [CrossRef]
162. Kang, M.H.; Wang, J.; Makena, M.R.; Lee, J.S.; Paz, N.; Hall, C.P.; Song, M.M.; Calderon, R.I.; Cruz, R.E.; Hindle, A.; et al. Activity of MM-398, nanoliposomal irinotecan (nal-IRI), in Ewing's family tumor xenografts is associated with high exposure of tumor to drug and high SLFN11 expression. *Clin. Cancer Res.* **2015**, *21*, 1139–1150. [CrossRef]
163. Pascual-Pasto, G.; Castillo-Ecija, H.; Unceta, N.; Aschero, R.; Resa-Pares, C.; Gomez-Caballero, A.; Vila-Ubach, M.; Munoz-Aznar, O.; Sunol, M.; Burgueno, V.; et al. SPARC-mediated long-term retention of nab-paclitaxel in pediatric sarcomas. *J. Control. Release* **2022**, *342*, 81–92. [CrossRef] [PubMed]
164. Amoroso, L.; Castel, V.; Bisogno, G.; Casanova, M.; Marquez-Vega, C.; Chisholm, J.C.; Doz, F.; Moreno, L.; Ruggiero, A.; Gerber, N.U.; et al. Phase II results from a phase I/II study to assess the safety and efficacy of weekly nab-paclitaxel in paediatric patients with recurrent or refractory solid tumours: A collaboration with the European Innovative Therapies for Children with Cancer Network. *Eur. J. Cancer* **2020**, *135*, 89–97. [CrossRef]
165. Oesterheld, J.E.; Reed, D.R.; Setty, B.A.; Isakoff, M.S.; Thompson, P.; Yin, H.; Hayashi, M.; Loeb, D.M.; Smith, T.; Makanji, R.; et al. Phase II trial of gemcitabine and nab-paclitaxel in patients with recurrent Ewing sarcoma: A report from the National Pediatric Cancer Foundation. *Pediatr. Blood Cancer* **2020**, *67*, e28370. [CrossRef] [PubMed]
166. Trucco, M.M.; Meyer, C.F.; Thornton, K.A.; Shah, P.; Chen, A.R.; Wilky, B.A.; Carrera-Haro, M.A.; Boyer, L.C.; Ferreira, M.F.; Shafique, U.; et al. A phase II study of temsirolimus and liposomal doxorubicin for patients with recurrent and refractory bone and soft tissue sarcomas. *Clin. Sarcoma Res.* **2018**, *8*, 21. [CrossRef] [PubMed]
167. Cai, L.; Li, J.; Zhang, X.; Lu, Y.; Wang, J.; Lyu, X.; Chen, Y.; Liu, J.; Cai, H.; Wang, Y.; et al. Gold nano-particles (AuNPs) carrying anti-EBV-miR-BART7-3p inhibit growth of EBV-positive nasopharyngeal carcinoma. *Oncotarget* **2015**, *6*, 7838–7850. [CrossRef] [PubMed]
168. Kovacs, D.; Igaz, N.; Marton, A.; Ronavari, A.; Belteky, P.; Bodai, L.; Spengler, G.; Tiszlavicz, L.; Razga, Z.; Hegyi, P.; et al. Core-shell nanoparticles suppress metastasis and modify the tumour-supportive activity of cancer-associated fibroblasts. *J. Nanobiotechnol.* **2020**, *18*, 18. [CrossRef]

169. De Jong, W.H.; Van Der Ven, L.T.; Sleijffers, A.; Park, M.V.; Jansen, E.H.; Van Loveren, H.; Vandebriel, R.J. Systemic and immunotoxicity of silver nanoparticles in an intravenous 28 days repeated dose toxicity study in rats. *Biomaterials* **2013**, *34*, 8333–8343. [CrossRef]
170. Naumann, J.A.; Widen, J.C.; Jonart, L.A.; Ebadi, M.; Tang, J.; Gordon, D.J.; Harki, D.A.; Gordon, P.M. SN-38 Conjugated Gold Nanoparticles Activated by Ewing Sarcoma Specific mRNAs Exhibit In Vitro and In Vivo Efficacy. *Bioconjug. Chem.* **2018**, *29*, 1111–1118. [CrossRef]
171. Da Silva Ferreira, V.; Eugenio, M.F.C.; Del Nery Dos Santos, E.; de Souza, W.; Sant'Anna, C. Cellular toxicology and mechanism of the response to silver-based nanoparticle exposure in Ewing's sarcoma cells. *Nanotechnology* **2021**, *32*, 115101. [CrossRef]
172. Alhaddad, A.; Adam, M.P.; Botsoa, J.; Dantelle, G.; Perruchas, S.; Gacoin, T.; Mansuy, C.; Lavielle, S.; Malvy, C.; Treussart, F.; et al. Nanodiamond as a vector for siRNA delivery to Ewing sarcoma cells. *Small* **2011**, *7*, 3087–3095. [CrossRef]
173. Kang, H.; Nagy, J.; Triche, T. Abstract 3705: Targeted anticancer drug delivery to Ewing's sarcoma using human anti-CD99 targeted hybrid polymerization liposomal nanoparticles. *Cancer Res.* **2018**, *78*, 3705. [CrossRef]
174. Sakpakdeejaroen, I.; Somani, S.; Laskar, P.; Mullin, M.; Dufes, C. Regression of Melanoma Following Intravenous Injection of Plumbagin Entrapped in Transferrin-Conjugated, Lipid-Polymer Hybrid Nanoparticles. *Int. J. Nanomed.* **2021**, *16*, 2615–2631. [CrossRef]
175. Kang, H.; Nagy, J.; Mitra, S.; Triche, T. Abstract 2875: Targeted therapy of Ewing's sarcoma by human anti CD99 targeted hybrid polymerized liposomal nanoparticles (HPLNs) encapsulating anticancer agents. *Cancer Res.* **2019**, *79*, 2875. [CrossRef]
176. Ginsberg, J.P.; Goodman, P.; Leisenring, W.; Ness, K.K.; Meyers, P.A.; Wolden, S.L.; Smith, S.M.; Stovall, M.; Hammond, S.; Robison, L.L.; et al. Long-term survivors of childhood Ewing sarcoma: Report from the childhood cancer survivor study. *J. Natl. Cancer Inst.* **2010**, *102*, 1272–1283. [CrossRef]
177. Casey, D.L.; Cheung, N.V. Immunotherapy of Pediatric Solid Tumors: Treatments at a Crossroads, with an Emphasis on Antibodies. *Cancer Immunol. Res.* **2020**, *8*, 161–166. [CrossRef]
178. Verduin, M.; Hoeben, A.; De Ruysscher, D.; Vooijs, M. Patient-Derived Cancer Organoids as Predictors of Treatment Response. *Front. Oncol.* **2021**, *11*, 641980. [CrossRef]

cancers

MDPI

Article

T Cells Directed against the Metastatic Driver Chondromodulin-1 in Ewing Sarcoma: Comparative Engineering with CRISPR/Cas9 vs. Retroviral Gene Transfer for Adoptive Transfer

Busheng Xue [1], Kristina von Heyking [1], Hendrik Gassmann [1], Mansour Poorebrahim [1], Melanie Thiede [1], Kilian Schober [2], Josef Mautner [3,4], Julia Hauer [1,5], Jürgen Ruland [4,6,7,8], Dirk H. Busch [2,4,5], Uwe Thiel [1,5,*,†] and Stefan E. G. Burdach [1,5,6,9,10,*,†]

1 Department of Pediatrics, Children's Cancer Research Center, Kinderklinik München Schwabing, School of Medicine, Technical University of Munich, 80804 Munich, Germany
2 Institute for Medical Microbiology, Immunology and Hygiene, School of Medicine, Technical University of Munich, 81674 Munich, Germany
3 Department of Gene Vectors, Helmholtz Centre Munich, 81377 Munich, Germany
4 DZIF, German Center for Infection Research, Partner Site Munich, Germany Institute of Clinical, 81675 Munich, Germany
5 Munich Childhood Health Alliance (CHANCE) e.V, 80337 Munich, Germany
6 DKTK German Cancer Consortium, Partner Site Munich, 81675 Munich, Germany
7 Institute of Chemistry and Pathobiochemistry, TUM School of Medicine, Technical University of Munich, 81675 Munich, Germany
8 Center for Translational Cancer Research (TranslaTUM), 81675 Munich, Germany
9 Translational Pediatric Cancer Research-Institute of Pathology, School of Medicine, Technical University of Munich, 81675 Munich, Germany
10 Department of Molecular Oncology, British Columbia Cancer Research Centre and Academy of Translational Medicine, University of British Columbia, Vancouver, BC V5Z 1L3, Canada
* Correspondence: uwe.thiel@tum.de (U.T.); stefan.burdach@tum.de (S.E.G.B.)
† These authors contributed equally to this work.

Citation: Xue, B.; von Heyking, K.; Gassmann, H.; Poorebrahim, M.; Thiede, M.; Schober, K.; Mautner, J.; Hauer, J.; Ruland, J.; Busch, D.H.; et al. T Cells Directed against the Metastatic Driver Chondromodulin-1 in Ewing Sarcoma: Comparative Engineering with CRISPR/Cas9 vs. Retroviral Gene Transfer for Adoptive Transfer. *Cancers* **2022**, *14*, 5485. https://doi.org/10.3390/cancers14225485

Academic Editor: Jaume Mora

Received: 8 October 2022
Accepted: 4 November 2022
Published: 8 November 2022

Simple Summary: The canonical methods of TCR gene delivery in pre-clinical and clinical applications are based on viral transduction of full-coding sequences, including α and β-chains recognizing tumor-specific antigens and tumor-associated antigens. As the transduced α- and β-chains may mispair with the endogenous α- and β-chains, the resultant new antigen specificities may cause auto-reactivity, potentially leading to graft-versus-host disease. The mispaired TCRs may also lose their function. We assessed the feasibility of endogenous TCR orthotopic replacement with a TCR containing a CHM1 targeting sequence via CRISPR/Cas9, evaluated tumor recognition and cytotoxicity function of the CRISPR/Cas9-engineered T cells; compared the prevention of endogenous TCR expression in CRISPR/Cas9 vs. retrovirally engineered T cells. We show that both engineered T cell products specifically recognize tumor cells and elicit cytotoxicity in vitro, with CRISPR/Cas9 engineered T cells providing a more prolonged cytotoxic activity.

Abstract: Ewing sarcoma (EwS) is a highly malignant sarcoma of bone and soft tissue with early metastatic spread and an age peak in early puberty. The prognosis in advanced stages is still dismal, and the long-term effects of established therapies are severe. Efficacious targeted therapies are urgently needed. Our previous work has provided preliminary safety and efficacy data utilizing T cell receptor (TCR) transgenic T cells, generated by retroviral gene transfer, targeting HLA-restricted peptides on the tumor cell derived from metastatic drivers. Here, we compared T cells engineered with either CRISPR/Cas9 or retroviral gene transfer. Firstly, we confirmed the feasibility of the orthotopic replacement of the endogenous TCR by CRISPR/Cas9 with a TCR targeting our canonical metastatic driver chondromodulin-1 (CHM1). CRISPR/Cas9-engineered T cell products specifically recognized and killed HLA-A*02:01+ EwS cell lines. The efficiency of retroviral transduction was higher compared to CRISPR/Cas9 gene editing. Both engineered T cell products specifically recognized tumor cells and elicited cytotoxicity, with CRISPR/Cas9 engineered T cells providing prolonged cytotoxic activity. In conclusion, T cells engineered with CRISPR/Cas9 could be feasible for immunotherapy of EwS and may have the advantage of more prolonged cytotoxic activity, as compared to T cells engineered with retroviral gene transfer.

Keywords: Ewing sarcoma; chondromodulin-1; immunotherapy; orthotopic TCR replacement; CRISPR/Cas9; retroviral transduction

1. Introduction

Ewing sarcoma (EwS) is a highly malignant bone and soft tissue cancer in children and adolescents characterized by early metastasis [1,2]. Primary EwS is treated by a combination of chemotherapy and surgery and/or radiation [3]. Patients with metastatic and refractory disease have been treated with extended radiation and high-dose chemotherapy, which is efficacious only in subgroups [4–6]. Moreover, current therapies are associated with acute and chronic adverse effects that compromise the quality of life in survivors [1] by chemotherapy-associated myeloid dysplastic syndrome, leukemia, and radiation-associated sarcoma. At the age of 50 years, 60% of childhood cancer survivors either died of long-term treatment effects or suffer from life threatening conditions [7]. No novel therapeutic modalities have been successfully introduced into the standard care of advanced EwS in the last 40 years. The overall survival, thus, remains unsatisfactory, especially in patients suffering metastasis or early relapse who have a 5-year overall survival < 30% [8]; novel therapeutic approaches are in urgent need.

The T cell receptor (TCR) can recognize peptide antigens presented on the cell membrane of the host cells by the histocompatibility complex (MHC)/human leukocyte antigen (HLA) system [9]. TCR is a heterodimer comprised most commonly of an α- and a β-chain [10], or alternatively of a γ and a δ chain [11]. TCR-based adoptive therapy allows the genetic redirection of the T cell specificity. Transduction with viral vectors is the conventional method of antigen-specific TCR insertion by either retro- or the lentivirus particles [12]. However, the random insertion of viruses into the genome raised safety concerns about insertional mutations and tumorigenesis, albeit the latter has been mainly observed in stem cells [13]. An unusual case has been described in T cells, where CAR-T cells originated from a single clone in which lentiviral vector-mediated insertion of the CAR transgene disrupted the TET2 gene and improved the expansion of the therapeutic clone to cause leukemia remission [14].

CRISPR/Cas9 engineered orthotopic TCR replacement leads to accurate α and β chain pairing, and regulation of the transgenic TCR is similar to that of physiological T cells [15]. The canonical methods of TCR gene delivery in pre-clinical and clinical applications are based on viral transduction of full-coding sequences, including α- and β-chains recognizing tumor-specific antigens (TSA) and tumor-associated antigens (TAA) [16,17]. As the transduced α- and β-chains may mispair with the endogenous α- and β-chains, the resultant new antigen specificities may cause auto-reactivity, potentially leading to graft-versus-host disease (GvHD). The mispaired TCRs may also lose their function. Albeit, we did not see relevant GvHD in our patients up to now, the number of patients is low [18], and the mechanisms are not fully understood. Schober et al. [15] at our university established a non-viral TRBC knock-out/TRAC knock-in model, which displayed a TCR regulation pattern very similar to that of a physiological T cell population [19–21].

The study reported here was initiated to compare T cells against the metastatic driver of EwS CHM1 engineered with CRISPR/Cas9 vs. retroviral gene transfer. In general, we asked whether CRISPR/Cas9 engineered T cell receptor insertion to the TRAC locus of CD3+ T cells preserves physiological properties and yields a therapeutic product that is at least as efficacious in immunotherapy of EwS as the product generated by retroviral gene transfer. Specifically, we assessed (1) the feasibility of an orthotopic replacement of the endogenous TCR with a TCR containing a CHM1 targeting sequence, (2) TCR expression, as well as tumor recognition and cytotoxicity function of CRISPR/Cas9-engineered T cells, (3) comparative prevention of endogenous TCR expression in CRISPR/Cas9 vs. retrovirally engineered T cells, and finally (4) characterization of CHM1 as a unique target. Figure 1 is a graphical abstract; the workflow of the study is depicted in Supplementary Figure S1.

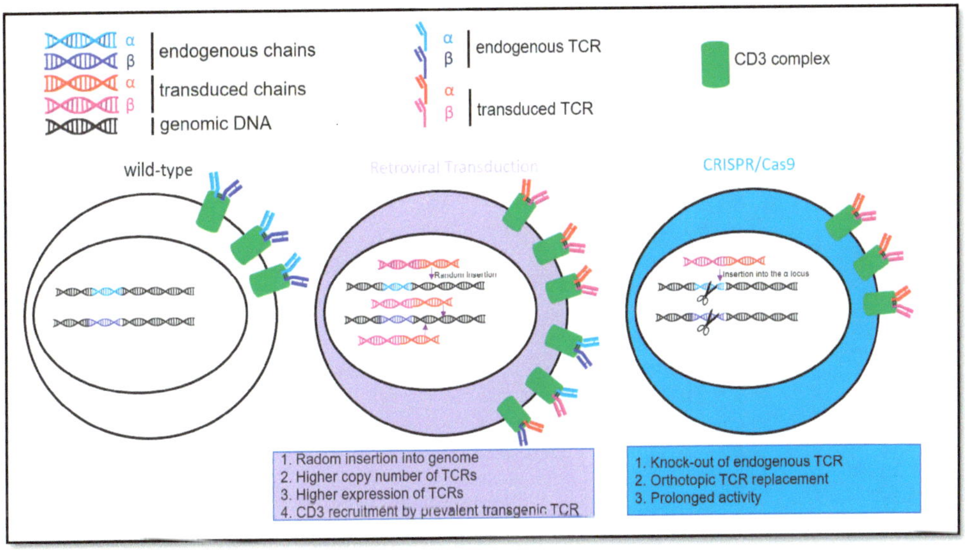

Figure 1. Graphical abstract illustrating the comparison of CRISPR/Cas9-engineered orthotopic TCR replacement or retrovirus transduced random TCR insertion into the T cell genome.

2. Materials and Methods

2.1. Cell Lines

The cell lines used in this work were described previously [22,23]. EwS cell lines and K562 were cultured in RPMI 1640 medium (Life Technologies Limited, Paisley, UK) with 10% fetal bovine serum (FBS) (Life Technologies Limited, Paisley, UK) and serum containing 100 U/mL penicillin and 100 ug/mL streptomycin (Life Technologies Corporation, Grand Island, NY, USA). For LCL and T2 cells, 1 mM Na-pyruvate and 1 mM non-essential amino acids (Life Technologies Limited, Paisley, UK) were also added into the medium. The 293Vec-RD114 packaging cells were cultured in DMEM (Life Technologies Limited, Paisley, UK) with 10% fetal bovine serum (FBS) (Life Technologies Limited, Paisley, UK) containing 100 U/mL penicillin and 100 ug/mL streptomycin (Life Technologies Corporation, Grand Island, NY, USA), and 1 mM Na-pyruvate and 1 mM non-essential amino acids (Life Technologies Limited, Paisley, UK). T cells were cultured in AIM-V (Life Technologies Limited, Darmstadt, Germany) medium with 5% human-AB serum (SIGMA-ALDRICH CHEMIE GmbH, Steinheim, Germany) containing 100 U/mL penicillin and 100 ug/mL streptomycin (Life Technologies Corporation, Grand Island, NY, USA). PBMCs were isolated by density-gradient centrifugation with Ficoll-Paque (GE Healthcare, Uppsala, Sweden), according to supplier's instructions from healthy donor, as described previously [24]. Healthy donor blood samples were purchased from DRK-Blutspendedienst (Baden-Wuerttemberg-Hessen, Ulm, Germany; obtained after IRB approval and informed consent).

2.2. Expansion of TCR-Transgenic T Cells

After being purified with anti-PE magnetic beads (Miltenyi Biotec, Bergisch Gladbach, Germany) coupled to PE-mTCR antibody (Biolegend, San Diego, CA, USA), CHM1^{319}-TCR-transgenic T cells were cultured in 25 cm^2 cell culture flasks (TPP, Trasasingen, Switzerland).

Cells were cultured with 25 mL T cell medium supplemented with OKT3 (50 ng/mL, Biolegend, San Diego, CA, USA), IL-2 (100 IU/mL, R&D Systems, Minneapolis, MI, USA), IL-7 (5 ng/mL, R&D Systems, Minneapolis, MI, USA), and IL-15 (2 ng/mL, R&D Systems, Minneapolis, MI, USA). Interleukins were added every other day. Irradiated LCL (100 Gy, 5×10^6) and PBMCs (peripheral blood mononuclear cells) (30 Gy, 2.5×10^7) pooled from at least three different donors are added as feeder cells.

2.3. Functional Characterization of CHM1^{319}/HLA-A*02:01-Specific TCR Transgenic T Cells

T cell specificity was confirmed with interferon-γ (IFNγ)-Elispot assay (Mabtech AB, Nacka Strand, Sweden), according to the manufacturer's information. A variable of 500 to 1×10^4 T cells were used targeting 2×10^4 EwS; K562 or T2 cells co-cultured for 20 h at 37 °C, 5% CO_2, Elispot Reader (machine and software version 5.0; Advanced Imaging Devices GmbH, Straßberg, Germany) were used for detection.

T cell-mediated cytotoxicity was monitored with the impedance xCELLigence assay (Roche Diagnostics, Penzberg, Germany), allowing continuous measurement of T cell activity against target cell lines, including A673 and SK-N-MC. The 1×10^4 A673 or 3×10^4 SK-N-MC cells were plated 48 h before the addition of 5×10^3 T cells.

2.4. Western Blot

After being washed once with PBS, the product was harvested and solubilized in RIPA lysis buffer (Abcam, ab156034, Waltham, MA USA) with protease inhibitor (Complete mini, Roche Diagnostic, Mannheim, Germany). Protein concentration was determined by BCA (Thermo Fisher Scientific, Ulm, Germany). Ten to fifty micrograms of protein extract were resolved on 10% SDS-PAGE and transferred onto PVDF membrane (Thermo Fisher Scientific, Dreieich, Germany). Primary antibody included PARP (Cell Signaling Technology, Danvers, MA, USA). GAPDH (Santa Cruz Biotechnology, Dallas, TX, USA) served as control. Second antibody used Anti-Rabbit (Santa Cruz Biotechnology, Dallas, TX, USA) and Anti-Mouse antibodies (Santa Cruz Biotechnology, Dallas, TX, USA). Detection was performed with ECL chemiluminescence reagent (Amersham Biosciences, Little Chalfont, UK).

2.5. TCR DNA Template Design

DNA template was synthesized by GeneArt (Life Technologies, Thermo Fisher Scientific, Dreieich, Germany). The DNA structure of CRISPR/Cas9-mediated HDR had 5′ homology arm (300–400 bp pairs), P2A, TCR-β, T2A, TCR-α, and bGHpA tail, as described previously [15]. The related sequences are available as Supplementary Table S1.

2.6. CRISPR/Cas9 Mediated TCR KI and Retrovirus Transduction

For fresh PBMC from buffy coat, CRISPR/Cas9-engineered endogenous TCR KO with or without exogenous TCR insertion was performed two days after T cell activation with CD3/28 Dynabeads™ (Thermo Fisher Scientific Baltics UAB, Vilnius, Lithuania) and 100 IU/mL IL-2, 5 ng/mL IL-7, and 5 ng/mL IL-15 for 2 days. For frozen PBMC, thaw the cells and culture them with T cell medium plus 50 IU/mL IL-2 for one day before activation. The murinized and codon-optimized TCR construct (pMP71_CHM1_mu_opt) in retrovirus group was used, as it was previously [22]. The packaging cell line 293Vec-RD114™ was seeded 24 h before transfection into six well plates. Transfection of plasmid for the production of retrovirus was performed using TransIT-293 (Mirus Bio LLC, Maison, WI, USA).

2.7. Analysis of Published Chip-Sequence Data and Microarray

ChIP-sequence data (GSE61944: GSM1517546, GSM1517547, GSM1517555, GSM1517556, GSM1517569, GSM1517570, GSM15175472, GSM1517573, GSM1517577, GSM1517581) were downloaded from the GEO database, and processed and displayed in the IGV browser [25]. Expression of CHM1 in EwS and bone marrow mesenchymal stem cell was mined from GEO database (GSE17618 and GSE6691), CCLE [26], and ProteomicsDB [27].

2.8. Statistical Analysis

GraphPad Prism (GraphPad Software, San Diego, CA, USA) was used to calculate mean and standard deviation of the mean (SD). Differences between groups were determined using the unpaired two-tailed Student's t-test with p-values < 0.05 being considered statistically significant (* $p < 0.05$; ** $p < 0.01$; *** $p < 0.001$; **** $p < 0.0001$).

3. Results

3.1. Feasibility of Orthotopic Replacement of the Endogenous T Cell Receptor with a T Cell Receptor Containing Chondromodulin-1 Targeting Sequence

3.1.1. CRISPR/Cas9-Engineered Orthotopic TCR Replacement

Based on our previous work [18] on immunotherapy of EwS, we focused on targeting the chondromodulin-1 peptide 319 (CHM1^{319}) VIMPCSWWV. The T cell receptor (TCR) DNA template containing the sequence targeting the CHM1^{319} peptide [22] was established for homology-directed repair (HDR) (Supplementary Figure S2A). We performed PCR to amplify the knock-in (KI) fragment from the right homology arm to the left homology arm (Supplementary Figure S2B) to generate an abundant PCR product.

Next, we accomplished CRISPR/Cas9-engineered knock-out (KO) of the endogenous T cell receptor (hTCR), combined with or without CHM1^{319}-TCR insertion into lymphocytes from peripheral blood mononuclear cells (PBMC). Single α- or β-strand, as well as double-strand KO, result in the loss of endogenous TCR surface expression (Figure 2A). Endogenous TCR KO combined with CHM1^{319}-TCR insertion leads to a T cell population containing a murinized TCR (mTCR), which is hTCR negative, indicating successful CRISPR/Cas9-engineered gene editing. The KO efficacy was approximately 98.5%. In contrast, the KI efficacy in T cells from thawed T cells ranged between 11–23% (21% in Figure 1A, corresponding isotype staining is shown in Supplementary Figure S3A), while the efficiency of KI in fresh T cells reached 45% (Figure 2B, corresponding isotype staining is shown in Supplementary Figure S3B).

The CD3 complex is a heterodimeric glycoprotein cooperating with the TCR to convey signal transduction upon interactions with the antigenic peptides [28]. Upon combined TCR KO and KI, we see CD3 surface expression only in the population, where KI was successful, as indicated by Vβ23 (Vβ23-PE, Figure 2C, right panel, Q2) while the KI negative population remains CD3 negative (Figure 2C, right panel, Q4, corresponding isotype is shown in Supplementary Figure S3C). This indicates that CD3 expression is linked to TCR expression, which reversely indicates a successful CRISPR/Cas9-engineered gene editing.

3.1.2. Tumor Recognition and Cytotoxicity by CRISPR/Cas9-Engineered T Cells

For functional analysis of CRISPR/Cas9-engineered T cells, we assessed T cell activation by IFNγ-Elispot and tumor cell apoptosis (cleaved-PARP) by Western blot. Selection of engineered T cells from six donors was initiated utilizing anti-murine TCR antibody coupled beads. The expression of the murinized TCR sequence (mTCR) was assessed by FACS analysis, yielding a 92.7% homogenous transgenic product in a representative experiment. (Figure 2D, right panel, Q2). CHM1^{319}-restricted TCR transgenic T cells, specifically recognized T2 cells loaded with CHM1^{319} peptide, while T2 cells loaded with an HLA-A*02:0- binding influenza control peptide (FLU) were not recognized (Figure 2E, left panel). Cl-PARP as a parameter of apoptosis was specifically induced by T cells with orthotopic TCR replacement in HLA-A*02:01+ A673 but not in the HLA-A*02:01- SK-N-MC line (Figure 2E, right panel and Supplementary Figure S3D). Some marginal and variable cl-PARP was still seen after co-culture with TCR KO cells.

CRISPR/Cas9-engineered T cells secreted IFNγ when co-cultured with the HLA-A*02:01+ A673 and TC-71 EwS cell lines. In contrast, when co-cultured with HLA-A*02:01- cell lines SB-KMS-KS or SK-N-MC, only marginal IFNγ release was observed (Figure 2F). These findings indicate that CRISPR/Cas9-engineered T cells caused specific HLA-restricted EwS cell line recognition.

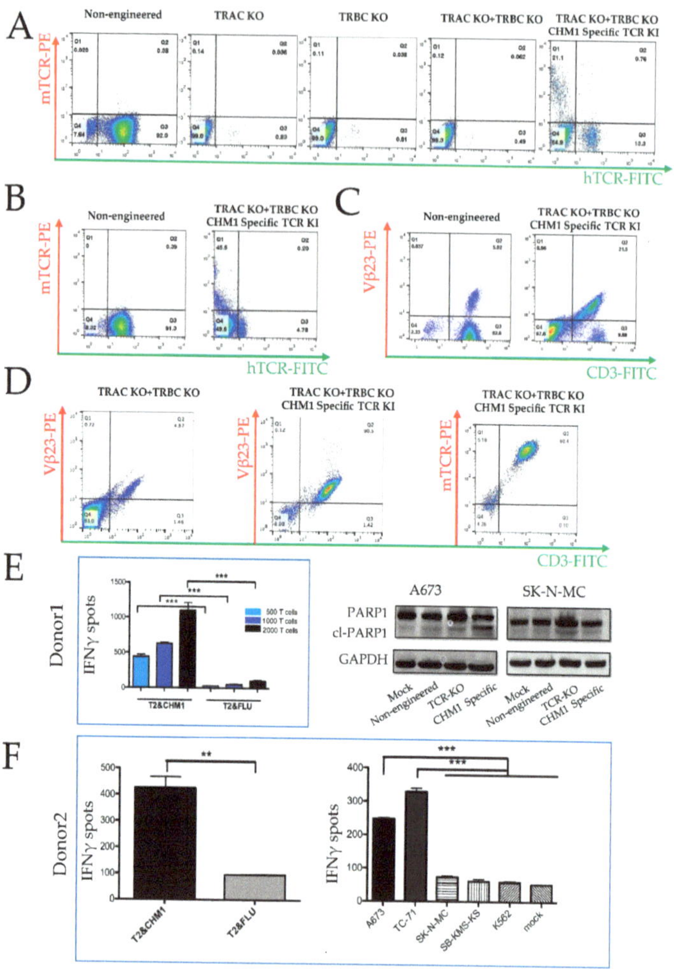

Figure 2. Feasibility of orthotopic TCR replacement by CRISPR/Cas9. (**A**) After knock-out (KO) of either α-chain (TRAC KO) or β-chain (TRBC KO) or of both chains, the endogenous TCR (hTCR) is undetectable by FACS analysis. After KO of both endogenous chains combined with knock-in (KI) of the CHM1^{319}-specific TCR, mTCR is detectable (the TRBC constant chain of the CHM1^{319}-specific TCR was murinized). Thawed T cells, KI efficiency: 21.3% (thawed T cells). Non-engineered: electroporation of CRISPR/Cas9 but without guide RNA. (**B**) KO of both chains combined with transgenic TCR knock-in: mTCR positive T cells are hTCR negative. Fresh T cells, KI efficiency: 45.5%. (**C**) Pentamer staining confirms transgenic TCR expression as panel B. (**D**) CHM1^{319}-TCR expression in the final product of CHM1^{319}-TCR insertion after enrichment. (**E**) Donor 1: IFNγ release with dose-dependent manner (500, 1000, and 2000) of transgenic T cells after exposure to T2 cells loaded with either CHM1^{319}-peptide or with control-peptide (FLU) for 20 h in donor 1; PARP cleavage (cl-PARP) analyzed by SDS-PAGE after co-culture of A673 (HLA-A*0201+) or SK-N-MC (HLA-A*0201-) with either no T cells (Mock), non-engineered T cells with TCR-KO, or T cells with orthotopic TCR replacement with CHM1^{319} TCR (CHM1-specific), all from donor 1. (**F**) Donor 2: IFNγ release to assess the specific reactivity against several tumor cell lines after co-culture with 1000 CHM1^{319}-TCR transgenic T cells in donor 2 (A673 and TC-71: HLA-A*0201+ EwS, SK-N-MC, and SB-KMS-KS: HLA-A*0201- EwS, K562: MHC- NK cell control). IFNγ release transgenic T cells after exposure to T2 cells loaded with either CHM1^{319}-peptide or with control-peptide (FLU) serving as positive or negative control. Error bars represent standard deviation of triplicates experiments. ** means $p < 0.01$, *** means $p < 0.001$.

3.2. *Higher Efficiency of Retroviral Transduction Compared to Gene Editing by CRISPR/Cas9*

3.2.1. TCR Transgenic T Cells Engineered by CRISPR/cas9 vs. Retroviral Gene Transduction

The experimental design is described in detail in the method section, as well as Supplementary Figure S1. In brief, after isolating PBMC from buffy coat, we stimulated T cells with CD3/CD28 Dynabeads for two days. Meanwhile, we amplified the KI-DNA fragment for CRISPR/Cas9 transduction and transfected 293Vec-RD114 packaging cells with the pMP71-CHM1-TCR plasmid for retrovirus production. Subsequently, we purified the KI-DNA fragment for CRISPR/Cas9 or harvested the retrovirus for transduction. Next, we isolated the transgenic T cells with anti-mTCR antibody (mTCR-PE) and expanded the transgenic T cells for further functional analysis and in vivo experiments.

3.2.2. Higher Efficiency of Retroviral Transduction Compared to Gene Editing by CRISPR/Cas9

We first assessed retroviral transduction efficacy. After transduction of T cells from two donors with GFP-containing control vector pMP71-GFP, we checked the GFP expression by fluorescence microscopy (Figure 3A) and by FACS analysis (Figure 3B). The cells from donor 1 were thawed, whereas cells from donor 2 were fresh. The transduction rate of thawed T cells was 78% (Figure 3B, upper panel), while fresh T cells reached 90% (Figure 3B, lower panel).

When comparing CRISPR/Cas9 orthotopic single gene replacement and multiple random insertions by retroviral transduction, we found retroviral transduction to be consistently higher. The efficacy of endogenous TCR orthotopic replacement with CHM1^{319}-TCR ranged from 11% to 45%, whereas efficacy of retroviral transduction mainly ranged from 70% to 90%. Figure 3C,D depict the results of non-engineered CRISPR/Cas9 and retrovirus transduced T cells derived from the same donor. Retrovirus transduction efficacy was 77% (Figure 3C, lower panel Q2 plus Q3), whereas CRISPR/Cas9 transduction efficacy was only 19% (Figure 3C, middle panel Q2 plus Q3). As expected, the replacement of the endogenous TCR was more efficient in the CRISPR/Cas9 (Figure 3D, middle panel Q5 plus Q6), as compared to the retrovirus (Figure 3D, lower panel Q5 plus Q6) group. Corresponding isotype staining is shown in Supplementary Figure S4A. Figure 3E shows the transduction efficacy via CRISPR/Cas9 or retrovirus. Although there is biological variability, possibly donor-dependent, the efficacy is significantly higher for retrovirus, as compared CRISPR/Cas9 ($p < 0.0001$).

3.3. *Prevention of Endogenous TCR Expression in CRISPR/Cas9 vs. Retrovirally Engineered T Cells*

3.3.1. Requirement of High Retroviral Gene Transduction Efficacy and High CRISPR/Cas9 KO Efficacy for Prevention of Endogenous TCR Expression and TCR Chain Mispairing

One of the advantages of orthotopic TCR replacement by CRISPR/Cas9 is that it avoids the mispairing of endogenous and exogenous TCR chains, and thus averts the generation of promiscuous TCRs recognizing off-target antigens. To gauge this postulated advantage of CRISPR/Cas9 vs. retroviral engineering, we compared the expression of endogenous TCR after CRISPR/Cas9 gene editing vs. retroviral transduction. We found that the decrease in endogenous TCR surface expression in the retrovirus group was similar to the expression in the CRISPR/Cas9 group (Figure 4A and Supplementary Figure S4B). After CHM1^{319}-TCR transduction via retrovirus, the transgene positive population (red curve, Figure 4B) shows less endogenous TCR transgene, as compared to the negative population (blue curve, Figure 4B).

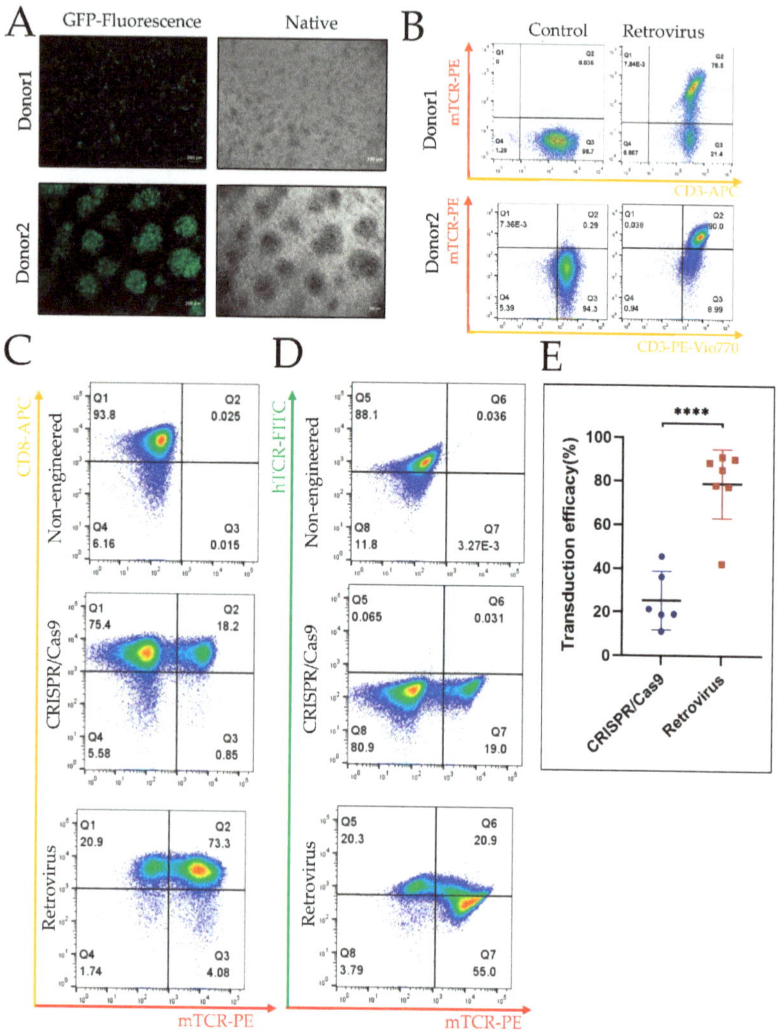

Figure 3. Efficacy of gene editing by CRISPR/Cas9 vs. retroviral transduction, as well as the phenotype of the T cell products. (**A**) A GFP sequence containing pMP71 vector was used to assess transduction efficiency of T cells in general. Transduction is performed twice at day 1 and day 2 of culture. Four days after the first transduction, representative fluorescence microscopy was performed to assess the transduction efficacy of GFP in T cells. Although the colony size is different between the two donors, transduction rates are comparable. Material from donor 1 was thawed; material from donor 2 was fresh. Scale bar: 200 µm. (**B**) The CHM1-TCR sequence containing the pMP71-CHM1-TCR vector was used to transfect T cells. Four days after the first transduction, representative FACS analysis was performed to access the transduction rates. FACS analysis of donor 1 was performed on a Becton Dickinson FACS while analysis of donor 2 was performed on a FACS from Miltenyi. This is the reason for the shift in gating used for donor 1 vs. donor 2. (**C**) T cells were stained with anti-CD8 (CD8-APC) and anti-mTCR (mTCR-PE) after culture to assess the efficacy of transduction. (**D**) T cells were stained with anti-endogenous TCR (hTCR-FITC) and anti-mTCR (mTCR-PE) after culture to assess the efficacy of transduction, the constant domain of the transgenic TCR beta chain being murinized. CRISPR/Cas9 and retrovirus (Retrovirus) transduced T cells from the same donor. (**E**) Transduction efficacy via CRISPR/Cas9 or retrovirus in T cells. **** means $p < 0.0001$.

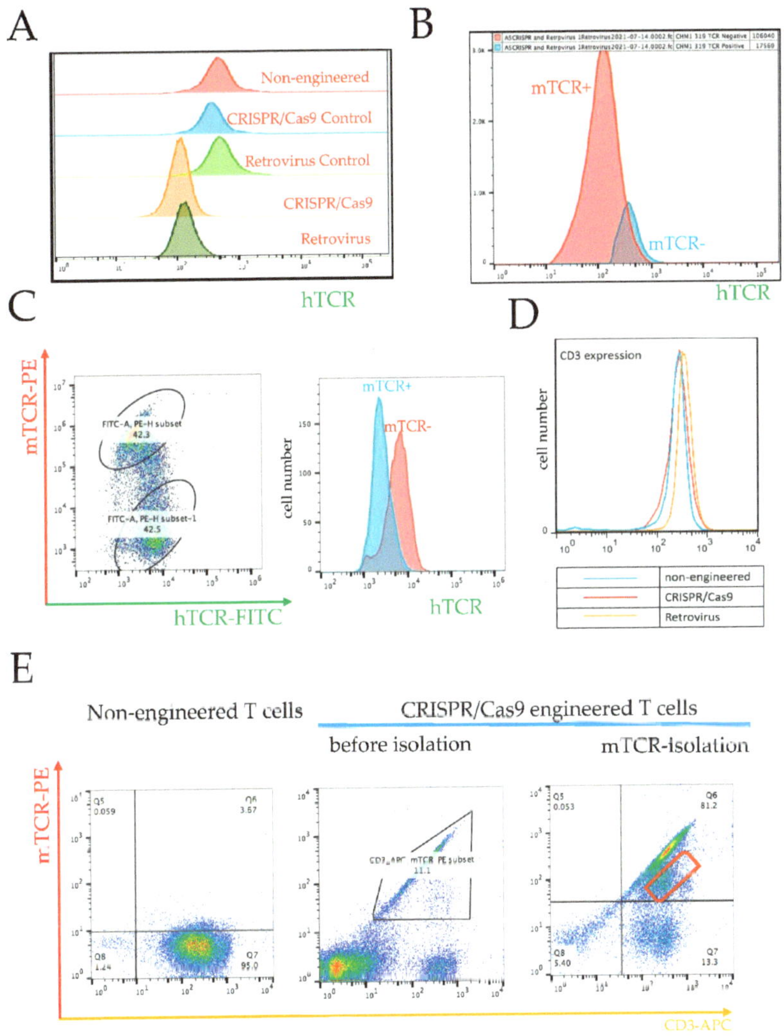

Figure 4. Endogenous TCR expression after gene editing by CRISPR/Cas9 vs. retroviral transduction. (**A**) FACS analysis of expression of endogenous TCR expression (hTCR) from a representative donor: "Non-engineered" designates a control non-engineered T cells; "CRISPR/Cas9 Control" designates a control containing T cells transduced by electroporation of CRISPR/Cas9 but without guide RNA; "Retrovirus Control" designates a control containing T cells exposed to centrifugation only with no retrovirus added; "CRISPR/Cas9" designates T cells with TCR replacement by CRISPR/Cas9, and "Retrovirus" designates T cells with TCR transfer by retrovirus. The "CRISPR/Cas9" and "Retrovirus" groups were analyzed after isolation with anti-mTCR antibody. (**B**) FACS analysis of expression of endogenous TCR expression (hTCR) after retrovirus-mediated TCR transfer (T cells were analyzed before isolation with anti-mTCR antibody). (**C**) FACS determines the expression of endogenous TCR expression (hTCR) on cell membrane after retrovirus-mediated TCR transfer with low efficiency (T cells were analyzed before isolation with mTCR antibody). (**D**) CD3 expression in non-engineered T cells, engineered T cells with orthotopic TCR replacement (CRISPR/Cas9), and engineered T cells retroviral transduction (Retrovirus). (**E**) CRISPR/Cas9-engineered transgenic T cells from thawed donor: left panel, non-engineered T cells; middle panel: CRISPR/Cas9-engineered transgenic T cells before mTCR selection with an anti-mTCR antibody (CRISPR/Cas9); right panel, CRISPR/Cas9-engineered transgenic T cells after mTCR selection with an anti-mTCR antibody. The red cloud in Q6 indicates the transgenic T cell products with a failure KO of endogenous β chain (CRISPR/Cas9).

We also noted that repression of the endogenous receptor after retroviral transduction depends on transduction efficacy (Figure 4C). In this experiment, the retroviral transduction efficacy was only 42%. In the setting of this low transduction efficacy, we can identify two distinct subpopulations (Figure 4C, left panel): the upper cloud represents a subpopulation with high mTCR expression, i.e., the transduced subpopulation, whereas the lower cloud represents a population with a low mTCR expression. The high mTCR expressing subpopulation has a lower hTCR expression, as compared to the subpopulation represented by the lower cloud, which is characterized by low mTCR and higher hTCR expression. This low mTCR/high hTCR subpopulation is comprised of non-transduced cells, as indicated by low mTCR expression. Although there is some overlap between both subpopulations, the peaks of transduced and non-transduced subpopulations are distinct (Figure 4C, right panel) Isotype control is depicted in Supplementary Figure S4D. This finding implicates that a low retroviral transduction efficacy will yield a heterogenous product containing a large subpopulation at risk for mispairing and causing autoimmune side effects. Moreover, retrovirally transduced T cells express slightly more TCR, as documented by CD3 expression (Figure 4D and Supplementary Figure S4E), due to multiple gene copies, possibly because CD3 and TCR work as a protein complex for intracellular signaling transduction.

3.3.2. Failure in KO of Endogenous β Chain Generates a Subpopulation with TCR Misparing

For clinical application, we have to ensure the KO of both endogenous TCR chains. Failure of β chain KO constitutes a risk of mispairing the transduced α with the endogenous β chain. The ratio of α to β chains in a single CRISPR/Cas9-engineered cell, where TRBC KO did not work, would be 2:1, since both transgenic chains are expressed from the α locus, in addition to the endogenous β chain expressed from the TRBC locus. When comparing the transduction efficacy of both procedures, in retrovirally transduced T cells the amount of transduced TCRs per cell are, by definition and by observation, higher ($n > 1$) than the single endogenous TCR ($n = 1$). On the other hand, if CRISPR/Cas9 KO is suboptimal, the risk of mispairing between the exogenous α with the endogenous β chains in a single cell, where TRBC KO did not work, would be X:1, with X being > 1. This implies that the risk of mispairing may be higher in CRISPR/Cas9, as compared to retrovirally transduced T cells, in case the KO is incomplete (Figure 4E). Failure in the knockout of the endogenous β chain leads to a subgroup of T cells expressing both the transgenic TCR and a mixed TCR with the endogenous β chain and the transgenic α chain. This is most likely the reason why we see some cells besides the main population in Figure 4E. We can also see this phenomenon in the right panel of Figure 2A. In Q2, we see a 0.76 population of the whole engineered T cells expressing both "mTCR" and "hTCR"; these cells are also a result of the failure of the knockout of the endogenous β chain.

3.4. *Specific Tumor Cell Recognition and Cytotoxicity In Vitro by Both T Cell Products with Better Prolonged Activity of CRISPR/Cas9-Engineered T Cells*

We next assessed the cytotoxic effects on and specific recognition of EwS cell lines by transgenic T cells obtained with either orthotopic TCR replacement by CRISPR/Cas9 or random TCR placement by retroviral transfer. After isolation of the transgenic T cell with mTCR antibody, we co-cultured the T cells with HLA-A*02:01+ A673. Here, we identified an increase in the cl-PARP in A673 after co-culture with both transgenic T cell products. Furthermore, cl-PARP1 was observed with retrovirally transduced T cells, compared to CRISPR/Cas9-engineered T cells (Figure 5A and Supplementary Figure S5). Of note, there was also an effect of the non-engineered T cells, probably due to non-specific alloresponse. Non-engineered cells retain their endogenous HLA TCR, recognizing the tumor HLA-disparate haplotype.

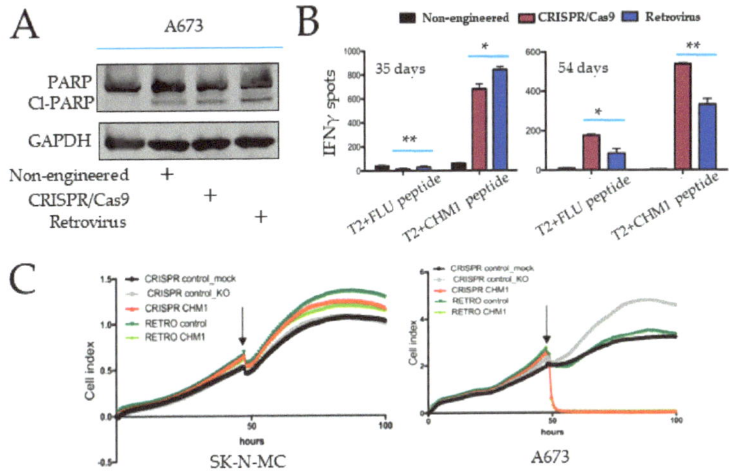

Figure 5. Adoptive transfer of CHM1[319]-TCR transgenic T cells-treated tumor-bearing Rag2-/-γC-/- mice. (**A**) Determination of cl-PARP in A673 (HLA-A*02:01+/CHM1+) cells by SDS-PAGE after co-culture of A673 cells with either no T cells, non-engineered T cells, engineered T cells with orthotopic TCR replacement, or retroviral transduction. (**B**) Evaluation of activation of T cells by IFNγ-ELISpot assay after co-culture with T2 cells plus CHM1 peptide on 35 days (left) and 54 days (right) after T cell isolation from PBMC. A total of 1000 T cells were used in the experiments. (**C**) xCELLigence detachment assays were performed with a different donor to compare the cytotoxic effect of T cells on HLA-A*02:01+/CHM1- SK-N-MC and HLA-A*02:01+/CHM1+ A673 cells. Treatment groups include T cells without CRISPR/Cas9 engineering but otherwise undergoing an identical procedure, including electroporation (CRISPR control_mock), T cells with CRISPR/Cas9-engineered TCR KO only (CRISPR control_KO), T cells with CRISPR/Cas9-engineered TCR KO and TCR replacement (CRISPR CHM1), T cells without retroviral TCR transduction but otherwise identical procedures, including ultracentrifugation (RETRO control) and T cells with retroviral TCR transduction (RETRO CHM1) on day 20. Error bars represent standard deviation of triplicates experiments. * means $p < 0.05$, ** means $p < 0.01$.

To assess specific recognition, we performed an IFNγ-Elispot on day 35 and day 54 after T cell culture by co-culture T cells with T2 cells loaded with either CHM1[319] peptide or with control peptide (FLU). On day 35, after T cell culture, we observed a higher IFNγ release with retrovirally transduced T cells, as compared to T cells with orthotopic TCR replacement (Figure 5B, left panel). In contrast, a higher IFNγ release was induced by T cells with orthotopic TCR replacement on day 54 (Figure 5B, right panel). This could be associated with or due to phenotypic differences (data not shown). Of note, we identified similar trends with the control peptide (Figure 5B), albeit with significantly weaker IFNγ signals.

We then performed xCELLigence assays to compare the cytotoxic effect on EwS cells with transgenic T cells on day 20. Both CRISPR/Cas9-engineered and retrovirus-transduced T cells lead to the significant killing, as measured by the detachment of the HLA- A*02:01+ A673 cell line without significant difference, but not to the killing of the HLA-A*02:01- SK-N-MC line, indicating HLA-A-restricted specificity (Figure 5C).

3.5. CHM1 as a Unique Immunotherapy Target in EwS

CHM1 Is a Direct Target of EWS-FLI1 Selectively Expressed in EwS

We have previously shown, EWS-FLI1 binds to the promotor and activates the transcription of CHM1 in EwS. Furthermore, CHM1 sustains the undifferentiated and invasive phenotype of EwS, which promotes lung metastasis of EwS [29]. It is highly expressed and required for metastasis [29] and serves as an EwS-specific antigen [18,22,30].

We first analyzed public CHIP-sequence data using the Integrative Genomics Viewer (IGV). CHIP-sequence data confirmed that EWS-FLI1 binds to two promotor sites of CHM1 and induces acetylation of H3K27 (H3K27ac) at both sites, which is associated with the activation of transcription (Supplementary Figure S6A). Forced expression of EWS-FLI1 in mesenchymal stem cell also enhances H3K27ac at the same sites (Supplementary Figure S6A).

We next mined the Cancer Cell Line Encyclopedia (CCLE) [31]. We found CHM1 mRNA is most highly expressed in EwS among all tumor cell lines (Supplementary Figure S6B). The Public Gene Expression Omnibus (GEO) database indicates that its expression in EwS tissues is significantly higher than in bone marrow mesenchymal stem cells (Supplementary Figure S6C). These results are in correspondence to our previous publications [29,32,33]. In addition, CHM1 expression does not correlate significantly with recurrence or metastasis (Supplementary Figure S6D).

Finally, we analyzed the expression of CHM1 at the protein level in normal tissue utilizing ProteomicsDB, developed by the Chair of Proteomics and Bioanalytics at the Technische Universität München and Cellzome GmbH [34,35]. We found only a very low expression in vitreous humor, lung, and heart and no expression in other tissues (Supplementary Figure S7).

4. Discussion

4.1. TCR-Based Immunotherapy of Ewing Sarcoma

There is an obvious medical need for novel therapies in advanced Ewing sarcoma (AES), i.e., in patients with bone or bone marrow metastasis or early relapse. High dose therapies with autologous hematopoietic stem-cell rescues have only been beneficial in selected subgroups [4,36]. Allogeneic hematopoietic stem-cell transplantation (allo-SCT) from healthy donors, as efficacious immunotherapy of leukemia, has offered hints for beneficial effects in solid tumors [37], possibly also in AES patients [4,36,38]. However, no difference in survival with reduced- versus high-intensity conditioning before allo-SCT [39] could be detected. There is also no difference in survival after HLA-mismatched versus HLA-matched allo-SCT [38]. These findings imply that allo-SCT is not sufficient for immunotherapy of AES, and novel therapeutic strategies are in urgent demand, such as TCR-based immunotherapy [40]. CAR T cells have also been developed for AES [41]. The limitation of this approach is that CARTs can only target membrane molecules. However, molecules that are indispensable for tumor metastasis do not necessarily happen to sit on the cell membrane. Tumor metastasis, however, is what immunotherapy has to address, for two reasons: (1) Most local tumors are not a therapeutic challenge anymore with modern precision therapies—including theranostics, functional imaging, and in situ biomarker-guided surgery—or proton or particle therapy (2) Most cancer patients do not die of the tumor but of metastasis. With TCR-based immunotherapy targeting tumor-associated antigens and metastatic drivers of EwS, such as CHM1 [22,33], STEAP1 [24], PAPPA [42], and PRAME [43], our group previously achieved efficacious in vitro and in vivo cytotoxic targeting of HLA-A*02:01+ EwS. TCR-based immunotherapy even led to partial regression without GvHD in refractory HLA-A2+ patients [18]. TCR-based adoptive therapy also showed promising anti-sarcoma effects by targeting NY-ESO-1, leading to objective clinical responses [44]. More than 600 clinical trials about TCR-based immunotherapy are in processing, according to ClinicalTrials.gov (https://clinicaltrials.gov, accessed on 28 April 2022).

Retrovirus- and lentivirus-based vectors are commonly used for TCR gene transfer in clinical trials [12]. Both viruses enable stable integration and efficient expression of exogenous TCRs in lymphocytes. However, the mispairing of endo- with exogenous TCRs limits the function of the transduced TCR and generates new antigens, which can cause autoreactivity or GvHD [45]. Nevertheless, there was no evidence of GvHD in the TCR-based adoptive therapy, targeting CHM1 in our treatment trials of EwS, including allogeneic donor lymphocyte infusions [15] or allogeneic transgenic T cells [18]. Several strategies, such as the murinization of TCR constant regions [46,47], codon optimization [48], and insertion of additional cysteine residues [49], have been proposed to prevent mispairing. We used codon optimization and murinization of TCR constant regions. However, these procedures cannot completely eliminate mispairing [50]. The random insertion of viruses into the genome also raises safety concerns, such as insertional mutations and tumorigenesis [13].

4.2. Orthotopic Replacement of TCR with Cytotoxic Functionality

To address the potential hazards of viral transduction, endogenous TCR KO with simultaneous non-viral orthotopic TCR replacement has been established. Orthotopic TCR gene replacement drives the translation of the transduced TCR gene sequence, and the activation via the endogenous TCR promoter provides functional results [15,51–53].

The comparison of orthotopic TCR gene replacement by CRISPR/Cas9 to random TCR gene insertion by retroviral gene transfer can be performed under different aspects. In clinical application, e.g., in established protocols for production of CAR T cells by retroviral transfer, a multiplicity of infections of the target cell with the retroviral vectors is used, yielding multiple gene copies in the genome of the target cell. In contrast, CRISPR/Cas9-engineered T cells contain only one TCR gene copy. We performed this study, prioritizing the clinical application. However, if a comparison of the biological effects of orthotopic TCR replacement vs. random TCR gene insertion is the aim of study, a MOI is to be chosen under which only a single retroviral insertion of the transgene into the genome of the target cell takes places. This consideration also illustrates that the term transduction efficacy, which is well established in the literature, may be somewhat misleading, since we are measuring yield rather than efficacy when comparing a multiplicity of infections of the target cell with the retroviral vectors to an orthotopic TCR gene replacement into the TRAC locus with a single gene copy by CRISPR/Cas9. Finally, it should be considered that the transduction procedure itself may change the phenotype and CD4/CD8 ratio of T cells.

This present work shows the feasibility of the orthotopic replacement of the endogenous T cell receptor (TCR) with CHM1^{319}-TCR targeting EwS by CRISPR/Cas9 and confirms previous publications with different TCRs [15,54]. Our CRISPR/Cas9-engineered T cell products demonstrated a strong specific cytotoxic effect towards HLA-A*02:01+ EwS, comparable to retrovirus-transduced T cells. Clinical application also requires the following considerations: Only patients with relapse or refractory disease after failure of standard therapy are eligible for TCR-based immunotherapy at present. These patients mostly have a short life expectancy and a narrow time window for immunotherapy, which has to be given when patients are, at least, in very good partial remission. In our clinical experience, the production process of therapeutic T cell products often takes too long to provide a clinical benefit within this narrow window. Thus, we were wondering which method yields enough engineered T cells in a short time. Thereby, we had to take into account the low KI efficacy with CRISPR/Cas9 and the high number of cells not surviving electroporation.

Comparing CRISPR/Cas9 with retrovirus-transduced T cells, our work indicates that high retroviral transduction efficacy can prevent endogenous TCR expression on the cell membrane, resembling CRISPR/Cas9-engineered T cells. This may indicate that a high efficiency of TCR transduction by the retrovirus is capable of competing with the endogenous TCR to form the heterocomplex with CD3 required for stable TCR membrane expression. This competition may help to avoid neo-antigen recognition due to TCR chain mispairing. However, high transduction rates may lead to abundant insertion of vector copy numbers (VCN) [55]. According to the reflection paper on clinical risk management, due to insertional mutations from the European Medicines Agency's Committee on Advanced Therapeutics [56], the risk of gene-modified cell therapies via insertional oncogenesis should be reduced by the restriction of VCN. Additionally, close-to-random transgene integration via viral transduction further limits the clinical application [57]. A low copy number is desired for safety reasons, and it also limits TCR expression, which ultimately compromises the functionality of the T cell product. Taken together, there are limitations by both high- and low-viral transduction rates.

On the other hand, gene editing by CRISPR/Cas9 generates structural defects of the nucleus, chromosomal truncations, micronuclei, and chromosome bridges, which initiate a mutational process and cause human congenital disease, even cancer [58,59]. Rare off-target effects were also identified when using TRAC guide RNA (gRNA) with wild-type Cas9, whereas no off-target effects were detected with the 'enhanced specificity' Cas9 variant eSp.Cas9 [53,60]. We performed our experiments by taking advantage of eSp.Cas9, which potentially avoids off-target effects. However, we did not evaluate the genome-wide editing specificity in the present work. For clinical application, we have to ensure the KO of both endogenous TCRs. If the KO of β chain fails, there is a possibility of the mispairing of the transduced α with the endogenous β chain. The ratio of α to β chain in a single cell would be 2:1, since both transgenic chains are expressed from the α locus. Although we do not present TCR sequencing data of CRISPR/Cas9-engineered T cells here, we published detection of transgenic TCRs using specific primers in retrovirally transduced T cells previously [22].

When we compared the transduction efficacy of both procedures, in retrovirally transduced T cells, the amount of transduced TCRs was significantly higher than the endogenous TCR in a single cell, as expected. Thus, the risk of mispairing between exogenous α to endogenous β chains in a single cell would be X:1, with X being >1, depending on the number of transduced gene copies. As expected, retrovirally transduced T cells express more TCR on their surface. However, to strictly compare the biology CRISPR/Cas9 vs. conventional editing, conditions need to be chosen that restrict integration to only one transgene copy. This would allow a strict comparison of orthotopic vs. random TCR placement into the genome. To this end, and to systematically compare CRISPR/Cas9 vs. conventional editing, some of us have recently analyzed expression and functionality of 51 CMV-specific TCRs to show that conventional genetic engineering leads to variable TCR expression and functionality as a result of variable copy numbers with untargeted integration. In contrast, the CRISPR/Cas9-mediated TCR replacement yields homogeneous TCR expression similar to physiological T cells. Product homogeneity after CRISPR/Cas9 TCR gene editing correlated with functionality. Such a well-defined product is obligatory in clinical application [53]. Since efficacies are never certain to be 100%, some of us have also established KI for both chains individually (TCR alpha into TRAC and TCR beta into TRBC), combined with HLA multimer sorting to enrich for the correctly edited population [15].

Finally, we identified a similar amount of CD3 expression on the T cell membrane after CRISPR/Cas9 engineering, compared to non-engineered T cells, while the CD3 expression in retroviral transduced cells was increased, compared to non-engineered T cells. This indicates variable TCR/CD3 complex expression likely depends on the number of TCR gene copies in the genome and the corresponding number of TCR molecules on the T cell membrane.

Several other limitations were also identified in our work, such as comparatively low transduction efficiency of CRISPR/Cas9, ranging from 10–45%, due to the imponderabilities of cultures or fresh vs. thawed status of the cultured lymphocytes. We could minimize cell death after electroporation, especially with TCR KI by directly culturing in Penicillin-Streptomycin (P/S)-free T cell medium after electroporation. We think that higher transduction rates with CRISPR/Cas9-engineered TCR KI are possible. Although cells with a high transduction rate after electroporation do not tolerate the antibiotics (P/S), an optimization of the protocol could involve culturing the cells for 24–48 h in T cell medium without P/S and changing back to the standard culture medium afterwards.

5. Conclusions

T cells engineered with CRISPR/Cas9 to address the metastatic driver, CHM1, are feasible for immunotherapy of EwS and may have the advantage of a more prolonged cytotoxic activity, as compared to T cells engineered with retroviral gene transfer, in addition to their homogeneity and functional predictability.

Supplementary Materials: The following supporting information can be downloaded at: https://www.mdpi.com/article/10.3390/cancers14225485/s1, Table S1: DNA sequence of knock-in CHM1[319]-TCR. Supplementary Figure S1: Procedures to compare transgenic T cells engineered by either CRISPR/Cas9 or retroviral gene transfer; Supplementary Figure S2: Schema and DNA template for gene editing by CRISPR/Cas9; Supplementary Figure S3: Isotype controls of FACS staining; Supplementary Figure S4: Isotype controls of FACS staining and endogenous TCR expression after gene editing by CRISPR/Cas9 vs. retroviral transduction; Supplementary Figure S5: Quantification of cl-PARP density after co-culture of A673cells with engineered T cells. *** means $p < 0.001$; Supplementary Figure S6: CHM1 is a direct target of EWS-FLI1 selectively expressed in EwS. **** $p < 0.0001$; Supplementary Figure S7: CHM1 protein expression in tissues according to ProteomicsDB.

Author Contributions: S.E.G.B. and U.T. conceived and designed the study. B.X., K.v.H. and M.T. performed the experimental work, K.v.H. and U.T. contributed xCELLigence assays, M.T. and B.X. performed in IFNγ-Elispot assays, H.G. helped in FACS analysis. K.v.H., K.S. and D.H.B. provided CRISPR/Cas9 method. S.E.G.B., U.T., J.H., J.R., K.v.H. and J.M. advised on the experimental design, data analysis, and/or interpretation. M.P., B.X. and S.E.G.B. reanalyzed the data with regard to compensation of the FACS channels. S.E.G.B. and B.X. co-wrote the manuscript. All authors have read and agreed to the published version of the manuscript. All experiments have been conducted in the Children's Cancer Research Center, Kinderklinik München Schwabing; FACS data analysis and compensation was performed at Translational Pediatric Cancer Research Institute of Chemistry and Pathobiochemistry, TUM School of Medicine, Technical University of Munich.

Funding: We are grateful to the China Scholarship Council (#201908210290) and the Cura Placida Children's Cancer Research Foundation (#06082021/#5110378) for their financial support to B.X., as well as to the St. Baldrick's Foundation (Martha's BEST Grant for all #663113) and the 'Du musst kämpfen' Wohltätigkeitsinitiative (Access to Systems Biology-Based Individualized Cell Therapies for Adolescents with Refractory Pediatric Cancer, AdoRe, #200310stb) for their grants to S.E.G.B. U.T. is funded by the Deutsche Forschungsgemeinschaft (DFG, German Research Foundation)—Project number 501830041, the "Zukunft Gesundheit e.V." and the Dr. Sepp and Hanne Sturm Memorial Foundation. U.T. and H.G. are funded by the Robert Pfleger Foundation and U.T., H.G., and S.E.G.B. by the Wilhelm Sander-Foundation (U.T. 2018.072.1. and 2021.007.1.; S.E.G.B. 2018.072.1). D.H.B. and K.S. received funding for the OTR development from SFB-TRR 338/1 2021—452881907 (project A01) and SFB 1054/3—210592381 (project B09). J.R. was supported by research grants from the Deutsche Forschungsgemeinschaft (DFG, German Research Foundation-Project-ID 360372040—SFB 1335), and the European Research Council (ERC) under the European Union's Horizon 2020 research and innovation programme (grant agreement number 834154). H.G. was also funded by the KKF program of TUM School of Medicine.

Institutional Review Board Statement: Not applicable.

Data Availability Statement: The data of this study can be found in Supplementary Materials.

Acknowledgments: Special thanks go to Sebastian J. Schober for experimental support and together with Poul H.B. Sorensen for expertise in the critical review of data and manuscript. Thanks also go to Wilko Weichert for the provision of work space and general support. This work constitutes part of the doctoral thesis of Busheng Xue. It has been presented in part at the 2022 American Society of Pediatric Hematology/Oncology (ASPHO) Conference, 4–7 May, Pittsburgh, PA, USA.

Conflicts of Interest: S.E.G.B. has an ownership interest in PDL BioPharma and served as a consultant to EOS Biotechnology Inc., Bayer AG and Swedish Orphan Biovitrum AB. S.E.G.B. held US and EU intellectual properties in gene-expression analysis. D.H.B. has a consulting contract with and receives sponsored research support from Juno Therapeutics/BMS. The other authors declare no conflict of interest.

References

1. Grünewald, T.G.; Cidre-Aranaz, F.; Surdez, D.; Tomazou, E.M.; de Álava, E.; Kovar, H.; Sorensen, P.H.; Delattre, O.; Dirksen, U. Ewing sarcoma. *Nat. Rev. Dis. Prim.* **2018**, *4*, 5. [PubMed]
2. Riggi, N.; Suvà, M.L.; Stamenkovic, I. Ewing's sarcoma. *N. Engl. J. Med.* **2021**, *384*, 154–164. [PubMed]
3. Thacker, M.M.; Temple, H.; Scully, S.P. Current treatment for Ewing's sarcoma. *Expert Rev. Anticancer Ther.* **2005**, *5*, 319–331. [CrossRef] [PubMed]

4. Burdach, S.; Jürgens, H. High-dose chemoradiotherapy (HDC) in the Ewing family of tumors (EFT). *Crit. Rev. Oncol.* **2002**, *41*, 169–189. [CrossRef]
5. Burdach, S.T.; Jürgens, H.; Peters, C.; Nürnberger, W.; Mauz-Körholz, C.; Körholz, D.; Paulussen, M.; Pape, H.; Dilloo, D.; Koscielniak, E. Myeloablative radiochemotherapy and hematopoietic stem-cell rescue in poor-prognosis Ewing's sarcoma. *J. Clin. Oncol.* **1993**, *11*, 1482–1488. [CrossRef]
6. Burdach, S.; Thiel, U.; Schöniger, M.; Haase, R.; Wawer, A.; Nathrath, M.; Kabisch, H.; Urban, C.; Laws, H.J.; Dirksen, U.; et al. Total body MRI-governed involved compartment irradiation combined with high-dose chemotherapy and stem cell rescue improves long-term survival in Ewing tumor patients with multiple primary bone metastases. *Bone Marrow Transpl.* **2009**, *45*, 483–489. [CrossRef] [PubMed]
7. Bhakta, N.; Liu, Q.; Ness, K.K.; Baassiri, M.; Eissa, H.; Yeo, F.; Chemaitilly, W.; Ehrhardt, M.J.; Bass, J.; Bishop, M.W.; et al. The cumulative burden of surviving childhood cancer: An initial report from the St Jude Lifetime Cohort Study (SJLIFE). *Lancet* **2017**, *390*, 2569–2582.
8. Gaspar, N.; Hawkins, D.S.; Dirksen, U.; Lewis, I.J.; Ferrari, S.; Le Deley, M.-C.; Kovar, H.; Grimer, R.; Whelan, J.; Claude, L.; et al. Ewing Sarcoma: Current Management and Future Approaches Through Collaboration. *J. Clin. Oncol.* **2015**, *33*, 3036–3046.
9. Burdach, S.; Kolb, H.J. The vigor of defense against non-self: Potential superiority of allorestricted T cells in immunotherapy of cancer? *Front. Oncol.* **2013**, *3*, 100. [CrossRef]
10. Rudolph, M.G.; Stanfield, R.L.; Wilson, I.A. How TCRs bind MHCs, peptides, and coreceptors. *Annu. Rev. Immunol.* **2006**, *24*, 419–466. [CrossRef]
11. Davis, M.M.; Bjorkman, P.J. T-cell antigen receptor genes and T-cell recognition. *Nature* **1988**, *334*, 395–402.
12. Manfredi, F.; Cianciotti, B.C.; Potenza, A.; Tassi, E.; Noviello, M.; Biondi, A.; Ciceri, F.; Bonini, C.; Ruggiero, E. TCR Redirected T Cells for Cancer Treatment: Achievements, Hurdles, and Goals. *Front. Immunol.* **2020**, *11*, 1689. [CrossRef] [PubMed]
13. Howe, S.J.; Mansour, M.; Schwarzwaelder, K.; Bartholomae, C.; Hubank, M.; Kempski, H.; Brugman, M.; Pike-Overzet, K.; Chatters, S.J.; De Ridder, D.; et al. Insertional mutagenesis combined with acquired somatic mutations causes leukemogenesis following gene therapy of SCID-X1 patients. *J. Clin. Investig.* **2008**, *118*, 3143–3150. [CrossRef] [PubMed]
14. Fraietta, J.A.; Nobles, C.L.; Sammons, M.A.; Lundh, S.; Carty, S.A.; Reich, T.J.; Cogdill, A.P.; Morrissette, J.J.D.; DeNizio, J.E.; Reddy, S.; et al. Disruption of TET2 promotes the therapeutic efficacy of CD19-targeted T cells. *Nature* **2018**, *558*, 307–312. [PubMed]
15. Schober, K.; Müller, T.R.; Gökmen, F.; Grassmann, S.; Effenberger, M.; Poltorak, M.; Stemberger, C.; Schumann, K.; Roth, T.L.; Marson, A.; et al. Orthotopic replacement of T-cell receptor alpha- and beta-chains with preservation of near-physiological T-cell function. *Nat. Biomed. Eng.* **2019**, *3*, 974–984. [CrossRef] [PubMed]
16. Sarukhan, A.; Garcia, C.; Lanoue, A.; von Boehmer, H. Allelic Inclusion of T Cell Receptor α Genes Poses an Autoimmune Hazard Due to Low-Level Expression of Autospecific Receptors. *Immunity* **1998**, *8*, 563–570. [CrossRef]
17. Coulie, P.G.; Karanikas, V.; Colau, D.; Lurquin, C.; Landry, C.; Marchand, M.; Dorval, T.; Brichard, V.; Boon, T. A monoclonal cytolytic T-lymphocyte response observed in a melanoma patient vaccinated with a tumor-specific antigenic peptide encoded by gene MAGE-3. *Proc. Natl. Acad. Sci. USA* **2001**, *98*, 10290–10295. [CrossRef]
18. Thiel, U.; Schober, S.J.; Einspieler, I.; Kirschner, A.; Thiede, M.; Schirmer, D.; Gall, K.; Blaeschke, F.; Schmidt, O.; Jabar, S.; et al. Ewing sarcoma partial regression without GvHD by chondromodulin-I/HLA-A*02:01-specific allorestricted T cell receptor transgenic T cells. *Oncoimmunology* **2017**, *6*, e1312239.
19. Graef, P.; Buchholz, V.R.; Stemberger, C.; Flossdorf, M.; Henkel, L.; Schiemann, M.; Drexler, I.; Höfer, T.; Riddell, S.R.; Busch, D.H. Serial Transfer of Single-Cell-Derived Immunocompetence Reveals Stemness of CD8+ Central Memory T Cells. *Immunity* **2014**, *41*, 116–126. [CrossRef]
20. Kaneko, S.; Mastaglio, S.; Bondanza, A.; Ponzoni, M.; Sanvito, F.; Aldrighetti, L.; Radrizzani, M.; La Seta-Catamancio, S.; Provasi, E.; Mondino, A.; et al. IL-7 and IL-15 allow the generation of suicide gene–modified alloreactive self-renewing central memory human T lymphocytes. *Blood* **2009**, *113*, 1006–1015. [CrossRef]
21. Buchholz, V.R.; Flossdorf, M.; Hensel, I.; Kretschmer, L.; Weissbrich, B.; Gräf, P.; Verschoor, A.; Schiemann, M.; Höfer, T.; Busch, D.H. Disparate Individual Fates Compose Robust CD8 + T Cell Immunity. *Science* **2013**, *340*, 630–635. [CrossRef] [PubMed]
22. Blaeschke, F.; Thiel, U.; Kirschner, A.; Thiede, M.; Rubio, R.A.; Schirmer, D.; Kirchner, T.; Richter, G.H.; Mall, S.; Klar, R.; et al. Human HLA-A*02:01/CHM1+ allo-restricted T cell receptor transgenic CD8+ T cells specifically inhibit Ewing sarcoma growth in vitro and in vivo. *Oncotarget* **2016**, *7*, 43267–43280. [CrossRef] [PubMed]
23. Schirmer, D.; Grünewald, T.G.P.; Klar, R.; Schmidt, O.; Wohlleber, D.; Rubio, R.A.; Uckert, W.; Thiel, U.; Bohne, F.; Busch, D.H.; et al. Transgenic antigen-specific, HLA-A*02:01-allo-restricted cytotoxic T cells recognize tumor-associated target antigen STEAP1 with high specificity. *OncoImmunology* **2016**, *5*, e1175795.
24. Schober, S.J.; Thiede, M.; Gassmann, H.; Prexler, C.; Xue, B.; Schirmer, D.; Wohlleber, D.; Stein, S.; Grünewald, T.G.P.; Busch, D.H.; et al. MHC Class I-Restricted TCR-Transgenic CD4+ T Cells Against STEAP1 Mediate Local Tumor Control of Ewing Sarcoma In Vivo. *Cells* **2020**, *9*, 1581.
25. Robinson, J.T.; Thorvaldsdóttir, H.; Wenger, A.M.; Zehir, A.; Mesirov, J.P. Variant Review with the Integrative Genomics Viewer. *Cancer Res.* **2017**, *77*, e31–e34. [CrossRef] [PubMed]

26. Barretina, J.; Caponigro, G.; Stransky, N.; Venkatesan, K.; Margolin, A.A.; Kim, S.; Wilson, C.J.; Lehár, J.; Kryukov, G.V.; Sonkin, D.; et al. The Cancer Cell Line Encyclopedia enables predictive modelling of anticancer drug sensitivity. *Nature* **2012**, *483*, 603–607. [CrossRef] [PubMed]
27. Wilhelm, M.; Schlegl, J.; Hahne, H.; Gholami, A.M.; Lieberenz, M.; Savitski, M.M.; Ziegler, E.; Butzmann, L.; Gessulat, S.; Marx, H.; et al. Mass-spectrometry-based draft of the human proteome. *Nature* **2014**, *509*, 582–587. [CrossRef] [PubMed]
28. Acuto, O.; Reinherz, E.L. The human T-cell receptor. Structure and function. *N. Engl. J. Med.* **1985**, *312*, 1100–1111.
29. von Heyking, K.; Calzada-Wack, J.; Göllner, S.; Neff, F.; Schmidt, O.; Hensel, T.; Schirmer, D.; Fasan, A.; Esposito, I.; Müller-Tidow, C.; et al. The endochondral bone protein CHM1 sustains an undifferentiated, invasive phenotype, promoting lung metastasis in Ewing sarcoma. *Mol. Oncol.* **2017**, *11*, 1288–1301.
30. Biele, E.; Schober, S.J.; Prexler, C.; Thiede, M.; Heyking, K.V.; Gassmann, H.; Eck, J.; Xue, B.; Burdach, S.; Thiel, U. Monocyte Maturation Mediators Upregulate CD83, ICAM-1 and MHC Class 1 Expression on Ewing's Sarcoma, Enhancing T Cell Cytotoxicity. *Cells* **2021**, *10*, 3070. [CrossRef]
31. Ghandi, M.; Huang, F.W.; Jané-Valbuena, J.; Kryukov, G.V.; Lo, C.C.; McDonald, E.R., III; Barretina, J.; Gelfand, E.T.; Bielski, C.M.; Li, H.; et al. Next-generation characterization of the Cancer Cell Line Encyclopedia. *Nature* **2019**, *569*, 503–508. [CrossRef] [PubMed]
32. Staege, M.S.; Hutter, C.; Neumann, I.; Foja, S.; Hattenhorst, U.E.; Hansen, G.; Afar, D.; Burdach, S.E.G. DNA Microarrays Reveal Relationship of Ewing Family Tumors to Both Endothelial and Fetal Neural Crest-Derived Cells and Define Novel Targets. *Cancer Res.* **2004**, *64*, 8213–8221. [CrossRef]
33. Thiel, U.; Pirson, S.; Müller-Spahn, C.; Conrad, H.; Busch, D.H.; Bernhard, H.; Burdach, S.; Richter, G.H.S. Specific recognition and inhibition of Ewing tumour growth by antigen-specific allo-restricted cytotoxic T cells. *Br. J. Cancer* **2011**, *104*, 948–956. [CrossRef] [PubMed]
34. Schmidt, T.; Samaras, P.; Frejno, M.; Gessulat, S.; Barnert, M.; Kienegger, H.; Krcmar, H.; Schlegl, J.; Ehrlich, H.C.; Aiche, S.; et al. ProteomicsDB. *Nucleic Acids Res.* **2018**, *46*, D1271–D1281. [PubMed]
35. Samaras, P.; Schmidt, T.; Frejno, M.; Gessulat, S.; Reinecke, M.; Jarzab, A.; Zecha, J.; Mergner, J.; Giansanti, P.; Ehrlich, H.-C.; et al. ProteomicsDB: A multi-omics and multi-organism resource for life science research. *Nucleic Acids Res.* **2019**, *48*, D1153–D1163. [CrossRef]
36. Koch, R.; Gelderblom, H.; Haveman, L.; Brichard, B.; Jürgens, H.; Cyprova, S.; Berg, H.V.D.; Hassenpflug, W.; Raciborska, A.; Ek, T.; et al. High-Dose Treosulfan and Melphalan as Consolidation Therapy Versus Standard Therapy for High-Risk (Metastatic) Ewing Sarcoma. *J. Clin. Oncol.* **2022**, *40*, 2307–2320.
37. Copelan, E.A. Hematopoietic stem-cell transplantation. *N. Engl. J. Med.* **2006**, *354*, 1813–1826.
38. Thiel, U.; Schober, S.J.; Ranft, A.; Gassmann, H.; Jabar, S.; Gall, K.; von Lüttichau, I.; Wawer, A.; Koscielniak, E.; Diaz, M.A.; et al. No difference in survival after HLA mismatched versus HLA matched allogeneic stem cell transplantation in Ewing sarcoma patients with advanced disease. *Bone Marrow Transpl.* **2021**, *56*, 1550–1557. [CrossRef]
39. Thiel, U.; Wawer, A.; Wolf, P.; Badoglio, M.; Santucci, A.; Klingebiel, T.; Basu, O.; Borkhardt, A.; Laws, H.-J.; Kodera, Y.; et al. No improvement of survival with reduced- versus high-intensity conditioning for allogeneic stem cell transplants in Ewing tumor patients. *Ann. Oncol.* **2011**, *22*, 1614–1621. [CrossRef]
40. Nicolini, A.; Rossi, G.; Ferrari, P.; Carpi, A. Minimal residual disease in advanced or metastatic solid cancers: The G0-G1 state and immunotherapy are key to unwinding cancer complexity. Semin. *Cancer Biol.* **2020**, *79*, 68–82.
41. Altvater, B.; Kailayangiri, S.; Lanuza, L.P.; Urban, K.; Greune, L.; Flügge, M.; Meltzer, J.; Farwick, N.; König, S.; Görlich, D.; et al. HLA-G and HLA-E Immune Checkpoints Are Widely Expressed in Ewing Sarcoma but Have Limited Functional Impact on the Effector Functions of Antigen-Specific CAR T Cells. *Cancers* **2021**, *13*, 2857. [CrossRef]
42. Kirschner, A.; Thiede, M.; Grünewald, T.G.; Alba Rubio, R.; Richter, G.H.; Kirchner, T.; Busch, D.H.; Burdach, S.; Thiel, U. Pappalysin-1 T cell receptor transgenic allo-restricted T cells kill Ewing sarcoma in vitro and in vivo. *Oncoimmunology* **2017**, *6*, e1273301. [CrossRef] [PubMed]
43. Brohl, A.S.; Sindiri, S.; Wei, J.S.; Milewski, D.; Chou, H.-C.; Song, Y.K.; Wen, X.; Kumar, J.; Reardon, H.V.; Mudunuri, U.S.; et al. Immuno-transcriptomic profiling of extracranial pediatric solid malignancies. *Cell Rep.* **2021**, *37*, 110047. [PubMed]
44. Robbins, P.F.; Kassim, S.H.; Tran, T.L.N.; Crystal, J.S.; Morgan, R.A.; Feldman, S.A.; Yang, J.C.; Dudley, M.E.; Wunderlich, J.R.; Sherry, R.M.; et al. A Pilot Trial Using Lymphocytes Genetically Engineered with an NY-ESO-1–Reactive T-cell Receptor: Long-term Follow-up and Correlates with Response. *Clin. Cancer Res.* **2015**, *21*, 1019–1027. [CrossRef] [PubMed]
45. van Loenen, M.M.; de Boer, R.; Amir, A.L.; Hagedoorn, R.S.; Volbeda, G.L.; Willemze, R.; van Rood, J.J.; Falkenburg, J.F.; Heemskerk, M.H. Mixed T cell receptor dimers harbor potentially harmful neoreactivity. *Proc. Natl. Acad. Sci. USA* **2010**, *107*, 10972–10977. [CrossRef] [PubMed]
46. Cohen, C.J.; Zhao, Y.; Zheng, Z.; Rosenberg, S.A.; Morgan, R.A. Enhanced Antitumor Activity of Murine-Human Hybrid T-Cell Receptor (TCR) in Human Lymphocytes Is Associated with Improved Pairing and TCR/CD3 Stability. *Cancer Res.* **2006**, *66*, 8878–8886. [CrossRef]
47. Sommermeyer, D.; Uckert, W. Minimal Amino Acid Exchange in Human TCR Constant Regions Fosters Improved Function of TCR Gene-Modified T Cells. *J. Immunol.* **2010**, *184*, 6223–6231. [CrossRef]
48. Scholten, K.B.; Kramer, D.; Kueter, E.W.; Graf, M.; Schoedl, T.; Meijer, C.J.; Schreurs, M.W.; Hooijberg, E. Codon modification of T cell receptors allows enhanced functional expression in transgenic human T cells. *Clin. Immunol.* **2006**, *119*, 135–145. [CrossRef]

49. Cohen, C.J.; Li, Y.F.; El-Gamil, M.; Robbins, P.F.; Rosenberg, S.A.; Morgan, R.A. Enhanced Antitumor Activity of T Cells Engineered to Express T-Cell Receptors with a Second Disulfide Bond. *Cancer Res.* **2007**, *67*, 3898–3903. [CrossRef]
50. Provasi, E.; Genovese, P.; Lombardo, A.L.; Magnani, Z.I.; Liu, P.-Q.; Reik, A.; Chu, V.; Paschon, D.E.; Zhang, L.; Kuball, J.; et al Editing T cell specificity towards leukemia by zinc finger nucleases and lentiviral gene transfer. *Nat. Med.* **2012**, *18*, 807–815 [CrossRef]
51. Eyquem, J.; Mansilla-Soto, J.; Giavridis, T.; van der Stegen, S.J.C.; Hamieh, M.; Cunanan, K.M.; Odak, A.; Gönen, M.; Sadelain, M Targeting a CAR to the TRAC locus with CRISPR/Cas9 enhances tumour rejection. *Nature* **2017**, *543*, 113–117. [PubMed]
52. Roth, T.L.; Puig-Saus, C.; Yu, R.; Shifrut, E.; Carnevale, J.; Li, P.J.; Hiatt, J.; Saco, J.; Krystofinski, P.; Li, H.; et al. Reprogramming human T cell function and specificity with non-viral genome targeting. *Nature* **2018**, *559*, 405–409. [PubMed]
53. Müller, T.R.; Jarosch, S.; Hammel, M.; Leube, J.; Grassmann, S.; Bernard, B.; Effenberger, M.; Andrä, I.; Chaudhry, M.Z.; Käuferle, T.; et al. Targeted T cell receptor gene editing provides predictable T cell product function for immunotherapy. *Cell Rep. Med.* **2021**, *2*, 100374. [CrossRef]
54. Moosmann, C.; Müller, T.R.; Busch, D.H.; Schober, K. Orthotopic T-cell receptor replacement in primary human T cells using CRISPR-Cas9-mediated homology-directed repair. *STAR Protoc.* **2022**, *3*, 101031.
55. Santeramo, I.; Bagnati, M.; Harvey, E.J.; Hassan, E.; Surmacz-Cordle, B.; Marshall, D.; Di Cerbo, V. Vector Copy Distribution at a Single-Cell Level Enhances Analytical Characterization of Gene-Modified Cell Therapies. *Mol. Ther.-Methods Clin. Dev.* **2020**, *17*, 944–956. [CrossRef] [PubMed]
56. Aiuti, A.; Cossu, G.; De Felipe, P.; Galli, M.C.; Narayanan, G.; Renner, M.; Stahlbom, A.; Schneider, C.K.; Voltz-Girolt, C. The Committee for Advanced Therapies' of the European Medicines Agency Reflection Paper on Management of Clinical Risks Deriving from Insertional Mutagenesis. *Hum. Gene Ther. Clin. Dev.* **2013**, *24*, 47–54. [PubMed]
57. Monjezi, R.; Miskey, C.; Gogishvili, T.; Schleef, M.; Schmeer, M.; Einsele, H.; Ivics, Z.; Hudecek, M. Enhanced CAR T-cell engineering using non-viral Sleeping Beauty transposition from minicircle vectors. *Leukemia* **2016**, *31*, 186–194. [CrossRef]
58. Leibowitz, M.L.; Papathanasiou, S.; Doerfler, P.A.; Blaine, L.J.; Sun, L.; Yao, Y.; Zhang, C.-Z.; Weiss, M.J.; Pellman, D. Chromothripsis as an on-target consequence of CRISPR–Cas9 genome editing. *Nat. Genet.* **2021**, *53*, 895–905. [CrossRef]
59. Cullot, G.; Boutin, J.; Toutain, J.; Prat, F.; Pennamen, P.; Rooryck, C.; Teichmann, M.; Rousseau, E.; Lamrissi-Garcia, I.; Guyonnet-Duperat, V.; et al. CRISPR-Cas9 genome editing induces megabase-scale chromosomal truncations. *Nat. Commun.* **2019**, *10*, 1136. [CrossRef]
60. Slaymaker, I.M.; Gao, L.; Zetsche, B.; Scott, D.A.; Yan, W.X.; Zhang, F. Rationally engineered Cas9 nucleases with improved specificity. *Science* **2016**, *351*, 84–88. [CrossRef]

cancers

MDPI

Article

Identification of Factors Driving Doxorubicin-Resistant Ewing Tumor Cells to Survival

Semyon Yakushov [1], Maxim Menyailo [2], Evgeny Denisov [2], Irina Karlina [1], Viktoria Zainullina [2], Kirill Kirgizov [3], Olga Romantsova [3], Peter Timashev [4] and Ilya Ulasov [1,*]

1 Group of Experimental Biotherapy and Diagnostics, Institute for Regenerative Medicine, World-Class Research Centre "Digital Biodesign and Personalized Healthcare", I.M. Sechenov First Moscow State Medical University, 119991 Moscow, Russia
2 Laboratory of Cancer Progression Biology, Cancer Research Institute, Tomsk National Research Medical Center, Russian Academy of Sciences, 634009 Tomsk, Russia
3 Research Institute of Pediatric Oncology and Hematology at N.N. Blokhin National Medical Research Center of Oncology, Ministry of Health of Russia, 115478 Moscow, Russia
4 World-Class Research Centre "Digital Biodesign and Personalized Healthcare", Sechenov First Moscow State Medical University, 119991 Moscow, Russia
* Correspondence: ulasov_i_v@staff.sechenov.ru; Tel.: +7-901-797-5406

Simple Summary: It is known that doxorubicin is one of the standards for chemotherapy treatment against Ewing sarcoma. Despite its widespread use, doxorubicin treatment initiates tumor escape mechanisms and disease relapse. Our study aims to identify the potential biomarkers of doxorubicin resistance in primary cultures of Ewing sarcoma cells using single-cell transcriptomic and proteomic analyses. To assess the specificity of identified gene biomarkers, we used publicly available datasets to represent mRNA profiles of patient samples and short-lived cultures of tumor cells, established earlier. Through our investigation, we confirmed that MGST1 and the new marker COL6A2 are both produced by doxorubicin-resistant cells and demonstrated clinical significance for the survival of patients with Ewing sarcoma.

Abstract: Background: Ewing sarcoma (ES) cells exhibit extreme plasticity that contributes to the cell's survival and recurrence. Although multiple studies reveal various signaling pathways mediated by the EWSR1/FLI1 fusion, the specific transcriptional control of tumor cell resistance to doxorubicin is unknown. Understanding the molecular hubs that contribute to this behavior provides a new perspective on valuable therapeutic options against tumor cells. Methods: Single-cell RNA sequencing and LC-MS/MS-based quantitative proteomics were used. Results: A goal of this study was to identify protein hubs that would help elucidate tumor resistance which prompted ES to relapse or metastasize. Several differentially expressed genes and proteins, including adhesion, cytoskeletal, and signaling molecules, were observed between embryonic fibroblasts and control and doxorubicin-treated tumor cell lines. While several cancer-associated genes/proteins exhibited similar expression across fibroblasts and non-treated cells, upregulation of some proteins belonging to metabolic, stress response, and growth pathway activation was uniquely observed in doxorubicin-treated sarcoma cells, respectively. The novel information on differentially expressed genes/proteins provides insights into the biology of ES cells, which could help elucidate mechanisms of their recurrence. Conclusions: Collectively, our results identify a novel role of cellular proteins in contributing to tumor cell resistance and escape from doxorubicin therapy and contributing to ES progression.

Keywords: Ewing sarcoma; scRNA-seq; proteomics; doxorubicin; cell resistances

Citation: Yakushov, S.; Menyailo, M.; Denisov, E.; Karlina, I.; Zainullina, V.; Kirgizov, K.; Romantsova, O.; Timashev, P.; Ulasov, I. Identification of Factors Driving Doxorubicin-Resistant Ewing Tumor Cells to Survival. *Cancers* **2022**, *14*, 5498. https://doi.org/10.3390/cancers14225498

Academic Editors: Stefan Burdach, Uta Dirksen and Poul H. Sorensen

Received: 6 October 2022
Accepted: 3 November 2022
Published: 9 November 2022

1. Background

Ewing sarcoma (ES) is a primary tumor that affects bones [1]—and much less often, soft tissues [2]—with a greater incidence in young people aged 10–15 years. The survival rate for ES patients has improved significantly with combined treatment, including surgery, chemotherapy, and radiation; however, this harsh treatment leaves patients with disabilities. Additionally, most of the time this disease occurs locally, with 5-year event-free survival up to 78% [3], but almost 20% of patients will experience localized disease relapse and later produce metastases (with an overall survival rate of around 28% [4,5]). Although the lung is the most prevalent location for ES metastases (85% of all patients), a recent investigation suggests that bone metastases become a valuable predictor for lung metastases during ES progression [6]. Therefore, understanding the mechanisms of tumor escape from therapy will lead to the development of better therapeutic options that are needed to elevate the therapeutic effect of current modalities and improve the outcome.

Progression and metastasis of ES is a complex biological process that involves bone lytic destruction and tumor cell propagation. It has been demonstrated that ES cells initiate osteolysis on the border between normal and neoplastic tissue via increased immobilization of macrophages, since they have been involved in the destruction of the bone extracellular collagen-containing matrix (ECCM) [7]. Along with the activation of osteoclast differentiation [8], ECCM destruction lays a foundation for tumor vascularization in the area of bone resorption and activation of angiogenesis [9]. The last step is critical for tumor cell spread and dissemination to the nearby tissue. The development of distant metastases requires the tumor cells to have an elevated level of matrix-cell adhesion proteins that allow tumor cells to attach and penetrate tissue. Several studies, particularly in the last decade, hypothesized that proteins of the carbonic anhydrase group, more specifically CA9 [10] and cadherin-11 [11], act as regulators of processes associated with sarcoma migration, invasion, and metastasis. Although recent studies provided further evidence for these [12] and other [13] proteins, such as MGST1 [14], to be involved in tumor dissemination, these proteins appeared to share an ability to regulate the formation of metastases in various cancers [15]. Despite scientific reports that various chemotherapeutic drugs [16,17] induce prometastatic proteins, therefore the tumor escape molecular mechanisms are not fully understood and have limited therapeutic efficacies against ES.

To understand the processes that are responsible for tumor cell survival, and compare our findings with data published in the field of ES resistance to DOX [18], we aim to identify the molecular basis for the cellular reaction in response to doxorubicin stress, which precedes most ES escapes from stress and generates tumor tissue growth. In this study, we used RNA-seq analysis to assess differentially expressed genes (DEGs) using the Gene Expression Omnibus (GEO) database. We observed the upregulation of 992 genes. To validate the specificity of gene expression and their attribution to the tumor cells, we performed single-cell RNA-seq and tandem mass tag (TMT) proteomic approaches to investigate the transcriptome and proteome changes of patient-derived ES culture in comparison to embryonic fibroblasts (M19). Through transcriptomic analysis, we found 45 (27 up- and 18 downregulated) DEGs, and at the protein level, we found 124 (44 up- and 78 downregulated) differentially expressed proteins (DEPs) in doxorubicin-treated ES cells vs. doxorubicin-treated M19 cells. Furthermore, a comparison of proteomic data of doxorubicin-treated fibroblasts and doxorubicin-treated ES cells identified a unique protein signature of 11 markers, among which COL6A2, PSME1 and FLNC demonstrated clinical significance by survival analysis GSE63155 for ES patients. Our results highlight key regulatory proteins that coordinate the response of ES cells to the doxorubicin treatment by integrating signaling from MGST1 cellular metabolism, and the COL6A2 molecule, whose interactions were underappreciated in the previous investigations. This adds to our knowledge of the molecular regulatory network of doxorubicin resistance and the transcriptional hub that governs tumor escape from doxorubicin, and might hold promise as a future therapeutic target.

2. Materials and Methods

2.1. Materials

DMEM media, glutamine, fetal bovine serums (FBS) and protease inhibitor cocktail tablets (EDTA-free) were purchased from Roche Diagnostics (GmbH, Mannheim, Germany).

2.2. Primary Tissue Samples

Our study included 14 tumor tissue specimens with a confirmed diagnosis of ES and treated in 2020–2021 at the clinical facility of NIMC N.N. Blokhin. All parents of sick children gave written informed consent to use their clinical data in agreement with approval by the institutional ethics committee (Research Institute of Pediatric Oncology and Hematology at N.N. Blokhin National Medical Research Center of Oncology). Based on an agreement between Sechenov First Moscow State Medical University and NIMC N.N. Blokhin, the clinical specimens were available to isolate tumor samples. Under this agreement, we established a few primary cultures from ES tumor samples; however, for the current study, we used only ES36 cells. The percentage of tumor cells in each resected tissue was more than 80% (pathologist assessment). The patient who was a donor for ES36 sarcoma cells developed a tumor mass around the tibia and later received pilot anti-relapse treatment under EE2008 protocol. However, around 1 year later this patient still relapsed.

2.3. Cell Cultures

Human primary Ewing sarcoma ES36 cells and human embryonic fibroblasts M19 were established as described [19]. All cell lines were routinely tested for purity and mycoplasma contamination. Human sarcoma cell line ES36 (from a 4-year-old male) and M19 (from a 9-year-old female) were grown in DMEM media, supplemented with 10% FBS and l-glutamine (1%) at 37°, 5% CO_2. M19 cells were received from colleagues at the National Research Center for Epidemiology and Microbiology named after Honorary Academician N.F. Gamaleya of the Ministry of Health of the Russian Federation. The patient-derived ES sarcoma was used in passage 3. ES36 and M19 were authenticated using STR protocol (24 markers) at Sistema Biotech (Moscow, Russia). The multiplex analysis was conducted using the 3500 Genetic Analyzer (Applied Biosystems, Waltham, MA, USA) and data were assessed via GeneMapper ID-X v1.4 (Applied Biosystems, Waltham, MA, USA). The received profiles of cells were matched against ATCC-based STR profiles. Additionally, EWSR1/FLI1 fusion was PCR-amplified using primers, visualized with agarose gel, and sequenced. ES36 and M19 cells were used to titrate Doxorubicin (10 Mm stock), and to obtain IC50. We further selected a dose of Dox 10-fold lower than IC50 at ES36, and with no impact on M19 proliferation. Under these conditions, we incubated ES36 and M19 for 7 days with Dox(0.125 mkM) before analysis.

2.4. cDNA Library Preparation and Single-Cell RNA-Seq

Single-cell cDNA libraries were prepared using Single Cell 3′ Reagent Kit v3.1 and a 10× Genomics Chromium Controller. The number of cells in each channel of the Single-Cell Chip G did not exceed 5000. The dsDNA High Sensitivity kit measured the concentration of cDNA libraries on a Qubit 4.0 fluorometer (Thermo Fisher Scientific, Waltham, MA, USA). The quality of cDNA libraries was assessed using High Sensitivity D1000 ScreenTape on a 4150 TapeStation (Agilent, Santa Clara, CA, USA). The ready cDNA libraries were pooled, denatured, and sequenced on NextSeq 2000 (Illumina, San Diego, CA, USA) using pair-end reads (28 cycles for reading 1, 91 cycles for reading 2, and 8 cycles for the i7 index). Cell Ranger software (version 6.1.1) provided by 10× Genomics (version 6.1.1) (Pleasanton, CA, USA) was used to perform sample demultiplexing, alignment to the GRCh38 transcriptome, filtering, barcode counting, and UMI counting. The resulting datasets, M19 and ES36, were subjected to quality control. The subset was produced at a mark of 25% for mitochondrial material and recorded duplicates for each of the samples. Functional annotation was performed using R gProfiler2 gene ontology analysis [20].

2.5. Proteomic Data Generation and Analysis

We extracted proteins from the live ES36 patient-derived ES adherent cells grown in DMSO (n = 3) or doxorubicin (n = 3) using the methanol–chloroform method. Then, each sample was treated with 100 mM $(NH4)_2CO_3$ (pH 7.8) in the presence of 0.2% ProteaseMax solution (Promega, Madison, WI USA), before sonication for 10 min, followed by treatment with 100 μM Tris(2-carboxyethyl)phosphine during 1 h incubation at 55 °C and 100 μM Iodoacetamide (1 h incubation at room temperature). A digestion mix of 0.1% of trypsin and 0.2% of Protease Mix were added to each reaction for a further 3 hours of incubation and then the reaction was terminated. The protein concentration was measured using a «BCA protein assay kit» (EMD Millipore, Billerica, MA, USA) and the rest of the sample was dry-vacuumed using Centrivap (Labconco, Kansas City, MO, USA) at +45 °C. The digested and enriched peptides were dried in a speed vacuum and reconstituted with 0.1% formic acid in water for liquid chromatography-mass spectrometry (LC-MS/MS) analysis. An Q Exactive HF-X Mass Spectrometer (Thermo Fisher Scientific, Waltham, MA, USA), coupled to a Dionex 3000 RS Nano (Thermo Fisher Scientific, Waltham, MA, USA), was used to analyze digested peptides. We performed MS/MS data acquisition, converted data using an mzML format (XML (eXtensible Markup Language)-based common file format for proteomics mass spectrometric data), and processed with ThermoRawFileParser software (Thermo Fisher Scientific, Waltham, MA, USA) based on two protocols: the first used Identipy + Biosaur + Scavager to distinguish hybrid spectra. In the second we used MSFragger + Philosopher(peptideprophet/proteinprophet) + IonQuant for data quantification. Spectra were searched against species-specific TrEMBL protein databases generated from the human genome and bovine serum proteins, detected early by us in fetal bovine serum samples.

The study of relative changes in protein quantities was carried out using the DEP software package, which is based on the LIMMA statistical package (an empirical Bayesian method developed initially for the analysis of microchip data). For protein quantification, counts from peptide spectrum matches (PSM) were normalized and filtered using the MNAR (missed-not-at-random) method, log2-transformed, and hierarchically clustered by expression patterns through the time series in R., using the entire protein-coding gene set as a background.

2.6. Data Source, Data Processing, and Data Distribution

The RNA-seq data and the corresponding clinical data of overall survival (OS) were downloaded from Gene Expression Omnibus (GEO) (https://www.ncbi.nlm.nih.gov/geo/ data portal. Gene expression profiles of 46 samples with primary ES deposed at GSE63155 (n = 46) were used (https://www.ncbi.nlm.nih.gov/geo/query/acc.cgi?acc=GSE63155 (accessed on 2 February 2021)) [21]. Due to the lack of data on relapses, information about the end of the experiment (treatment) for patients with the status of "survivor" was regarded as definitively true. For the analysis, gene expression data from all 46 patients was used without censorship. The normalized data were uploaded by the authors and retrieved from the data portal mentioned above. The LIMMA R package was used to identify differentially expressed genes (DEGs). Gene probe IDs for which an "-/NA" notation was not available were removed. To obtain differential expression profiles, the GEO data were divided into two groups with respect to survivorship (n = 32, survivors) and non-survivors (n = 14) after receiving treatment.

2.7. Functional GO Enrichment Analysis

Relative DEG lists were analyzed by a web server for functional enrichment analysis. g:Profiler Datasets with adjusted $p < 0.05$ (Bonferroni correction) were deemed meaningful enrichment pathways. The "Cluster Profiler" R package was used to combine gene annotation and gene expression analysis results. The results are visualized in a dot-plot.

2.8. Experimental Validation of Fold-Change Values Using Semiquantitative RT-PCR Analysis

QIAzol lysis reagent (Qiagen, Germantown, MD, USA) was used to isolate total RNA from 14 ES patients (6 females and 8 males). RNAase-free DNAase1 (Invitrogen, Waltham, MA, USA) was used to remove residual genomic DNA in isolated total RNA. Total RNA was quantified using a Nanodrop spectrophotometer (Beckman Coulter, Inc., Brea, CA USA). Once quantified, 10 ng of total RNA per sample was used to synthesize single-strand cDNA with the iScript cDNA synthesis kit (Bio-Rad, Hercules, CA USA) for one-step RT-PCR analysis. The synthesized single-strand cDNA of 100 and 50 ng per sample was utilized as a template for semi-quantitative PCR analysis with an Encyclo Plus PCR kit (Evrogen, Moscow, Russia) following the manufacturer's instructions (https://evrogen.com/products/Encyclo-PCR-kit/Encyclo-PCR-kit.shtml (accessed on 2 February 2021). Primer pairs for PCR analyses were selected from previous studies [22]. All PCRs were performed on three replicates per cDNA sample. To obtain the fold change, MGST1 and COL6A2 values of analyzed samples were normalized to β-Actin mRNA expression, as an endogenous reference gene.

2.9. Statistical Analyses

GSE63155 includes an analysis of whole-genome expression, whereas GSE146221 is a single-cell analysis of three established ES cell lines (TC71, CHLA9, and CHLA10). The GSE146221 data were generated by 10× Genomics single-cell RNA-sequencing, whereas GSE63155 was generated using Affymetrix platforms. Cox regression analysis was performed for GSE63155 and gene expression correlation analysis for GSE146221. DEGs analysis was performed between groups of survivors and non-survivors, and between individuals with UMAP clusters for GSE63155 and GSE146221 datasets, respectively. Using proteomic profiles of ES36, ES36 DOX, M19, and M19-DOX, we compared DEPs between the following groups: ES36 vs. M19, ES36 vs. ES36-DOX, and M19-DOX vs. ES36-DOX samples, respectively. Our single-cell transcriptomic dataset was used to compare DEGs between clusters 11 and 15 of ES36 vs. M19, ES36 vs. M19 cells and individual UMAP clusters. All analysis was conducted with a statistical significance limit of $p = 0.05$. DEGs analysis of scRNA clusters was performed by the Seurat package R with a statistical significance limit of p adjusted = 0.05. All statistical analyses were performed using GraphPad Prism 8.3.0 and R stat Version 4.0.2 (R-project.org (accessed on 25 January, 2022)), and R packages were obtained from the Bioconductor project (www.bioconductor.org (accessed on 25 January 2021)). Statistical significance was set at $p < 0.05$.

2.10. Availability of Data and Materials

The datasets (SRA transcriptomic access data: PRJNA846952 and proteomic access data: in progress) used and/or analyzed during the current study are available.

3. Results

A significant number of genes are elevated in ES tumor tissue.

To identify DEGs that are relevant to tumor cell survival, we analyzed the GSE63155 microarray database after normalization. Based on the patient survival status (survivors ($n = 32$) vs. non-survivors ($n = 16$)), we applied the univariate and multivariate Cox proportional hazards regression methods for 48 tumor tissues available for analysis (Table 1).

Both analyses showed that an event-free survival (EFS) event was a risk factor, while gender, age, primary tumor site, and tumor content did not correlate with OS. Keeping the cutoff value strict among 17,634 genes, genes with p-value < 0.05 of their expression between the two groups produced DEGs and the data were depicted in a volcano plot (Figure 1A). Using one-way analysis of variance, we found that multiple transcripts were upregulated in the non-survivor tumor samples. Overall, the Cox regression revealed 812 and 194 genes that were up- or downregulated during the analysis, respectively. Assuming that benefits for ES patient survival occurred from the inhibition of neoplastic signaling inside tumors, we performed GO term analysis. As shown in Figure 1B,C, the

GO analysis of upregulated genes revealed an association of DEGs with blood vessel formation, angiogenesis (PECAM1/TGFBI/CD34/THY1/SEMA6A) and cell migration (MMP9/CCL21/MMP2/AKT3/ITGA6/FLT1). At the same time, GO analysis indicates that DEGs mainly regulate cellular metabolism, including cytochrome P450 and glutathione transferase (Table 2).

Table 1. Univariate and multivariate Cox analysis of the correlation clinical factors with OS in patients with ES.

	Univariate			Multivariate		
	Beta	**HR (95% CI)**	***p*.Value**	**Beta**	**HR (95% CI)**	***p*.Value**
Sex	−0.47	0.62 (0.21–1.9)	0.4	−2.358	0.095 (0.005–1.84)	0.120
Age_at_enrollment_days	0.00033	1 (1–1)	0.07	−0.004	0.996 (0.98–1.0)	0.465
Age	0.11	1.1 (0.99–1.3)	0.075	1.637	5.14 (0.085–309.385)	0.433
Primary_tumor_site	−0.28	0.76 (0.35–1.6)	0.47	1.502	4.492 (0.377–53.53)	0.235
Efs_event	2	7.7 (3.4–17)	9.5e-07	3.941	51.476 (3.736–709.32)	0.003
Tumor_content	0.77	2.2 (0.57–8.2)	0.26	−0.562	0.57 (0.088–3.676)	0.554

Abbreviations: ES, Ewing sarcoma; EFS event; HR, hazard ratio; 95% CI, 95% confidence interval.

To investigate further whether some of these genetic traits are related to clinical prognosis, clinical information was extracted from the GSE63155 database. Based on that information, the top 30 genes overexpressed in the non-survivor group were CCL18, LUM, SPP1, VCAN, LPL, MMP9, FAP, TSPAN7, THBS1, CHN1, LYZ, HBB, EFEMP1, CTSC, COL12A1, VCAM1, C8orf4, NETO2, FAM198B, LAPTM5, LOX, COL3A1, TM4SF18, PAX3, GPNMB, AHR, DOCK11, RASGRP3, CDH2, and COL11A1, mostly responsible for extracellular matrix organization (GO:0030198) and cell adhesion (GO:0007155). Among the abundant genes with high mRNA expression, significantly worse OS was detected for *SPP1* (Hazard Ratio, log-rank 8.3), VCAM1 (7.9), CCL18 (5.1), CTSC (4.6), THBS1 (4.5), and the log-rank p-values were: 7×10^{-5}, 12×10^{-5}, 5×10^{-4}, 1×10^{-4} and 1×10^{-4} relatively. No statistically significant difference in prognosis was obtained in GEO-deposed patient tumor samples with high and low expressions for LUM (0.056), LYZ (0.18), HBB (0.095), EFEMP1 (0.127), and COL12A1 (0.25) markers. Although previously for ES patients, SPP1 [23], MGST1 [18], and miR34a [24] were considered as markers of ES progression and ES cell sensitivity towards doxorubicin, no additional prognostic factors have been proposed based on individualized treatment options. As chemotherapeutic stress in normal and neoplastic cells results in the production of stress signals, we aimed to establish markers for pathological state and doxorubicin resistance using patient-derived ES cells and embryonic fibroblasts as controls and determine the clinical significance of biomarkers using scRNA-seq and mass-spectroscopy.

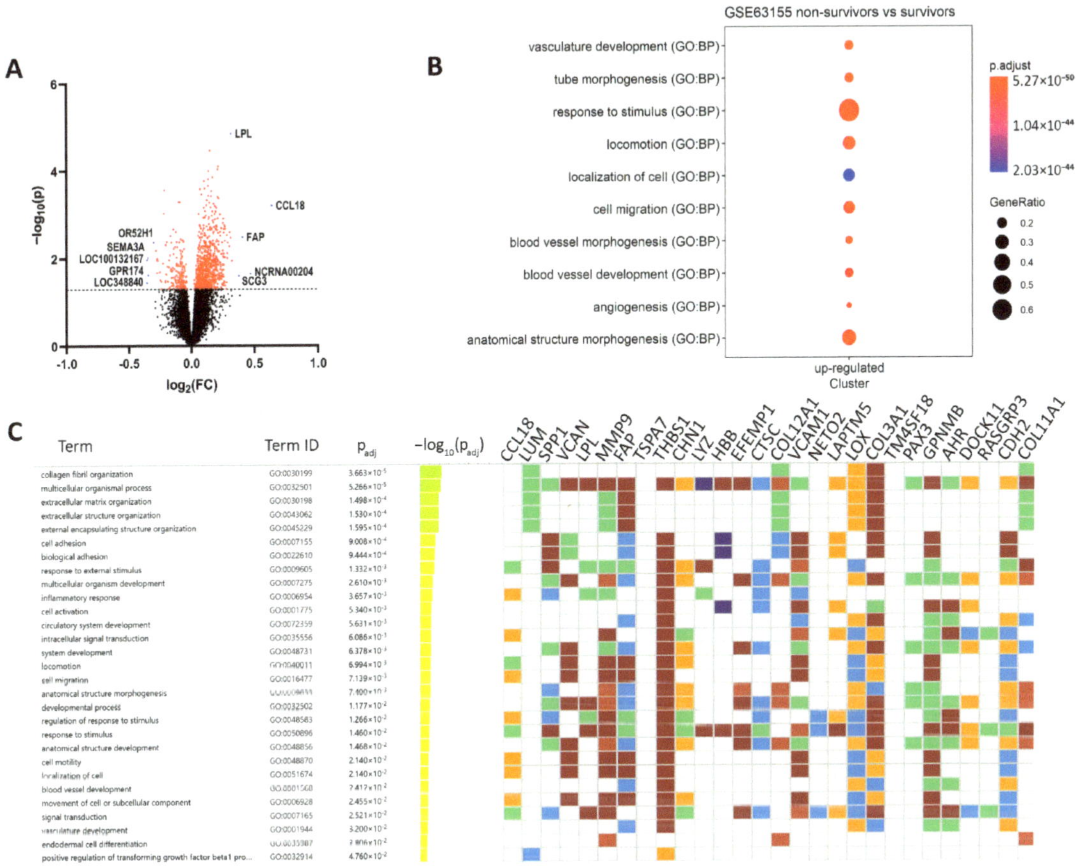

Figure 1. Differentially regulated genes (DEGs) in ES patients with various survival statuses. (**A**) Volcano plot diagram of differentially regulated genes in the tumor tissue from survivor (n = 32) vs. nonsurvivor (n = 14) patient samples (data were extracted from GSE63155). The data were extracted from GSE63155 and resulted in 1006 genes. Differentially regulated genes with $p \leq 0.05$ are labeled red. The x-axis represents Log2 fold change; the y-axis represents negative log10-transformed p-values. Dots on the left side represent transcripts upregulated in survivors, and dots on the right side represent transcripts upregulated in non-survivors. Blue dots represent most and downregulated genes by log2fold change. (**B**) a GO analysis of DEGs associated with ES patient survival. The node sizes indicate gene counts enriched in the pathway and the color shows the p-value from low (red) to high (blue) level. Significance was set at $p < 0.05$. DEGs, differentially expressed genes; GO, gene ontology. The plot was created using the R gProfiler2. (**C**) Significant enrichment of molecular function of interacting proteins; gene-ontology-based GSEA analysis indicates the top 30 genes most expressed in non-survivor vs. survivor patient samples. The heatmap indicates a log2 fold change in mRNA expression. The colors for different evidence codes in the table: red: inferred from experiment, direct assay, mutant phenotype, genetic interaction, physical interaction; purple: inferred from experiment, direct assay, mutant phenotype, genetic interaction, physical interaction; green: traceable author, non-traceable author, inferred by a curator; orange: expression pattern, sequence or structural similarity, genomic context, sequence model, sequence alignment, sequence ontology, the biological aspect of the ancestor, rapid divergence; blue: reviewed computational analysis, electronic annotation.

Table 2. The biological process and pathways regulated by DEGs.

	Description	ID	*p* Value	Count/Gene ID
Downregulated	locomotion	GO:0040011	1.14×10^{-26}	CCL18/VCAN/MMP9/FAP/THBS1/CHN1/VCAM1/LOX/COL3A1/GPNMB/CDH2/SULF1/GJA1/ACTA2/TGFBR2/SMOC2/CCL21/MMP2/DCLK1/AKT3/CD200/MAP1B/RND3/KDR/MCC/LEF1/SGK1/SDC2/ENPP2/PECAM1/KITLG/CCR1/ADAMTS1/ITGA6/EPS8/ADAMTS9/EFNB2/CD34/THY1/SEMA6A/LYN/ALX1/S100A8/ITGA9/MTUS1/NRP1/PLVAP/PRKD1/ETS1/LYVE1/CDH5/PLXNC1/PODXL/ITGA2/FCER1G/TLR4/TEK/NRP2/PDGFRB/LAMB1/ELMO1/EMP2/MCTP1/FLT1/DOCK2/PTPRC/XBP1/CHL1/SATB2/PLXNA2/DLC1
	angiogenesis	GO:0001525	4.39×10^{-26}	FAP/THBS1/GPNMB/SULF1/THBS2/PTPRB/TGFBR2/SMOC2/MMP2/CALCRL/CYBB/ADAM12/AKT3/ANPEP/KDR/LEF1/ENPP2/COL15A1/CEMIP2/TGFBI/APLNR/ADAMTS1/ADAMTS9/EFNB2/CD34/THY1/SEMA6A/SAT1/NRP1/PRKD1/ETS1/COL4A1/CDH5/EMCN/EPAS1/TEK/COL4A2/NRP2/PDGFRB/EMP2/FLT1/XBP1
	positive regulation of cell migration	GO:0030335	2.52×10^{-19}	MMP9/THBS1/GPNMB/ACTA2/TGFBR2/SMOC2/CCL21/MMP2/AKT3/KDR/LEF1/ENPP2/PECAM1/KITLG/CCR1/ADAMTS1/ITGA6/THY1/SEMA6A/LYN/NRP1/PLVAP/PRKD1/ETS1/LYVE1/CDH5/PODXL/ITGA2/TLR4/TEK/NRP2/PDGFRB/LAMB1/FLT1/PTPRC/XBP1
	cell differentiation	GO:0030154	6.24×10^{-16}	SPP1/VCAN/LPL/MMP9/CHN1/EFEMP1/COL12A1/VCAM1/LOX/COL3A1/DOCK11/CDH2/COL11A1/KRT10/SULF1/STEAP4/GJA1/A2M/ACTA2/TGFBR2/CCL21/MMP2/CD53/GPM6B/DCLK1/ADAM12/PLEK/MAP1B/ANPEP/KDR/ESRP1/LEF1/SGK1/SDC2/PECAM1/COL15A1/KITLG/CCR1/TGFBI/APLNR/ITGA6/MAP2/ADAMTS9/MTSS1/EFNB2/CD34/CD36/THY1/SEMA6A/LYN/HEY2/RAI14/ALX1/S100A8/RPS6KA2/SPRY4/NRP1/PRKD1/TMEM119/ETS1/PRICKLE1/COL4A1/CDH5/PLXNC1/PODXL/ITGA2/SLC6A6/EPAS1/FCER1G/TLR4/TEK/COL4A2/FRMD6/NRP2/PDGFRB/LAMB1/EMP2/FLT1/FARP1/DOCK2/PTPRC/XBP1/CHL1/RPS6KA3/SATB2/PLXNA2/TAGLN/BHLHE40
	extracellular matrix organization	GO:0030198	8.07×10^{-14}	LUM/MMP9/FAP/COL12A1/LOX/COL3A1/COL11A1/SULF1/SMOC2/MMP2/GPM6B/COL14A1/MMP16/COL15A1/TGFBI/ADAMTS1/ITGA6/ADAMTS9/ITGA9/FBLN5/COL4A1/ITGA2/COL4A2/LAMB1/NID2
	cell-substrate adhesion	GO:0031589	7.69×10^{-11}	THBS1/VCAM1/COL3A1/EDIL3/CCL21/GPM6B/VWF/KDR/ITGA6/ADAMTS9/CD34/CD36/THY1/SPRY4/NRP1/FBLN5/LYVE1/ITGA2/TEK/LAMB1/EMP2/NID2/DLC1
	chemotaxis	GO:0006935	7.74×10^{-11}	CCL18/THBS1/CHN1/VCAM1/LOX/GPNMB/SMOC2/CCL21/KDR/LEF1/ENPP2/CCR1/EFNB2/SEMA6A/LYN/S100A8/ITGA9/MTUS1/NRP1/PRKD1/PLXNC1/ITGA2/FCER1G/NRP2/PDGFRB/FLT1/DOCK2/CHL1/PLXNA2
	cell-matrix adhesion	GO:0007160	1.55×10^{-9}	THBS1/VCAM1/COL3A1/CCL21/GPM6B/KDR/ADAMTS9/CD34/CD36/THY1/NRP1/FBLN5/LYVE1/ITGA2/TEK/EMP2/NID2/DLC1
	cell-substrate junction assembly	GO:0007044	0.0002	THBS1/GPM6B/KDR/ITGA6/THY1/NRP1/ITGA2/TEK/DLC1

Table 2. *Cont.*

Description	ID	p Value	Count/Gene ID
collagen-containing extracellular matrix	GO:0062023	3.43×10^{-21}	LUM/VCAN/MMP9/THBS1/EFEMP1/COL12A1/CTSC/COL3A1/CDH2/COL11A1/EDIL3/SULF1/A2M/THBS2/SMOC2/MMP2/VWF/COL14A1/SDC2/SPON1/PCOLCE/COL15A1/TGFBI/ADAMTS1/SPARCL/ADAMTS9/S100A8/BGN/HAPLN1/FBLN5/COL4A1/COL4A2/LAMB1/NID2
extracellular exosome	GO:0070062	6.20×10^{-10}	LUM/SPP1/MMP9/THBS1/LYZ/HBB/EFEMP1/COL12A1/CTSC/VCAM1/PPIC/EDIL3/KRT10/STEAP4/A2M/ACTA2/CD53/VWF/ANPEP/MAN1A1/PECAM1/PCOLCE/COL15A1/CEMIP2/TGFBI/EPS8/PRSS23/THY1/LYN/S100A8/CD14/BGN/PLVAP/MYO1B/ENTPD1/FBLN5/LYVE1/RFTN1/PODXL/NT5E/FCGR3A/COL4A2/LAMB1/DOCK2/RAB27B/PTPRC/MARCKS/CHL1/NID2/TNFSF10/PYGL
collagen trimer	GO:0005581	1.90×10^{-6}	LUM/COL12A1/LOX/COL3A1/COL11A1/COL14A1/COL15A1/CD36/COL4A1/COL4A2
collagen binding	GO:0005518	1.53×10^{-8}	LUM/MMP9/THBS1/LOX/VWF/COL14A1/PCOLCE/TGFBI/SPARCL1/ITGA2/NID2
glycosaminoglycan binding	GO:0005539	3.92×10^{-8}	VCAN/LPL/THBS1/GPNMB/COL11A1/SULF1/THBS2/TGFBR2/SMOC2/PCOLCE/ADAMTS1/NRP1/BGN/HAPLN1/LYVE1/NRP2/PTPRC
Degradation of the extracellular matrix	REAC:R-HSA-1474228	6.00×10^{-8}	SPP1/MMP9/COL12A1/COL3A1/COL11A1/A2M/MMP2/COL14A1/MMP16/COL15A1/ADAMTS1/ADAMTS9/COL4A1/COL4A2/LAMB1
Collagen formation	REAC:R-HSA-1474290	4.10×10^{-6}	MMP9/COL12A1/LOX/COL3A1/COL11A1/COL14A1/PCOLCE/COL15A1/ITGA6/COL4A1/COL4A2
Collagen degradation	REAC:R-HSA-1442490	2.99×10^{-5}	MMP9/COL12A1/COL3A1/COL11A1/MMP2/COL14A1/COL15A1/COL4A1/COL4A2

Table 2. *Cont.*

Description	ID	p Value	Count/Gene ID
Collagen chain trimerization	REAC:R-HSA-8948216	0.0004	COL12A1/COL3A1/COL11A1/COL14A1/COL15A1/COL4A1/COL4A2

3.1. Hypoxia and Metastatic Traits Distinguish Tumor Cells from Non-Neoplastic Cells

To further investigate the function and significance of differentially expressed genes in the ES cell culture, we conducted a scRNA-seq analysis of ES36 cells (patient-derived ES cells) vs. M19 embryonic fibroblasts. We produced samples' cDNA libraries using a commercial droplet-based system and sequenced the libraries to obtain transcriptomes covering 500–5000 genes per cell. In total, 5000 cells from each sample were included in the analysis (Figure 2A,L). Overall, we detected an average of 5000 (range 0–10,000) as highly variable genes in the ES36 sample and 5000 (range 0–10,000) in the M19 cells. The GO annotations showed that the functions of genes mainly enriched in tumor cells were related to cell proliferation (GO: 0008283, 37 counts, $p = 1 \times 10^{-6}$) and cellular adhesion (GO: 0007155, 17 counts, $p = 7.1\times 10^{-6}$). Inhibited genes were mostly represented by respiratory chain complex I (gamma subunit) mitochondrial (CORUM: 2919, 2 counts, $p = 0.004$) signaling. Our results found that the top 20 highly variable genes in the tumor populations were: CLSPN, CDCA8, CDC20, ISG15, KIF2C, ALPL, STMN1, IFI6, HMGN2, NASP, FABP3, DHRS3, HSPB7, ID3, PDPN, ERRFI1, SRARP, COL8A2, LINC00337, and SERINC2, and in the M19 populations they were: RHBDL2, CDC20, CDCA8, CLSPN, KIF2C, FABP3, ISG15, STMN1, SFN, CDA, DRAXIN, SNHG12, ERRFI1, E2F2, AL590434.1, PDPN, CCDC30, EPHB2, KLF17, and SERINC2. Of note, M19 and ES36 shared some of the mRNA signatures for CLSPN, CDCA8, CDC20, ISG15, KIF2C, STMN, FABP3, PDPN, ERRFI1, and SERINC2 genes. After filtering the sample's data (Figure 2A through 2E, and 2F through 2J), we applied the Seurat algorithm to further gene clustering. Fibroblasts and ES cell transcriptomes were visualized with a Uniform Manifold Approximation Projection (UMAP) employing ten principal components.

The multiple sequencing runs from M19 and ES36 cells were combined into a single uniform manifold approximation and projection (UMAP) for visualization (Figure 3A) and clusterization (Figure 3B). Comparable trends of subset colocalization between M19 and ES36 cultured cells remained, reinforcing the conclusion that common gene expression by each cell type is a shared phenomenon regardless of cell type. Across the UMAP plot, we found a distinct number of clusters in which a majority of cells consisted of ES36 (clusters 11, and 15), or mixed (clusters 0, 1, 2, 3, 4, 5, 6, 7, 8, 9, 10, 12, 13, 14, and 15) cells. Furthermore, analysis of C11 and C15 revealed a unique distribution of upregulated genes (Figure 3C,D). Thus, C11 mostly accumulates cells with genes responsible for cell motility, cell adhesion, and migration; in C15, most upregulated genes participate in HIF1 signaling, and downregulated genes are involved in cell adhesion. Overall, in ES cells, MALAT1, FTH1, FTL, VIM, MT-CO1, TMSB4X, FN1, MT-CO2, NEAT1, S100A6, EEF1A1, CALD1, MT-CYB, TPT1, RPL41, TAGLN, RPLP1, LGALS1, TMSB10, CD63, RPS2, ACTB, RPS18, RPS8, RPL37A, RPL13, ANXA2, COL1A2, RPL10, and ACTG1 were most upregulated compared to their expression in M19 cells.

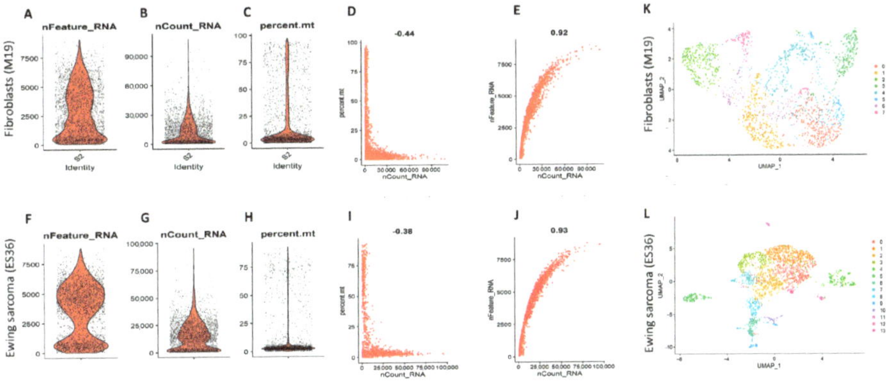

Figure 2. UMAPs of single-cell transcriptome data. Quality control filtering of embryonic fibroblasts (**A–E**) and ES36 tumor sample (**F–J**). Comparison of 10× Genomics-annotated doublets, singlets, and negatives was assessed using gene expression that was log-transformed using Seurat and presented as violin-box plots with highlighted median values. Violin plots showing the counts of each gene in each cell (**A**,**F**). Violin plots of the sum of the expression levels of all genes in each cell (**B**,**G**). Violin plots of the percentage of mitochondrial genes (**C**,**H**). Scatter plots for the percentage of mitochondrial genes in the sum of the expression levels of all genes in each cell (**D**,**I**). Scatter plots for the counts of genes (**E**,**J**). Embryonic human fibroblasts (**K**) and patient-derived ES (**L**) color-coded by cell identity. Colors show different cell clusters resulting from UMAP clustering using the SEURAT algorithm.

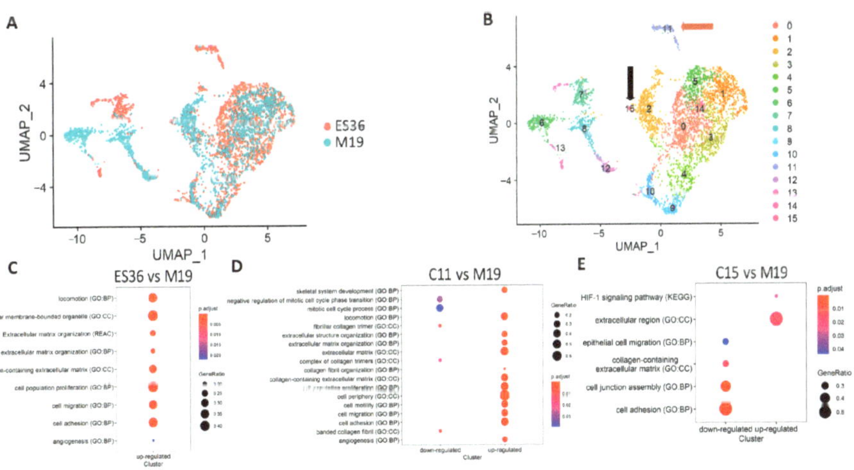

Figure 3. A single-cell ecosystem of embryonic fibroblasts and ES 36 cells. (**A**) Datasets for 10× Genomics analysis were derived from M19 fibroblasts and short-lived patient-derived ES culture (ES36). The 10× Genomics plots identify groups of cells with similar expression patterns. The average of 5000 cultured M19 cells and ES36 cells were green or red color-coded. (**B**) Distribution of single-cell population for each type of cell in the UMAP plot (first two dimensions) in accordance with the clusters made using the Seurat algorithm. Arrows point to the unique clusters that present in ES-patient-derived ES36 tumor culture; GO term analysis that summarizes the biological function and cellular component was performed for DEGs in ES36 tumor cells cluster 11 ($n = 131$ cells) (**C**) or cluster 15 ($n = 18$ cells) (**D**) vs. M19 cells, and also between these ES36 clusters (11 vs. 15), respectively (**E**). The specificity of gene expression value has been normalized and visualized as a circle that corresponds with the ratio of gene expression in ES36 cells vs. M19. The size of nodes indicates the number of cells in each cluster with that ratio.

3.2. Embryonic Fibroblasts (M19) and ES (ES36) Cells Are Genetically Close

To investigate whether transcriptional regulation in ES cells cooperates with protein expression, we performed a proteomic analysis of fibroblasts and ES tumor cells (Figure 4A) Specifically, we analyzed protein expression in mock or DOX-treated M19 and ES36- tumor cells to identify potential signature proteins and pathways that distinguish two types of cells and identify potential targets for future therapeutic interventions. We identified a total of 1069 (47.5%) and 1238 (55%) proteins in ES36 and M19, respectively, out of 2250 possible proteins (Figure 4A,B). For the identification of DEPs, we used a p-value ≤ 0.05. This proteome coverage was highly reproducible across the three biological replicates. Proteins that were identified in only one or two biological replicates (1415 (62.9%) of all identified proteins in ES36 and 1454 (64.6%) in M19) were excluded from downstream analysis. Comparative proteomic analysis revealed a subset of 979 proteins common to ES36 (out of 1765) and M19 (out of 1642). A protein signature associated with the cellular amide metabolic process (GO: 0043603) and translation (GO: 0006412) was revealed by the GO term category.

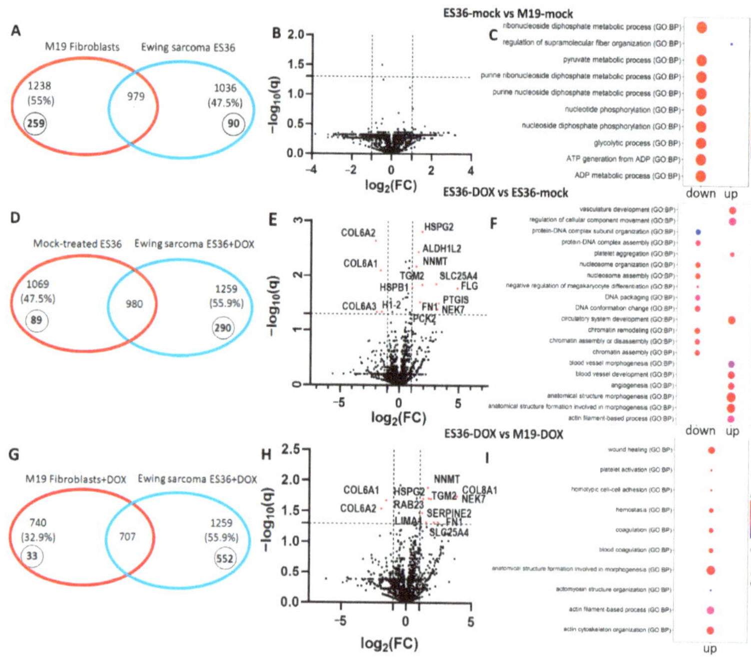

Figure 4. Human fibroblasts and ES cells ES36 display distinct protein compositions. (**A**,**D**,**G**) Venn diagram showing common and unique proteins in mock- or doxorubicin (DOX)-treated M19 and patient-derived ES36 cells. Overlapped proteins were identified in three replicates per each sample. Unique proteins for each type of cell and/or treatment are marked in bold and italics inside small circles. Created using Microsoft Office 2010. (**B**,**E**,**H**) Differential protein expressions between M19 and ES36 samples. Volcano plots of protein abundance between samples showing the distribution of proteins identified in ES36 and M19 during mock or doxorubicin treatment. Data presented as Log2 fold ratio between samples. The most significant genes are highlighted in red and we used an absolute threshold q(FDR) ≤ 0.05 and $|\log2(FC)| \geq 1$ (**C**,**F**,**I**) Gene ontology (GO) enrichment analysis for 60 potential targets (30 upregulated and 30 downregulated) of ES36 vs. M19, ES36-DOX vs. ES36-mock treated, and ES36-DOX vs.M19-DOX samples with $p < 0.05$ and most value of $|\log2(FC)|$ for more complete enrichment. The dot-plot of top-upregulated and downregulated biological functions was performed using p adjusted < 0.05. Hierarchical clustering of the protein molecular signature of ES36 and M19 was generated by the computational tool R gProfiler2. Supplemental Tables S1–S3 provide details of all molecular signatures.

From a unique protein perspective, M19 and ES36 cells possess 259 and 90 proteins, respectively. Further comparative pathway analysis revealed significant activation of proteins which regulate the enzyme activity (hydrolase, proteolysis, and peptidases) pathways in ES36-cultured cells, while organic acid and carbohydrate metabolism signaling pathways were found to be mostly regulated in M19-cultured cells. In M19 cells, we observed that of 130 proteins, most were involved in cellular catabolism (Figure 5A,D). Remarkably, whereas the extracellular matrix and cytoplasm become locations for the majority of upregulated proteins in ES36 and M19, the downregulated proteins in these types of cells reside mostly in the cytoplasm. Of note, the majority of most regulated proteins (GAPDH, LDHA, PGK1, TPI1, ENO1, ALDOA, GPI, PGAM1, ALDOC) responsible for the purine, glycolytic, pyruvate or ribonucleoside diphosphate metabolism of ES36 cells were inhibited (Figure 4C). For the top upregulated proteins of ES36 cells, R gProfiler2 reported "protein metabolism" (GO: 0051246), "response to stress" (GO: 0006950), and "regulation of cell death" (GO: 0010941) as unique GO term enrichments, and for the most downregulated proteins, the only GO term enrichments were "catabolic process" (GO: 0008104) and "response to growth factor stimulus" (GO: 0090287). At M19, the top 979 proteins with high expression were involved in "program cell death" (GO: 0012501) and "response to stress" (GO: 006950), and with low expression, the GO terms were "translation" (GO: 0006412) and "intracellular transport" (GO: 0046907). Using the analysis approach, we identified several proteins with differential expression between M19 and ES36. All these proteins are of particular interest since their roles in promoting metastases [25,26] and sarcoma progression [27,28] have been demonstrated.

3.3. Proteomic Profiling of DOX-Treated ES36 Cells Provides a Unique Signature

By proteome-based MS analysis, the composition of cytoplasmic proteins from fibroblasts and ES cells after doxorubicin treatment was investigated. First, we analyzed protein signatures between DOX-treated ES36 and mock-treated ES36 (Figure 4D,E). Compared with control cells, 980 common and 164 ($p < 0.05$) variable proteins were detected in DOX-treated ES cells. Among the 164 variable proteins in DOX-treated ES36 cells, 107 were upregulated and 57 were downregulated. The GO term analysis for downregulated proteins in the tumor-treated cells with DOX suggests suppression of chromatin remodeling and assembly, and vascular development and angiogenesis process, as top upregulated biological functions, can dictate the behavior of tumor cells in the presence of DOX (Figure 4F). We also looked for proteins that demonstrated a negative or positive trend in the presence of DOX in ES36 cells vs. mock-treated ES36. The top 30 of these DEPs with high expression were ACTG1, ACTC1, FN1, HSPB1, NNMT, GLS, ACTN1, SLC25A3, HTRA1, RTN4, PLS3, TGM2, ATP5F1A, SLC25A6, FHL2, MYH9, CNN2, VDAC1, MDH2, ALDOA, CKAP4, LMA1, ATP5F1B, VDAC2, FLNA, CLIC4, CNN3, GARS1, and DSTN; and the top 30 with low expression were: H2AC11, H4C1, H2BC12, H1-2, AHNAK, TFRC, HSPA8, ENO1, CFL1, COL6A1, COL1A1, COL6A3, TUBB4B, COL6A2, COL1A2, COTL1, CAPG, FASN, COL3A1, HNRNPM, PCBP1, KHSRP, LRP1, MTHFD1, FLNC, CUTA, HNRNPH3, SLC1A5, SEPTIN9, and RAB5C. The cellular components for several DEPs, such as HSPA8, YWHAZ, CAPG, COTL1, PDCD6IP, ACLY, ARPC5, PYGB, AK1, SLC25A3, CAPNS1, ARHGAP1, ATP6V1E1, TLN1, LAP3, DCTN2, NAGK, and USP14, were extracellular exosomes, with molecular components such as protein-containing complex binding (GO:MF 0044877) and cadherin binding (GO:MF 0045296). The differential expression suggests a decrease in complementary binding mediated by disordered protein domains of neoplastic cells to the other globular domains, including the core binding region of E-cad [29].

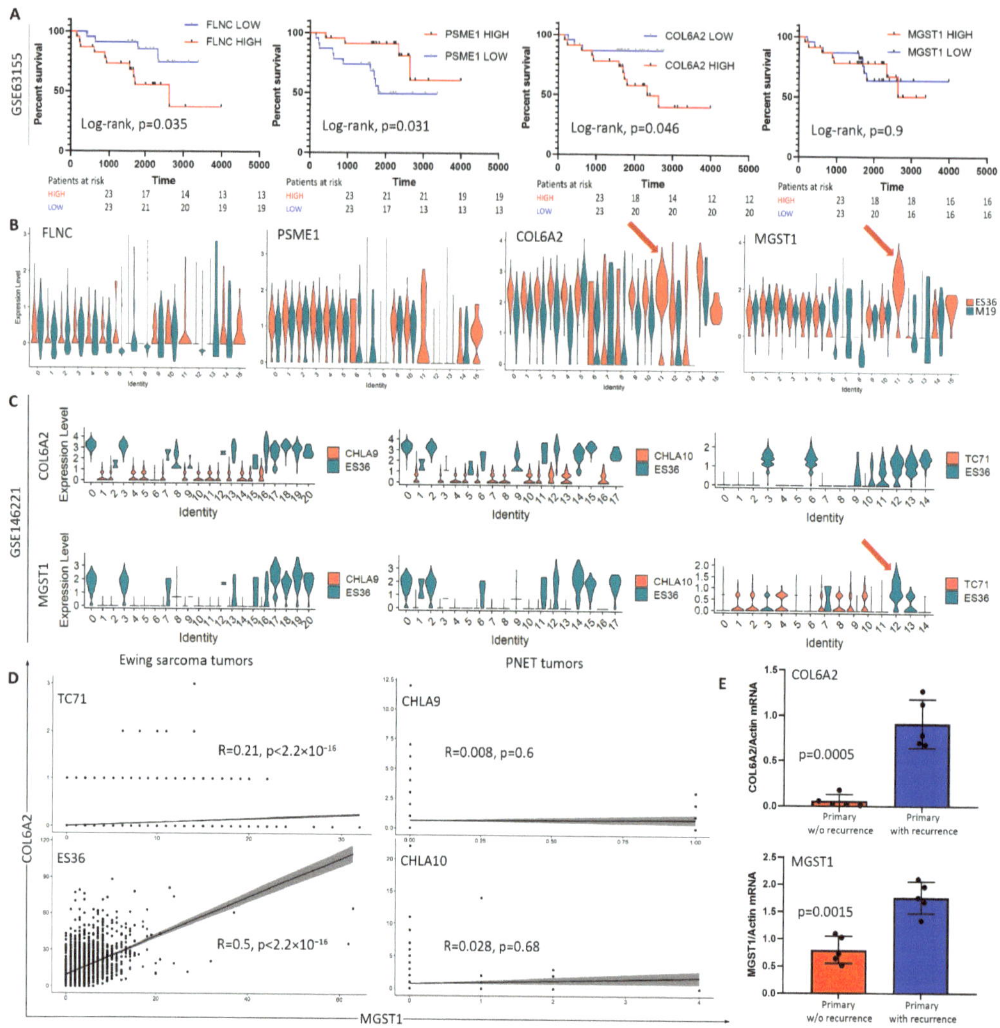

Figure 5. Validation of COL6A2 and MGST1 as markers for ES resistance and progression. (**A**) Group of 46 patients with ES was divided based on their marker expressions. Kaplan–Meier plot of OS for ES patients (GEO datasets GSE63155) stratified by transcriptomic data. The *p*-value was calculated using a two-sided log-rank test. (**B**,**C**) Single-cell RNA sequence discriminates a subset of tumor cells enriched with COL6A2 and MGIST1 markers across subclusters at ES36 and M19 cells (**B**) and in comparison with cellular subsets from TC71, CHLA9, and CHLA10 (**C**) (GEO dataset GSE146221). Data presented as violin plots are based on relative mRNA expression and clusterization of TC71, CHLA9, and CHLA10 with ES36 cells. Blue and black arrows indicate the association of MGST1 and COL6A2 with cluster 11 of ES36 cells. (**D**) Correlations between COL6A2 and MGIST1 expression in various ES cells. A significant positive correlation between COL6A2 and MGIST1 mRNA expression was detected in relapsed ES-based patient-derived cells and no such correlation was seen in PNET was considered significant. (**E**) mRNA levels of MGST1 and COL6A2 in selected primary patient samples (with/without recurrence, based on clinical information available) obtained from Federal State Budgetary Institution National Medical Research Center of Oncology named after N.N. Blokhin of the Ministry of Health of Russia. Semiquantitative PCR analysis of copy numbers of MGST1 and COL6A2 mRNAs obtained from absolute quantification using after normalization to b-actin. An unpaired t-test with Welch's correction, *p*-value difference between groups is presented.

We performed comparative proteomics to distinguish the metabolic and functional changes between embryonic fibroblasts and ES ES36 cells during DOX treatment. DOX exposure induces the expression of 707 common proteins in addition to 33 and 552 unique to fibroblasts or tumor cells, respectively (Figure 4G,H). Among 153 DEPs in the DOX-treated ES36 cells vs. M19, 100 proteins were upregulated and 53 were downregulated. The top 30 proteins with the most differences in means were: ACTB, ACTG1, TAGLN, ACTC1, FN1, HBA, TPM1, LDHA, HBB, HSPB1, NNMT, GLS, CSRP1, PLS3, ACTN1, HTRA1, CALU, CNN2, SLC25A3, MYH9, TGM2, FHL2, ANXA2, LIMA1, SLC25A6, MDH2, EEF1G, PEA15, CNN3, and CLIC4, and the top 30 with decreased expression were: AHNAK, TFRC, HSPA8, COL6A1, COL6A3, PDIA6, TUBB4B, KRT1, KRT10, COL6A2, COL1A1, CTTN, NAMPT, PDIA4, YWHAQ, P4HA2, GANAB, PSME1, PSME2, ITGB1, RANGAP1, PRKCSH, SEPTIN9, LRP1, COL3A1, FLNC, CPNE3, PLOD1, ATP1A1, and ANXA11. The GO term analysis for downregulated proteins in ES36 cells treated with DOX showed peptide cross-linking (GO:0018149); in particular, fibrillar collagen trimer (GO:0005583), collagen chain trimerization (REAC: R-HSA-8948216) and focal adhesion (KEGG:04510). The GO term analysis for upregulated proteins was: platelet aggregation (GO:0070527), platelet activation (GO:0030168), homotypic cell–cell adhesion (GO:0034109), actin cytoskeleton organization (GO:0030036), coagulation (GO:0050817), hemostasis (GO:0007599), cytoskeleton organization (GO:0007010), cell adhesion (GO:0007155) (Figure 4I). Of the regulated proteins, eight out of eleven proteins such as SLC25A6, EEF1G, PEA15, TUBB4, SEPTIN9, COL3A1, HBB, and HBA did not have any probes in the GEO dataset (GSE 63155). At the same time, PSME1 expression was associated with the improvement of ES patient survival (Log-rank test, $p = 0.031$), and COL6A2 and FLNC (Figure 5A) showed a negative impact on the survival of ES patients (Log-rank test, $p = 0.046$, and $p = 0.035$ relatively), and involvement in the regulation of integrin and ERK pathways (www.genecards.com (accessed on 1 February 2021)). Of note, the MGST1 marker was detected in two out of three samples in ES36-DOX and three out of three samples in M19-DOX. Although our strict policy on replicates excludes MGST1 from proteomic analysis, we compared COL6A2, PSME1, FLCN, and MGST1 for their clinical significance, correlation with clinical signs, and correlation against each other. We noticed that MGST1 expression was unique to ES36 Cluster 11 (C11), the same cluster where the gene's subset enriched with COL6A2 (Figure 5B). Furthermore, cluster assignment and overlay of the gene profiles between ES36 and CHLA10 patient-derived ES cells suggest sharing of the small gene population, including MGST1 and COL6A2 genes (Figure 5C). Therefore, mRNA expression of these biomarkers is straightforward in GEO datasets using TC71 and ES36 profiles (ES cells). Using randomly selected ES primary specimens with known clinical history, we analyzed the expression of COL6A2 and MGST1. Validation of scRNA-seq differentially expressed genes (DEGs) through semiquantitative real-time-polymerase chain reaction (RT-PCR) suggests the direct association of COL6A2 and MGST1 mRNAs with ES relapse, demonstrating a direct link between ES36 resistance, tumor regrowth, and increased tumor cell survival due to tumor resistance.

4. Discussion

We provide fresh evidence for tumor-cell-signaling activity generated from the interplay of metabolism, stress, and growth pathways that control ES treatment (Figure 4C). Our findings identify key regulatory transcription factors and kinases, as well as novel interactions between these pathways that drive distinct phases of this response in fibroblasts and tumor cells, and link well-characterized signaling mechanisms across cells of the same origin (embryonic fibroblasts and primary short-lived tumor cell cultures of ES36). To avoid the impact of interpatient heterogeneity of the cell signaling, our findings also suggest that it may be necessary to obtain fibroblasts and tumor cells from the same patient with Ewing sarcoma and carefully assess their transcription programs to validate differences in cellular signaling.

The treatment of ES is complicated by heterogeneity both internally and between tumors. To develop an effective anticancer therapeutic, a better understanding of tumor response is necessary. In this regard, reliable biomarkers may represent a source of therapy efficacy against tumor cells. Previous genomic analysis indicates several biomarkers relevant to glutathione metabolism (MGST1) [18] and cell-cycle upregulation mediated by ectopic mir34a [24] overexpression, supporting our GEO analysis of surviving patients with resected tumor samples enriched with cellular metabolism, including cytochrome P450 and glutathione transferase DEGs. Although these studies allow us to explore the intertumoral response to the therapeutic options, mainly chemotherapy, they also utilize responses from normal healthy cells that are present among tumor cell populations. To distinguish cells, we compared the first basal level of DEG expression between patient-derived ES36 and M19 embryonic fibroblasts and then we clustered the single-cell transcriptome which allowed us to detect 14 and 8 clusters, respectively. As a result of M19 and ES transcriptomics mixing, we detected a few unique clusters in ES36 cells (C11 and C15). Using the GO term biological functions, a member of a gene family that is relevant to migration was associated with the C11 cluster. Among genes, the highest expression detected in C11 and C15 of ES cells were: FTL, SERPINE2, CCDC80, CALD, and COL1A2 or FTL, TIMP1, TAGLN, SH3BGRL3, and CALD1 relatively. Despite the involvement of migration for the hub of eight genes, only COL1A2 was previously related to the OS of patients with ES [30]. On the molecular level, a reduced level of COL1A2 was inversely correlated with osteosarcoma proliferation and migration under cisplatin restriction [31]. Besides COL1A2, the ferritin light chain (FTL) expression was shown in association with the response to chemotherapy [32] and contributed to the proliferation, migration, and invasion [33] of osteosarcoma cells. SERPINE2 may drive self-proliferation and drug resistance in osteosarcoma [34], and has poor prognosis for patients with bladder [35] and ovarian [36] cancer with high COL1A2 expressions.

It is noteworthy that in the case of ES36 DOX vs. ES36 MOCK, we found a significant decrease in the expression of all three alpha-helices of the COL6 protein (Figure 4E), including COL6A2. This effect means a more complex effect of doxorubicin therapy on ES36, mainly, changing the content and 3D structure of the extracellular collagen matrix. At the same time, the expression of COL6A1 and COL6A2 is also significantly reduced in ES36 DOX versus M19 DOX (Figure 4H), which also indicates a specialized and unique functional nature of the decreasing COL6 level in ES36 cells in response to DOX. This may indicate the involvement of COL6 in the formation of vulnerability to doxorubicin, similar to breast cancer cells [37]. Whether the expression of MGST1 alone or in combination with COL6A2 is required for metastatic development has yet to be seen. However, our working hypothesis is that ES metastatic development might require multiple gene activation depending on the tumor stage. For instance, whereas MSGT1 and COL6A2 expression is required for ES cell survival during chemotherapeutic stress mediated by Doxorubicin, SOX2, KL4 and/or Oct4 [38] are vital for maintaining the stemness of ES cells and for the ES seeding and survival, as second-nodule activation of ZEB2 [39] and IGF-1R [40] transcriptional programs are most helpful. Among those, at some points of ES, such as seeding and cells making decisions to colonize bone or lung, additional proteins become influential, such as neuropeptide Y [41] or Caveolin-1 [42]. However, some studies [43] demonstrated the involvement of ES-based cellular proteins in the ability of ES to metastasize in vitro; the ultimate test for metastatic progression has required the presence of a microenvironment to support the invasion, migration, internalization and seeding of ES tumor cells. Therefore, in vivo tail injection [42,44] or injection into a muscle such as a calf [45] or gastrocnemius muscle [41,46] with tumor cells, or the use of the metastatic Ewing sarcoma mouse model [47] may hold the answers on ES seeding and metastasis development.

DOX-mediated stress is a complex cellular reaction produced by sarcoma cells. It is unclear why tumor cells express two proteins, one of each encoded alpha chain of collagen type VI and the second a protein with high glutathione transferase activity. It is generally agreed that MGST1 possesses a protein homodimerization activity, suggesting that stress response employs the MGST1 expression signal as a stressor signal for collagen filament re-engagement. It is also plausible that COL6A2 serves as an adapt messenger function to unify the cellular response during DOX stress and disrupt cellular adaptation via cooperation with other proteins. A possible degree of cooperation has been demonstrated previously for DSCAM and COL6A2 in the H9C2 cardiac cell line. In these cells, the transcriptional analysis of that interaction points to genes involved in adhesion and cardiac hypertrophy [48] that cause severe physiological and morphological defects in the heart of Drosophila. Further evidence suggests that glutathione and glutathione-related enzymes are at the forefront of the adaptive detoxification cellular response [49] to stress during DOX-mediated ROS production. Considering that DOX-resistant cervical cancer cells also display activation of GSH signaling, including MGST1 gene expression isoforms [50], activation of such proteins might represent a DOX-specific stress network that elaborates cellular function during the stress and becomes a common feature for various tumor cells. Crosstalk between various cellular functions mediated by COL6A2 and MGST1 coordinates tumor cell vulnerability and might define the severity of the ES cells' damage.

5. Conclusions

Based on the results of our study, by clustering scRNA-seq and subsequent DEG analysis, we were able to perform genotypic profiling of ES cells. Further proteomic and DEGs of GSE63155 analysis identified and showed the main differences in the expression of metastatic and primary-tumor markers that can be considered for targeted chemotherapy. Our data suggest that doxorubicin treatment mediates the induction of collagen remodeling and activation of glutathione metabolism, which allows premetastatic phenotype changes in ES tumor cells.

Supplementary Materials: The following supporting information can be downloaded at: https://www.mdpi.com/article/10.3390/cancers14225498/s1, Figure S1: Transcriptomic comparison of ES36 versus TC71, CHLA9, and CHLA10 patient-derived ES cells. Supplemental Tables S1–S3: Go terms of DEG expression in ES36 vs M19 fibroblasts, ES36-DOX vs ES36 primary tumor cells, and ES36-DOX (Doxorubicin-treated Ewing's sarcoma cells ES36) vs M19-DOX doxorubicin-treated human fibroblasts M19

Author Contributions: S.Y., M.M., I.U., I.K. and E.D. designed the research; S.Y., M.M., I.K., V.Z., I.U. and E.D. performed the research; S.Y., M.M., I.U., I.K., O.R. and E.D. analyzed the data; S.Y., M.M., I.U., I.K., O.R. and E.D. wrote the paper; P.T. and K.K. provided administrative and lab support; all authors edited the manuscript. All authors have read and agreed to the published version of the manuscript.

Funding: The study was supported by the Russian Science Foundation (#21-15-00213, IU, collecting primary samples, establishing primary culture, STR characterization).

Institutional Review Board Statement: Parents gave their consent to utilize the ES tumor tissue for research studies. All studies were approved by the Federal State Budgetary Institution National Medical Research Center of Oncology named after N.N. Blokhin of the Ministry of Health of Russia. A general protocol for tumor resection has been applied for the Research Institute of Pediatric Oncology and Hematology protocol since 2010, Day of approval 09.09.2010.

Informed Consent Statement: Our study included tumor tissue specimens with a confirmed diagnosis of ES (ES) and treated in 2021 at the clinical facility of NIMC N.N. Blokhin. All parents of sick children gave written informed consent to use the clinical data in agreement with approval of the institutional ethics committee (Research Institute of Pediatric Oncology and Hematology at N.N. Blokhin National Medical Research Center of Oncology.). Based on the agreement between Sechenov First Moscow State Medical University and NIMC N.N. Blokhin, the clinical specimens were available to isolate tumor samples.

Data Availability Statement: The RNA-seq datasets presented in this study can be found in the online repository (SRA transcriptomic access data: PRJNA846952). The names of the repository/repositories used in the study and their accession number(s) can be found below: https://www.ncbi.nlm.nih.gov/geo/, GSE63155, and GSE146221. While we are working to submit the proteomic files to the online repository, please use the following link to download the data (https://kapitul.ru/owncloud/index.php/s/UwSbISThjNXQyWP, password sarcoma); alternatively, data are available upon sending a request to Dr. Ilya Ulasov, ulasov75@yahoo.com.

Acknowledgments: All authors have read the journal's policy on disclosure of potential conflicts of interest and have disclosed no financial or personal relationship with organizations that could potentially be perceived as influencing the described research. We are also grateful to Dmitry Korzhenevsky and Aleksey M. Nesterenko (both from the Federal Center of Brain Research and Neurotechnology, Federal Medical Biological Agency) for their help with proteomic analysis, and Anna A. Khozyainova (Laboratory of Cancer Progression Biology, Cancer Research Institute, Tomsk National Research Medical Center, Russian Academy of Sciences) for her help with uploading files to the public database. Single-cell RNA-sequencing was carried out on the equipment of the Core Facility "Medical Genomics" (Tomsk NRMC).

Conflicts of Interest: The authors declare no conflict of interest.

References

1. Koelsche, C.; Kriegsmann, M.; Kommoss, F.K.F.; Stichel, D.; Kriegsmann, K.; Vokuhl, C.; Grunewald, T.G.P.; Romero-Perez, L.; Kirchner, T.; de Alava, E.; et al. DNA methylation profiling distinguishes Ewing-like sarcoma with EWSR1-NFATc2 fusion from Ewing sarcoma. *J. Cancer Res. Clin. Oncol.* **2019**, *145*, 1273–1281. [CrossRef] [PubMed]
2. Sankar, S.; Lessnick, S.L. Promiscuous partnerships in Ewing's sarcoma. *Cancer Genet.* **2011**, *204*, 351–365. [CrossRef]
3. Shulman, D.S.; Whittle, S.B.; Surdez, D.; Bailey, K.M.; de Alava, E.; Yustein, J.T.; Shlien, A.; Hayashi, M.; Bishop, A.J.R.; Crompton, B.D.; et al. An international working group consensus report for the prioritization of molecular biomarkers for Ewing sarcoma. *NPJ Precis. Oncol.* **2022**, *6*, 65. [CrossRef] [PubMed]
4. Miser, J.S.; Goldsby, R.E.; Chen, Z.; Krailo, M.D.; Tarbell, N.J.; Link, M.P.; Fryer, C.J.; Pritchard, D.J.; Gebhardt, M.C.; Dickman, P.S.; et al. Treatment of metastatic Ewing sarcoma/primitive neuroectodermal tumor of bone: Evaluation of increasing the dose intensity of chemotherapy—A report from the Children's Oncology Group. *Pediatr. Blood Cancer* **2007**, *49*, 894–900. [CrossRef]
5. Leavey, P.J.; Mascarenhas, L.; Marina, N.; Chen, Z.; Krailo, M.; Miser, J.; Brown, K.; Tarbell, N.; Bernstein, M.L.; Granowetter, L.; et al. Prognostic factors for patients with Ewing sarcoma (EWS) at first recurrence following multi-modality therapy: A report from the Children's Oncology Group. *Pediatr. Blood Cancer* **2008**, *51*, 334–338. [CrossRef] [PubMed]
6. Li, W.; Hong, T.; Liu, W.; Dong, S.; Wang, H.; Tang, Z.R.; Li, W.; Wang, B.; Hu, Z.; Liu, Q.; et al. Development of a Machine Learning-Based Predictive Model for Lung Metastasis in Patients With Ewing Sarcoma. *Front. Med.* **2022**, *9*, 807382. [CrossRef]
7. Margulies, B.S.; DeBoyace, S.D.; Damron, T.A.; Allen, M.J. Ewing's sarcoma of bone tumor cells produces MCSF that stimulates monocyte proliferation in a novel mouse model of Ewing's sarcoma of bone. *Bone* **2015**, *79*, 121–130. [CrossRef] [PubMed]
8. Taylor, R.; Knowles, H.J.; Athanasou, N.A. Ewing sarcoma cells express RANKL and support osteoclastogenesis. *J. Pathol.* **2011**, *225*, 195–202. [CrossRef] [PubMed]
9. Neve, A.; Cantatore, F.P.; Maruotti, N.; Corrado, A.; Ribatti, D. Extracellular matrix modulates angiogenesis in physiological and pathological conditions. *BioMed Res. Int.* **2014**, *2014*, 756078. [CrossRef] [PubMed]
10. Perut, F.; Carta, F.; Bonuccelli, G.; Grisendi, G.; Di Pompo, G.; Avnet, S.; Sbrana, F.V.; Hosogi, S.; Dominici, M.; Kusuzaki, K.; et al. Carbonic anhydrase IX inhibition is an effective strategy for osteosarcoma treatment. *Expert Opin. Ther. Targets* **2015**, *19*, 1593–1605. [CrossRef] [PubMed]
11. Hatano, M.; Matsumoto, Y.; Fukushi, J.; Matsunobu, T.; Endo, M.; Okada, S.; Iura, K.; Kamura, S.; Fujiwara, T.; Iida, K.; et al. Cadherin-11 regulates the metastasis of Ewing sarcoma cells to bone. *Clin. Exp. Metastasis* **2015**, *32*, 579–591. [CrossRef]
12. Supuran, C.T.; Alterio, V.; Di Fiore, A.; D'Ambrosio, K.; Carta, F.; Monti, S.M.; De Simone, G. Inhibition of carbonic anhydrase IX targets primary tumors, metastases, and cancer stem cells: Three for the price of one. *Med. Res. Rev.* **2018**, *38*, 1799–1836. [CrossRef] [PubMed]
13. Vieira, G.M.; Roberto, G.M.; Lira, R.C.; Engel, E.E.; Tone, L.G.; Brassesco, M.S. Prognostic value and functional role of ROCK2 in pediatric Ewing sarcoma. *Oncol. Lett.* **2018**, *15*, 2296–2304. [CrossRef]
14. Hetland, T.E.; Nymoen, D.A.; Emilsen, E.; Kaern, J.; Trope, C.G.; Florenes, V.A.; Davidson, B. MGST1 expression in serous ovarian carcinoma differs at various anatomic sites, but is unrelated to chemoresistance or survival. *Gynecol. Oncol.* **2012**, *126*, 460–465. [CrossRef] [PubMed]
15. van Kuijk, S.J.; Yaromina, A.; Houben, R.; Niemans, R.; Lambin, P.; Dubois, L.J. Prognostic Significance of Carbonic Anhydrase IX Expression in Cancer Patients: A Meta-Analysis. *Front. Oncol.* **2016**, *6*, 69. [CrossRef] [PubMed]
16. Li, Z.K.; Liu, J.; Chen, C.; Yang, K.Y.; Deng, Y.T.; Jiang, Y. Locally advanced malignant solitary fibrous tumour successfully treated with conversion chemotherapy, operation and postoperative radiotherapy: A case report. *J. Int. Med. Res.* **2021**, *49*, 300060521996940. [CrossRef]

17. Ulanet, D.B.; Ludwig, D.L.; Kahn, C.R.; Hanahan, D. Insulin receptor functionally enhances multistage tumor progression and conveys intrinsic resistance to IGF-1R targeted therapy. *Proc. Natl. Acad. Sci. USA* **2010**, *107*, 10791–10798. [CrossRef] [PubMed]
18. Scotlandi, K.; Remondini, D.; Castellani, G.; Manara, M.C.; Nardi, F.; Cantiani, L.; Francesconi, M.; Mercuri, M.; Caccuri, A.M.; Serra, M.; et al. Overcoming resistance to conventional drugs in Ewing sarcoma and identification of molecular predictors of outcome. *J. Clin. Oncol.* **2009**, *27*, 2209–2216. [CrossRef]
19. Yarovinsky, T.O.; Gorlina, N.K.; Cheredeev, A.N.; Kozlov, I.G.; Zorin, N.A.; Zorina, R.M. Alpha2-Macroglobulin Modulates Interactions between Lymphocytes and Fibroblasts. *Russ. J. Immunol.* **2001**, *6*, 1–8.
20. Raudvere, U.; Kolberg, L.; Kuzmin, I.; Arak, T.; Adler, P.; Peterson, H.; Vilo, J. g:Profiler: A web server for functional enrichment analysis and conversions of gene lists (2019 update). *Nucleic Acids Res.* **2019**, *47*, W191–W198. [CrossRef] [PubMed]
21. Volchenboum, S.L.; Andrade, J.; Huang, L.; Barkauskas, D.A.; Krailo, M.; Womer, R.B.; Ranft, A.; Potratz, J.; Dirksen, U.; Triche, T.J.; et al. Gene Expression Profiling of Ewing Sarcoma Tumors Reveals the Prognostic Importance of Tumor-Stromal Interactions: A Report from the Children's Oncology Group. *J. Pathol. Clin. Res.* **2015**, *1*, 83–94. [CrossRef] [PubMed]
22. Wang, P.P.; Ding, S.Y.; Sun, Y.Y.; Li, Y.H.; Fu, W.N. MYCT1 Inhibits the Adhesion and Migration of Laryngeal Cancer Cells Potentially Through Repressing Collagen VI. *Front. Oncol.* **2020**, *10*, 564733. [CrossRef] [PubMed]
23. Zhou, J.; Chen, Z.; Zou, M.; Wan, R.; Wu, T.; Luo, Y.; Wu, G.; Wang, W.; Liu, T. Prognosis and Immune Infiltration of Chromobox Family Genes in Sarcoma. *Front. Oncol.* **2021**, *11*, 657595. [CrossRef]
24. Nakatani, F.; Ferracin, M.; Manara, M.C.; Ventura, S.; Del Monaco, V.; Ferrari, S.; Alberghini, M.; Grilli, A.; Knuutila, S.; Schaefer, K.L.; et al. miR-34a predicts survival of Ewing's sarcoma patients and directly influences cell chemo-sensitivity and malignancy. *J. Pathol.* **2012**, *226*, 796–805. [CrossRef]
25. Krueger, S.; Kellner, U.; Buehling, F.; Roessner, A. Cathepsin L antisense oligonucleotides in a human osteosarcoma cell line: Effects on the invasive phenotype. *Cancer Gene Ther.* **2001**, *8*, 522–528. [CrossRef] [PubMed]
26. Yeh, L.T.; Lin, C.W.; Lu, K.H.; Hsieh, Y.H.; Yeh, C.B.; Yang, S.F.; Yang, J.S. Niclosamide Suppresses Migration and Invasion of Human Osteosarcoma Cells by Repressing TGFBI Expression via the ERK Signaling Pathway. *Int. J. Mol. Sci.* **2022**, *23*, 484. [CrossRef] [PubMed]
27. Zhou, Y.; Mu, T. LncRNA LINC00958 promotes tumor progression through miR-4306/CEMIP axis in osteosarcoma. *Eur. Rev. Med. Pharmacol. Sci.* **2021**, *25*, 3182–3199. [CrossRef]
28. Park, I.C.; Lee, S.Y.; Jeon, D.G.; Lee, J.S.; Hwang, C.S.; Hwang, B.G.; Lee, S.H.; Hong, W.S.; Hong, S.I. Enhanced expression of cathepsin L in metastatic bone tumors. *J. Korean Med. Sci.* **1996**, *11*, 144–148. [CrossRef]
29. Wiggers, F.; Wohl, S.; Dubovetskyi, A.; Rosenblum, G.; Zheng, W.; Hofmann, H. Diffusion of a disordered protein on its folded ligand. *Proc. Natl. Acad. Sci. USA* **2021**, *118*, e2106690118. [CrossRef]
30. Tang, M.; Liu, P.; Wu, X.; Gong, J.; Weng, J.; Gao, G.; Liu, Y.; Gan, L. COL3A1 and Its Related Molecules as Potential Biomarkers in the Development of Human Ewing's Sarcoma. *BioMed Res. Int.* **2021**, *2021*, 7453500. [CrossRef]
31. Zhang, J.; Huang, J.; Liu, W.; Ding, L.; Cheng, D.; Xiao, H. Identification of Common Oncogenic Genes and Pathways Both in Osteosarcoma and Ewing's Sarcoma Using Bioinformatics Analysis. *J. Immunol. Res.* **2022**, *2022*, 3655908. [CrossRef] [PubMed]
32. Fellenberg, J.; Bernd, L.; Delling, G.; Witte, D.; Zahlten-Hinguranage, A. Prognostic significance of drug-regulated genes in high-grade osteosarcoma. *Mod. Pathol.* **2007**, *20*, 1085–1094. [CrossRef] [PubMed]
33. Yu, G.H.; Fu, L.; Chen, J.; Wei, F.; Shi, W.X. Decreased expression of ferritin light chain in osteosarcoma and its correlation with epithelial-mesenchymal transition. *Eur. Rev. Med. Pharmacol. Sci.* **2018**, *22*, 2580–2587. [CrossRef]
34. Mao, M.; Wang, W. SerpinE2 promotes multiple cell proliferation and drug resistance in osteosarcoma. *Mol. Med. Rep.* **2016**, *14*, 881–887. [CrossRef] [PubMed]
35. Brooks, M.; Mo, Q.; Krasnow, R.; Ho, P.L.; Lee, Y.C.; Xiao, J.; Kurtova, A.; Lerner, S.; Godoy, G.; Jian, W.; et al. Positive association of collagen type I with non-muscle invasive bladder cancer progression. *Oncotarget* **2016**, *7*, 82609–82619. [CrossRef]
36. Yang, L.; Jing, J.; Sun, L.; Yue, Y. Exploring prognostic genes in ovarian cancer stage-related coexpression network modules. *Medicine* **2018**, *97*, e11895. [CrossRef]
37. Lovitt, C.J.; Shelper, T.B.; Avery, V.M. Doxorubicin resistance in breast cancer cells is mediated by extracellular matrix proteins. *BMC Cancer* **2018**, *18*, 41. [CrossRef]
38. Guzel Tanoglu, E.; Ozturk, S. miR-145 suppresses epithelial-mesenchymal transition by targeting stem cells in Ewing sarcoma cells. *Bratisl. Lek. Listy* **2021**, *122*, 71–77. [CrossRef]
39. Wiles, E.T.; Bell, R.; Thomas, D.; Beckerle, M.; Lessnick, S.L. ZEB2 Represses the Epithelial Phenotype and Facilitates Metastasis in Ewing Sarcoma. *Genes Cancer* **2013**, *4*, 486–500. [CrossRef]
40. Garofalo, C.; Manara, M.C.; Nicoletti, G.; Marino, M.T.; Lollini, P.L.; Astolfi, A.; Pandini, G.; Lopez-Guerrero, J.A.; Schaefer, K.L.; Belfiore, A.; et al. Efficacy of and resistance to anti-IGF-1R therapies in Ewing's sarcoma is dependent on insulin receptor signaling. *Oncogene* **2011**, *30*, 2730–2740. [CrossRef]
41. Hong, S.H.; Tilan, J.U.; Galli, S.; Izycka-Swieszewska, E.; Polk, T.; Horton, M.; Mahajan, A.; Christian, D.; Jenkins, S.; Acree, R.; et al. High neuropeptide Y release associates with Ewing sarcoma bone dissemination—In vivo model of site-specific metastases. *Oncotarget* **2015**, *6*, 7151–7165. [CrossRef] [PubMed]
42. Sainz-Jaspeado, M.; Lagares-Tena, L.; Lasheras, J.; Navid, F.; Rodriguez-Galindo, C.; Mateo-Lozano, S.; Notario, V.; Sanjuan, X.; Garcia Del Muro, X.; Fabra, A.; et al. Caveolin-1 modulates the ability of Ewing's sarcoma to metastasize. *Mol. Cancer Res.* **2010**, *8*, 1489–1500. [CrossRef] [PubMed]

43. Garcia-Dominguez, D.J.; Hajji, N.; Lopez-Alemany, R.; Sanchez-Molina, S.; Figuerola-Bou, E.; Moron Civanto, F.J.; Rello-Varona, S. Andres-Leon, E.; Benito, A.; Keun, H.C.; et al. Selective histone methyltransferase G9a inhibition reduces metastatic development of Ewing sarcoma through the epigenetic regulation of NEU1. *Oncogene* **2022**, *41*, 2638–2650. [CrossRef] [PubMed]
44. Yang, Y.; Ma, Y.; Gao, H.; Peng, T.; Shi, H.; Tang, Y.; Li, H.; Chen, L.; Hu, K.; Han, A. A novel HDGF-ALCAM axis promotes the metastasis of Ewing sarcoma via regulating the GTPases signaling pathway. *Oncogene* **2021**, *40*, 731–745. [CrossRef] [PubMed]
45. Lopez-Alemany, R.; Tirado, O.M. Metastasis Assessment in Ewing Sarcoma Using Orthotopic Xenografts. *Methods Mol. Biol.* **2021** *2226*, 201–213. [CrossRef]
46. Choo, S.; Wang, P.; Newbury, R.; Roberts, W.; Yang, J. Reactivation of TWIST1 contributes to Ewing sarcoma metastasis *Pediatr. Blood Cancer* **2018**, *65*, e26721. [CrossRef]
47. Cerra, C.; Harris, M.A.; Hawkins, C.J. Establishment and Characterisation of Metastatic Extraskeletal Ewing Sarcoma Mouse Models. *In Vivo* **2021**, *35*, 3097–3106. [CrossRef]
48. Grossman, T.R.; Gamliel, A.; Wessells, R.J.; Taghli-Lamallem, O.; Jepsen, K.; Ocorr, K.; Korenberg, J.R.; Peterson, K.L.; Rosenfeld, M.G.; Bodmer, R.; et al. Over-expression of DSCAM and COL6A2 cooperatively generates congenital heart defects. *PLoS Genet.* **2011**, *7*, e1002344. [CrossRef]
49. McIlwain, C.C.; Townsend, D.M.; Tew, K.D. Glutathione S-transferase polymorphisms: Cancer incidence and therapy. *Oncogene* **2006**, *25*, 1639–1648. [CrossRef]
50. Drozd, E.; Krzyszton-Russjan, J.; Marczewska, J.; Drozd, J.; Bubko, I.; Bielak, M.; Lubelska, K.; Wiktorska, K.; Chilmonczyk, Z.; Anuszewska, E.; et al. Up-regulation of glutathione-related genes, enzyme activities and transport proteins in human cervical cancer cells treated with doxorubicin. *Biomed. Pharmacother.* **2016**, *83*, 397–406. [CrossRef]

cancers

MDPI

Article

C-Reactive Protein Pretreatment-Level Evaluation for Ewing's Sarcoma Prognosis Assessment—A 15-Year Retrospective Single-Centre Study

Sarah Consalvo [1,*], Florian Hinterwimmer [2], Norbert Harrasser [1], Ulrich Lenze [1], Georg Matziolis [3], Rüdiger von Eisenhart-Rothe [1] and Carolin Knebel [1]

1 Department of Orthopaedics and Sports Orthopaedic, Klinikum Rechts der Isar, Technical University of Munich, 81675 Munich, Germany
2 Institute for AI and Informatics in Medicine, Technical University of Munich, 81675 Munich, Germany
3 Orthopaedic Department, University Hospital Jena, Campus Eisenberg, 81675 Eisenberg, Germany
* Correspondence: sarah.consalvo@mri.tum.de; Tel.: +49-89-4140-8012

Simple Summary: The importance of chronic inflammation in favouring a "cancer-friendly" microenvironment in most types of tumours has been analysed, except for in the case of Ewing's sarcoma. Ewing's sarcoma is a highly malignant blue round cell tumour that affects 2.9 in 1,000,000 children worldwide. The aim of this retrospective study was to analyse the role of C-reactive protein (CRP) as a prognostic factor. Serum CRP levels were significantly different in patients with a poorer prognosis and in patients who presented distant metastasis, whereas CRP levels were not significantly different in patients with local recurrence. In Ewing's sarcoma cases, we believe we can consider a CRP pretreatment value of >0.5 mg/dL as a sensitive prognostic risk factor indication for distant metastasis and poor prognosis.

Abstract: Background: A pathological/inflamed cellular microenvironment state is an additional risk factor for any cancer type. The importance of a chronic inflammation state in most diffuse types of tumour has already been analysed, except for in Ewing's sarcoma. It is a highly malignant blue round cell tumour, with 90% of cases occurring in patients aged between 5 and 25 years. Worldwide, 2.9 out of 1,000,000 children per year are affected by this malignancy. The aim of this retrospective study was to analyse the role of C-reactive protein (CRP) as a prognostic factor for Ewing's sarcomas. Methods: This retrospective study at Klinikum rechts der Isar included 82 patients with a confirmed Ewing's sarcoma diagnosis treated between 2004 and 2019. Preoperative CRP determination was assessed in mg/dL with a normal value established as below 0.5 mg/dL. Disease-free survival time was calculated as the time between the initial diagnosis and an event such as local recurrence or metastasis. Follow-up status was described as death of disease (DOD), no evidence of disease (NED) or alive with disease (AWD). The exclusion criteria of this study included insufficient laboratory values and a lack of information regarding the follow-up status or non-oncological resection. Results: Serum CRP levels were significantly different in patients with a poorer prognosis (DOD) and in patients who presented distant metastasis ($p = 0.0016$ and $p = 0.009$, respectively), whereas CRP levels were not significantly different in patients with local recurrence ($p = 0.02$). The optimal breakpoint that predicted prognosis was 0.5 mg/dL, with a sensitivity of 0.76 and a specificity of 0.74 (AUC 0.81). Univariate CRP analysis level >0.5 mg/dL revealed a hazard ratio of 9.5 (95% CI 3.5–25.5). Conclusions: In Ewing's sarcoma cases, we consider a CRP pretreatment value >0.5 mg/dL as a sensitive prognostic risk factor indication for distant metastasis and poor prognosis. Further research with more data is required to determine more sensitive cutoff levels.

Keywords: Ewing's sarcoma; CRP; prognosis; metastasis; local recurrence

Citation: Consalvo, S.; Hinterwimmer, F.; Harrasser, N.; Lenze, U.; Matziolis, G.; von Eisenhart-Rothe, R.; Knebel, C. C-Reactive Protein Pretreatment-Level Evaluation for Ewing's Sarcoma Prognosis Assessment—A 15-Year Retrospective Single-Centre Study. *Cancers* **2022**, *14*, 5898. https://doi.org/10.3390/cancers14235898

Academic Editor: Shinji Miwa

Received: 1 November 2022
Accepted: 27 November 2022
Published: 29 November 2022

1. Introduction

Ewing's sarcoma is a highly malignant blue round cell tumour, with 90% of cases occurring in patients between the age of 5 and 25 years. Worldwide, 2.9 out of 1,000,000 children per year are affected by this malignancy, with a slightly higher incidence in male patients (1.5 male for every −1 female) [1].

The 5-year survival has improved over the years from 40% to 70% because of multidisciplinary treatment strategies using chemotherapy and surgical therapy [2–4]. Unfortunately, the recurrence rate, even in initially localised tumours, is 25% [5]. More accurate and reliable factors need to be determined to distinguish high-risk patients. Yet, in a recent SEER database study, tumour size, older age and primary site were established as associated with poor prognosis and the presence of distant metastasis at diagnosis [6,7]. EWING 2008 is the current European protocol in which patients are divided into three groups according to clinical risk factors. The 3-year survival rate changes from 7578% for the standard risk group, to 51–55% for the high-risk group [8,9].

In order to identify more parameters for "high-risk" patients the focus of the International Cancer Society was on analysing the carcinoma microenvironment for the most common cancer entities (e.g., breast or lung cancer) [10,11]. The microenvironment consists of an extracellular matrix, immune-inflammatory cells, blood and lymphatic vascular networks. These cells and this matrix can increasingly develop the function of abnormal tissue and play a crucial role in metastatic spread and growth of malignancies [12]. Indeed, a healthy microenvironment may help to promote protection against tumourigenesis [13,14]. Conversely, the pathological state of that environment can be an additional risk factor for any type of cancer. C-reactive protein (CRP) is one of the proteins of the acute phase response and can be easily measured in routine blood sampling. The purpose of the acute phase reaction is to cause tissue damage locally through a local reaction and simultaneously prevent the damage from spreading too far. They are therefore relevant diagnostic markers of the extent of the body's response. Acute phase proteins are produced within 24 h; during which time, an increase in their concentration in the blood of 25% is measured [15,16].

Protein concentration can thus increase up to 1000-fold. The prognostic relevance of several inflammatory factors must be analysed. In Ewing's sarcoma, a high white blood cell (WBC) count is demonstrably associated with decreased event-free survival (EFS) [17]. However, the prognostic role of CRP in patients with Ewing's sarcoma has not been evaluated. This retrospective study's aim was to evaluate an accessible, worldwide and inexpensive means of assessing a patient's risk of distant metastases or reduced life expectancy at diagnosis, and thereby assess a further standardised risk factor for treatment evaluation.

2. Methods

This retrospective study included 82 patients who were treated at Klinikum rechts der Isar (Munich, Germany) between 2004 and 2019, and had a confirmed diagnosis of Ewing's sarcoma.

For all patients, diagnoses were validated through a histopathological examination as the reference standard. The local institutional review and ethics board (Klinikum rechts der Isar, Technical University of Munich) approved this retrospective study (N°48/20S). The exclusion criteria included insufficient laboratory values, a lack of information regarding follow-up and Rx or R1/R2 resection status. In total, 40 patients were included in this study.

2.1. *Preevaluation*

We analysed the medical records of all patients enrolled in this study. Histology was obtained via either incision or CT scan-guided biopsy performed in our institution, and confirmed at an interdisciplinary sarcoma board following the WHO guidelines.

Laboratory data were collected during pretreatment between one and up to a maximum of seven days before biopsy or the first surgical treatment at the first appointment at our clinic. All patients were screened for bacterial infection with urinary samples, lung

X-ray and total body clinical examination (e.g., urinary infection, pulmonary disease or other possible systemic bacterial infections) and were excluded if the screenings were positive. CRP levels were reported in milligrams per decilitre (mg/dL) with a normal value considered as below 0.5 mg/dL. The measurement was performed using the Cobas® 8000 modular analyser C702 (Fa. Roche, Basel, Switzerland). The follow-up investigations were performed in line with EWING 2008, International Guideline Harmonisation Group for Late Effects of Childhood Cancer, Late Effects Surveillance System [18].

All patients were followed up in our department every 3 months for the first 2 years, every 6 months from the third to the fifth year and at 12-month intervals thereafter in accordance with guidelines. Disease-free survival (DFS) time was calculated as the time between the initial diagnosis and an event such as local recurrence or distant metastasis. All metastasis locations were included. The follow-up time was defined as the time between the initial diagnosis and the last follow-up, or the last (unscheduled) presentation of the patient for a check-up in our clinic. The follow-up status of patients was classified as death of disease (DOD), no evidence of disease (NED) or, for patients who are currently alive but had a diagnostically confirmed distant metastasis or a proven local recurrence, or alive with disease (AWD).

2.2. Statistical Analysis

Data were processed and analysed using StatPlus:mac Pro 2020 (Fa. AnalystSoft, Walnut, CA, USA).

The national standard of 0.5 mg/dL was used to distinguish between 'increased' and 'decreased' CRP. The correlation between CRP value and overall survival (OS) was explored through a Pearson correlation model. The following were selected as guide values for the interpretation of the correlation coefficient: CC = 0 no linear correlation, CC = 0.3 weak positive linear correlation and CC = 0.5 positive linear correlation.

Statistically significant values were calculated with a p-value < 0.01. Sensibility and specificity based on the optimal identified cut point were calculated along the 95% interval.

The survival curves were generated using Kaplan–Meier analysis and evaluated using the log-rank test. The quantitative accuracy of the CRP measurements was measured by the area under the curve (ROC Curve/AUC). An AUC value >0.8 was considered a good predictive test ability. Two sample t-tests were performed to estimate the association between CRP levels and prognosis status, the presence of distance metastasis and local recurrence.

3. Results

A total of, 82 patients with a confirmed diagnosis of Ewing's sarcoma were included. In total, 42 patients were excluded: 14 patients presented insufficient laboratory values, 26 cases had a lack of information regarding the follow-up status and 2 cases had non-oncological resection (R1/R2). The inclusion criteria were met by 40 patients. Their clinical characteristics are summarised in Table 1. The mean age of the patients was 21.5 years with a minimum age of 4 years and a maximum age of 62 years. The majority of the cohort was male (70% vs. 30%). The mean female age was 23 years (12–62 years), whereas that of males was 21.5 years (4–54 years). With regard to sarcoma position, 47.5% of patients were affected at lower extremity localisations (27.5% femur and 20% tibia), pelvis 20%, humerus 5%, clavicle 5%, forearm 5%, foot 5% and spine 2.5%. At initial diagnosis, 10% (n = 4) showed multifocal bone involvement.

The follow-up status was defined as DOD in 17 patients (42%), NED in 21 patients (53%) and AWD in 2 patients (5%). No surgical therapy was performed in 13 patients (32.5%). In all other patients (67.5%), R0 resection was confirmed by the pathology department (n = 27). Of the 13 patients who had no surgical treatment, 11 had proven multiple metastases in CT staging (CT thorax/abdomen/pelvis). Chemotherapy and local radiotherapy were performed in 73% of the aforementioned cases. Palliative radiotherapy was performed on only one patient based on his/her request and in 18% only systemic therapy was performed. No surgical therapy was performed despite the absence of distant metastasis in two patients. In one case, the primary tumour was localised in the distal tibia with a skip lesion in the proximal fibula in the first patient. An oncological resection would have caused the complete loss of foot function with necessary tibial nerve resection In agreement with the parents, surgery was not performed and only chemotherapy with local radiation was conducted. The other case had a localised Ewing's sarcoma in the calcaneus and after completing neoadjuvant chemotherapy, he/she demonstrated completely regressive radiological findings, so in agreement with the parents no surgical resection was performed to avoid a functionally mutilating calcanectomy.

Table 1. Clinical characteristics and descriptive statistics of cohort.

Spalte1	*n*	Percentage	Mean	Range	SD
Age	40		21.5 years	4–62 years	13.54 years
Sex					
Male	28	70%	-	-	-
Female	12	30%	-	-	-
Site					
Femur	11	27.50%	-	-	-
Tibia	8	20%	-	-	-
Humerus	2	5%	-	-	-
Clavicula	2	5%	-	-	-
Radius	2	5%	-	-	-
Hip	8	20%	-	-	-
Foot	2	5%	-	-	-
Spine	1	2.50%	-	-	-
Multifocal	4	10%	-	-	-
Follow-up					
DOD	17	42.50%	-	-	-
NED	21	52.50%	-	-	-
AWD	2	5%	-	-	-
Metastasis					
Pulmonary	7	17.50%	-	-	-
Skeletal	2	5%	-	-	-
Multifocal	10	25%	-	-	-
At diagnosis	15	37.50%	-	-	-
None	6	15%			
Local recurrence	6	15%	-	-	-
DFS	40	-	3.5 years	0–17 years	54.5 years
Follow-up	40	-	4.79 years	0–17 years	53.6 years
CRP					
>0.5 mg/dL	27	67.50%	1.6 mg/dL	0.6–32.8 mg/dL	9 mg/dL
<0.5 md/dL	13	32. 50%	0.1 mg/dL	0.1–0.2 mg/dL	0.05 mg/dL

The median CRP value of the entire cohort was 1.6 mg/dL (0.1–32.8 mg/dL).

In the deceased patients, the median CRP level was 4.6 mg/dL (0.1–32.8 mg/dL). In those patients who remained alive but showed distant metastasis or local recurrence, the median CRP level was 0.6 mg/dL (0.1–0.6 mg/dL), whereas the median CRP value of the NED group was 0.7 mg/dL (0.1–5.8 mg/dL) (Figure 1). It can be shown that those patients in our cohort who ultimately died as a result of Ewing's sarcoma had higher preoperative CRP values on average on the basis of the aforementioned values.

In the group with an elevated CRP (>0.5 mg/dL) value, 59% had already died, 3% was alive but with metastases or a local recurrence and 37% was alive without any disease manifestation. In the low-CRP-value group (<0.5 mg/dL), 8% had already died, 8% were alive but with metastases or a local recurrence and 84% were alive without any manifestation of the disease. Of the entire cohort, 37% had pulmonary metastases, 52% had multifocal metastases (more than two organs) and 10% had metastases in organs other than the lungs. Metastases were already present in 37.5% of patients at initial diagnosis.

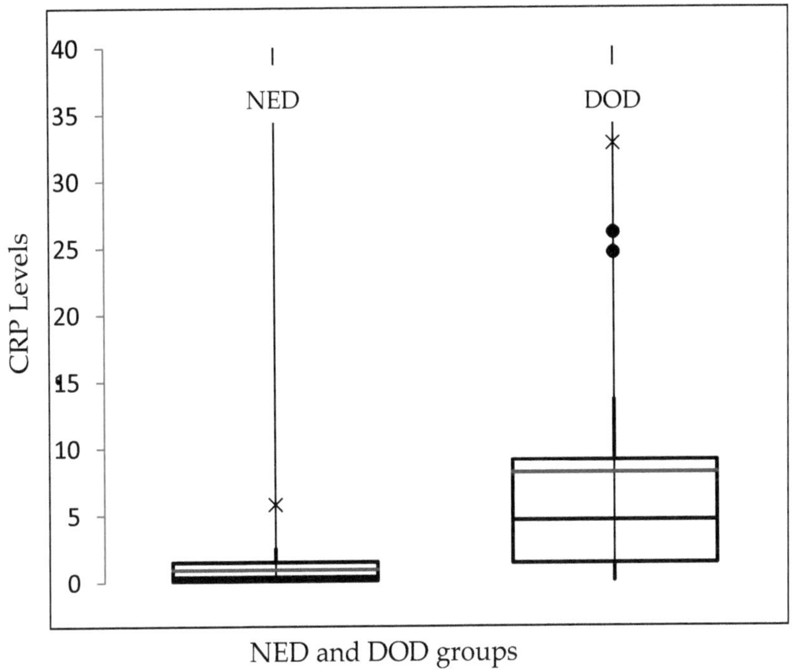

Figure 1. Box plot of the median CRP values (red line) depending on the follow-up status of the patients: no evidence of disease (NED) and death of disease (DOD). For the alive with disease (AWD) status, no diagram was created because only two patients were involved. ×: extreme outliners, •: mild outliners.

Survival Curves

The median follow-up time was 4.8 years (0–17 years). In contrast, the median DFS time was 3.5 years (0–17 years). Different survival curves could be established on the basis of the predefined cutoff value of 0.5 mg/dL depending on the CRP value.

Patients with a preoperative CRP value of >0.5 mg/dL had a median follow-up time of 4.5 years (2 months–13 years). In contrast, patients with a CRP value below the cutoff value of 0.5 mg/dL showed a median follow-up time of 5 years (0–17 years), as shown in Figure 2.

Univariate analysis of CRP level > 0.5 mg/dL revealed a hazard ratio (HR) of 9.5 (95% CI 3.5 to 25.5). A 5-year survival rate of 92% and 46% was calculated in the CRP < 0.5 mg/dL and >0.5 mg/dL groups, respectively.

The Ewing cohort showed a median DFS time of 7 months (0–17 years) with a preoperative CRP above 0.5 mg/dL, whereas patients with a CRP below 0.5 mg/dL showed a median DFS time of 5.5 years (0–11 years), as shown in Figure 3.

The risk estimation of the recurrence of an event in terms of local recurrence or distant metastasis was 8.3 times higher in the patients with an CRP level > 0.5 mg/dL compared to the group of patients with a CRP level below 0.5 mg/dL (8.3 HR and 95% CI 3–22.7).

The optimal breakpoint was confirmed to be 0.5 mg/dL (AUC 0.812, sensitivity = 0.76 and specificity = 0.74) (Figure 4).

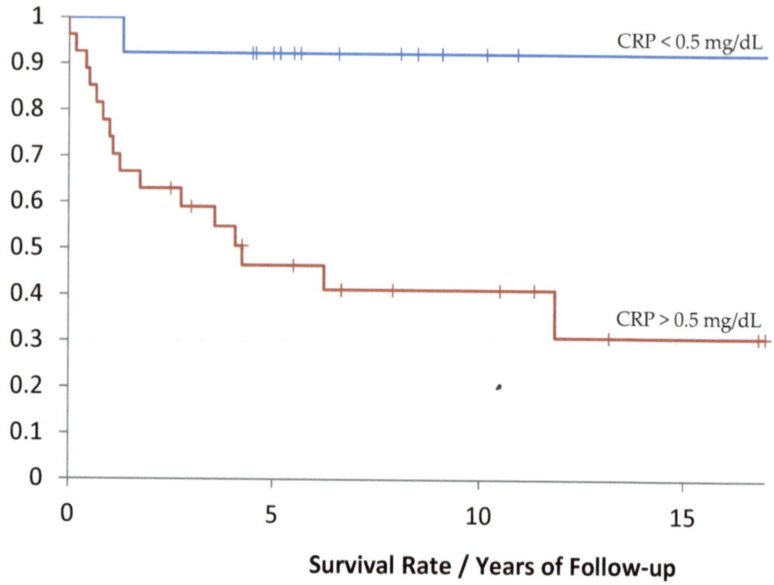

Figure 2. Kaplan-Meier curve of the survival rate depending on the CRP value: red corresponds to the group with an increased CRP value and blue corresponds to the group with a lower CRP value. The follow-up time is evaluated in years (*p*-value: 0.005).

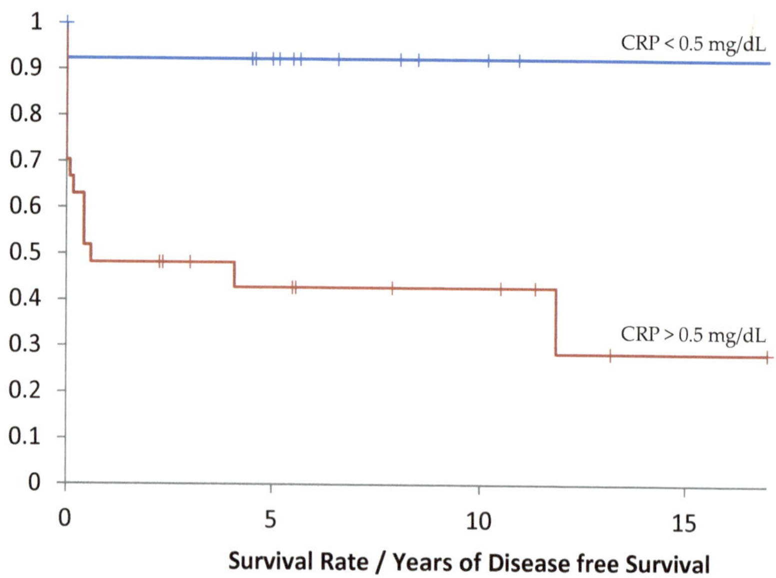

Figure 3. Kaplan-Meier curve of the recurrence/remote metastasis-free time as a function of the CRP value: red corresponds to the group with an increased CRP value andblue corresponds to the group with a lower CRP value. The DFS time is evaluated in years (*p*-value: 0.006).

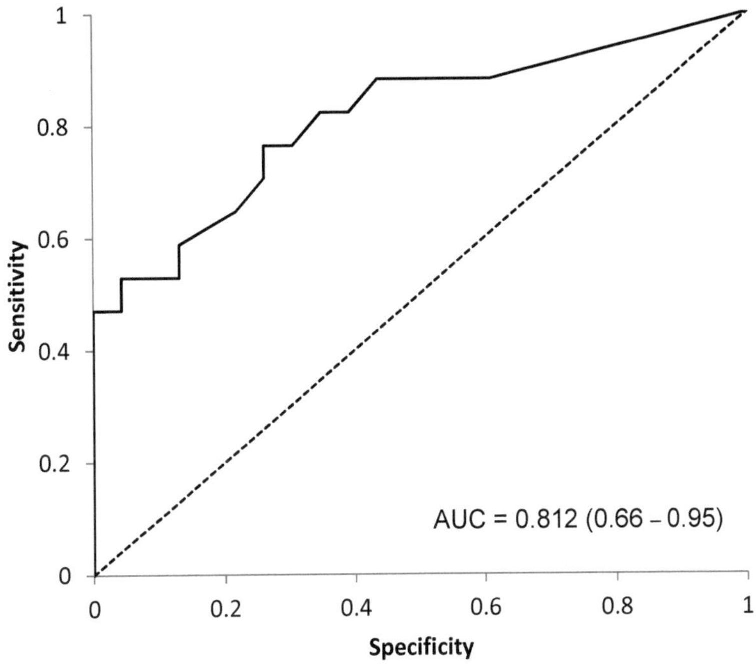

Figure 4. ROC Curve (AUC = 0.812). Optimal breakpoint analysis of CRP as a prognostic factor of OS.

The correlation between the CRP level and a reduced DFS as well as poor prognosis in Ewing sarcomas showed a CC (Correlation Coefficient) of 0.51, with a p value < 0.0005.

Two sample t-tests confirmed that CRP levels were a significant prognostic factor for poor prognosis ($p < 0.001$) and risk of the presence of distant metastasis ($p = 0.009$). No significantly relevant difference was found between CRP level and occurrence or local recurrence ($p = 0.02$) (Table 2).

Table 2. Two sample t-tests of CRP level as a predictor of the presence of metastasis, local recurrence and poor prognosis.

	CRP Levels n (%) 95% CI		CRP Levels n (%) 95% CI		p-Value
Presence of			Absence of		
Metastasis	19 (47.5%)	2.3 to 11.9	21 (52.5%)	0.3 to 1.8	0.009
LR	9 (22.5%)	−0.2 to 17.7	31 (77.5%)	0.6 to 4.5	0.02
Prognosis	DOD		NED/AWD		
	17 (42.5%)	2.9 to 13.3	23 (57.5%)	0.3 to 1.4	<0.001

4. Discussion

This study aimed to determine the prognostic value of a pretreatment serum CRP analysis. After evaluating the statistical relevance of CRP values in correlation with prognosis, we can prudently confirm the involvement of this protein in the acute phase response in the Ewing sarcoma microenvironment.

In accordance with our findings, Aggerholm-Pendersen et al. analysed a group of 172 patients with bone sarcoma (consisting of 63 chondrosarcomas and 109 between Ewing's sarcomas and osteosarcomas) and demonstrated that elevated CRP levels were associated with increased overall mortality [19]. In contrast to our study, osteosarcomas and Ewing's sarcomas were analysed in one group without any entity distinction. Likewise, in 2013, a poorer DFS was statistically confirmed ($p = 0.02$) in patients with Ewing's sarcoma and chondrosarcoma with an elevated CRP value. Here, no cutoff value was calculated and no distinction per entity was performed. Furthermore, patients with metastatic spread at diagnosis were excluded [20]. The calculated 5-year survival with elevated and reduced CRP levels was 57% and 79% ($p < 0.0001$), respectively [20]. In our study, we calculated a 5-year survival of 46% and 92% in the elevated and reduced CRP level groups ($p = 0.006$), respectively. Survival was not completely comparable because of the inclusion of chondrosarcomas in the Nakamura et al. study.

Two recent studies analysed the CRP–albumin ratio (CAR). In a cohort of 122 Ewing's sarcomas, Yong-Jiang et al. demonstrated that the CAR had a significantly larger AUC compared with the neutrophil–leucocyte ratio (NLR), platelet–leucocyte ratio, leucocyte–monocyte ratio and neutrophil–platelet ratio ROC curves. Therefore, higher levels of CAR were correlated with poor prognosis (HR 2.4, $p = 0.005$), and the calculated ratio was the most robust prognostic factor of all the aforementioned factors [21]. An optimal CAR cutoff value of 1.5 as a prognostic factor was calculated two years later. Therefore, the presence of metastasis and a CAR value under <1.5 were significantly associated with a reduced OS ($p < 0.05$). In particular, this study was performed only on spinal Ewing's sarcomas [22]. Even if the analysed prognostic factor was a ratio (CRP/albumin), a correlation and confirmation of our findings can be made.

Considering other inflammation agents, Biswas et al. predicted inferior EFS ($p = 0.009$) and local control ($p = 0.02$) rates for patients without metastases in a cohort of 60 extraosseous Ewing's sarcomas with a WBC count of $>11 \times 10^9$/L [23].

Pretreatment levels of 224 localised Ewing's sarcomas were analysed and a count of $>11 \times 10^9$/L WBC predicted an inferior EFS ($p = 0.003$) [24]. In the second study, 35 cases of head and neck Ewing's sarcoma were analysed that were diagnosed in the same institution and were treated with a uniform chemotherapy protocol. Multivariate analysis showed that baseline the WBC count independently predicted the EFS rate ($p = 0.04$). Patients with WBC $\leq 11.000/\mu$L had superior EFS, although no difference for OS was observed [25]. Although only for osteosarcomas, there are currently encouraging results regarding the better prognosis of these patients with addition of the conventional chemotherapy of mifamurtide. This is an immunomodulator that stimulates the immune response against tumour cells, e.g., in the lung, by activating macrophages and releasing proinflammatory or antitumour cytokines. This represents clear evidence of the immune system's central role in carcinogenesis even if it is currently only for osteosarcoma. The role of tumour-associated macrophages (TAMs) is also becoming a central question in the OS (Overall Survival) of cancer entities. Fujiwara et al. found that a high extent of TAM infiltration, a substantial microvascular density, elevated WBC counts (>6800 cells/μL) and CRP values > 0.2 mg/dL were significantly associated with worse prognosis [26]. In these cases, the CRP cut-off was set lower than in our study.

In consideration of our findings, we can confirm the role of pretreatment CRP values in prognosis, local recurrence and of the presence of distant metastasis in patients with Ewing's sarcoma (Table 2). The advantages of CRP blood testing over the above options are its lower cost and ubiquitous use in any clinic or outpatient practice. Sadly, it is very difficult to clearly confirm the clinical implication of the findings because of the rarity of this cancer subtype, and the small size of the analysed cohort. Even with a small sample size, we think that a confirmed R0 resection of Ewing's sarcoma can help eliminate an important bias in local recurrence and distant metastasis. Unfortunately, numerous patients were lost in the 15-year follow-up range, and complete follow-up could not be calculated.

Mifamurtide

We would like to put some focus on the new drug mifamurtide, to better understand the clinical implications of chronic inflammation regarding the "cancer-friendly" state.

Mifamurtide/L-MTP-PE (liposomal muramyl tripeptide phosphatidylethanolamine) or MAPACT (Takeda Pharmaceutical Company) is a synthetic drug which stimulates the immune response, thereby activating macrophages and monocytes [27]. MAPACT is currently approved in Europe for the treatment of non-metastatic Osteosarcoma in addition to standard chemotherapy [28,29].

Studies in vitro demonstrated that human macrophages can be induced by Mifamurtide to stimulate anti-tumor activity against Osteosarcoma cells. Punzo et al. treated macrophages obtained from peripheral blood mononucleated cells of healthy donors and MG63 cells (cells that have fibroblast morphology and are isolated from the bone patient with osteosarcoma) with Mifamurtide. MG63 cells co-cultured with Mifamurtide-activated macrophages showed a significant decrease in the metastasis, prognosis and inflammation markers compared to those co-cultured with macrophages which were not activated [30].

Mifamurtide is pharmacological proof that an immunomodulatory drug given together with standard adjuvant chemotherapy can improve the prognosis of cancer patients.

5. Conclusions

It can be confirmed that a preoperative evaluation of the CRP value in patients with Ewing's sarcoma represents a valuable prognosis indicator. Thus, the follow-up for patients with elevated preoperative CRP value >0.5 mg/dL could be modified for this higher-risk group. Further research and the central collection and analysis of data is required to determine the most sensitive cutoff values of pretreatment CRP levels in predicting poor survival or distant metastasis.

Author Contributions: S.C.: Project administration, Conceptualization, Writing—Original Draft, Data Curation; F.H.: Writing—Review & Editing, Formal analysis; N.H.: Writing—Review & Editing; U.L.: Writing—Review & Editing; G.M.: Supervision, Validation; R.v.E.-R.: Supervision, Validation and C.K.: Methodology, Supervision, Validation. All authors have read and agreed to the published version of the manuscript.

Funding: This research was funded by Department of Orthopaedics and Sports Orthopaedic, Klinikum rechts der Isar, Technical University of Munich.

Institutional Review Board Statement: The study was conducted according to the guidelines of the Declaration of Helsinki, and approved by the Institutional Ethics Committee of Klinikum rechts der Isar, Technical University of Munich (N°48/20S).

Informed Consent Statement: Not applicable. The paper is a retrospective study.

Data Availability Statement: The data that support the findings of this study are available on request from the corresponding author (Dott. S. Consalvo). The data are not publicly available due to restrictions e.g., their containing information that could compromise the privacy of research participants.

Conflicts of Interest: The authors declare no conflict of interest.

Abbreviations

DOD: death of disease; AWD, alive with disease; NED, no evidence of disease; CRP, C-reactive protein; DFS, disease-free survival; OS, overall survival; NLR, neutrophil–leucocyte ratio; EFS, event-free survival; TAMs, tumour-associated macrophages and OS, Overall Survival.

References

1. Picci, P.; Manfrini, M.; Fabbri, N.; Gambarotti, M.; Vanel, D. *Atlas of Musculoskeletal Tumors and Tumorlike Lesions: The Rizzoli Case Archive*; Springer International Publishing: Cham, Switzerland, 2016.
2. Ahmed, S.K.; Witten, B.G.; Harmsen, W.S.; Rose, P.S.; Krailo, M.; Marcus, K.J.; Randall, R.L.; DuBois, S.G.; Janeway, K.A.; Womer, R.B.; et al. Analysis of local control outcomes and clinical prognostic factors in localized pelvic Ewing sarcoma patients treated with radiation therapy: A Report from the Children's Oncology Group. *Int. J. Radiat. Oncol. Biol. Phys.* **2022**. [CrossRef]

3. Indelicato, D.J.; Vega, R.B.M.; Viviers, E.; Morris, C.G.; Bradfield, S.M.; Gibbs, C.P.; Bradley, J.A. Modern Therapy for Chest Wall Ewing Sarcoma: An Update of the University of Florida Experience. *Int. J. Radiat. Oncol. Biol. Phys.* **2022**, *113*, 345–354. [CrossRef] [PubMed]
4. Indelicato, D.J.; Vega, R.B.M.; Viviers, E.; Morris, C.G.; Bradfield, S.M.; Ranalli, N.J.; Bradley, J.A. Modern Therapy for Spinal and Paraspinal Ewing Sarcoma: An Update of the University of Florida Experience. *Int. J. Radiat. Oncol. Biol. Phys.* **2022**, *113*, 161–165. [CrossRef] [PubMed]
5. Durer, S.; Shaikh, H. Ewing Sarcoma. In *StatPearls*; StatPearls Publishing: Treasure Island, FL, USA, 2022.
6. Shi, J.; Yang, J.; Ma, X.; Wang, X. Risk factors for metastasis and poor prognosis of Ewing sarcoma: A population based study. *J. Orthop. Surg. Res.* **2020**, *15*, 88. [CrossRef] [PubMed]
7. Windsor, R.; Hamilton, A.; McTiernan, A.; Dileo, P.; Michelagnoli, M.; Seddon, B.; Strauss, S.J.; Whelan, J. Survival after high-dose chemotherapy for refractory and recurrent Ewing sarcoma. *Eur. J. Cancer* **2022**, *170*, 131–139. [CrossRef] [PubMed]
8. Anderton, J.; Moroz, V.; Marec-Berard, P.; Gaspar, N.; Laurence, V.; Martin-Broto, J.; Sastre, A.; Gelderblom, H.; Owens, C.; Kaiser, S.; et al. International randomised controlled trial for the treatment of newly diagnosed EWING sarcoma family of tumours—EURO EWING 2012 Protocol. *Trials* **2020**, *21*, 96. [CrossRef] [PubMed]
9. Dirksen, U.; Brennan, B.; Le Deley, M.C.; Cozic, N.; van den Berg, H.; Bhadri, V.; Brichard, B.; Claude, L.; Craft, A.; Amler, S.; et al. High-Dose Chemotherapy Compared with Standard Chemotherapy and Lung Radiation in Ewing Sarcoma with Pulmonary Metastases: Results of the European Ewing Tumour Working Initiative of National Groups, 99 Trial and EWING 2008. *J. Clin. Oncol.* **2019**, *37*, 3192–3202. [CrossRef]
10. Winther-Larsen, A.; Aggerholm-Pedersen, N.; Sandfeld-Paulsen, B. Inflammation-scores as prognostic markers of overall survival in lung cancer: A register-based study of 6210 Danish lung cancer patients. *BMC Cancer* **2022**, *22*, 63. [CrossRef]
11. Crassini, K.; Mulligan, S.P.; Best, O.G. Targeting chronic lymphocytic leukemia cells in the tumor microenviroment: A review of the in vitro and clinical trials to date. *World J. Clin. Cases* **2015**, *3*, 694–704. [CrossRef]
12. Noruzinia, M.; Tavakkoly-Bazzaz, J.; Ahmadvand, M.; Azimi, F.; Dehghanifard, A.; Khakpour, G. Young Breast Cancer: Novel Gene Methylation in WBC. *Asian Pac. J. Cancer Prev.* **2021**, *22*, 2371–2375. [CrossRef]
13. Danilova, A.B.; Baldueva, I.A. Neutrophils as tumor microenviroment member. *Vopr. Onkol.* **2016**, *62*, 35–44. [PubMed]
14. Wen, Y.; Zhu, Y.; Zhang, C.; Yang, X.; Gao, Y.; Li, M.; Yang, H.; Liu, T.; Tang, H. Chronic inflammation, cancer development and immunotherapy. *Front. Pharmacol.* **2022**, *13*, 1040163. [CrossRef] [PubMed]
15. Janciauskiene, S.; Wrenger, S.; Gunzel, S.; Grunding, A.R.; Golpon, H.; Welte, T. Potential Roles of Acute Phase Proteins in Cancer: Why Do Cancer Cells Produce or Take up Exogenous Acute Phase Protein Alpha1-Antitrypsin? *Front. Oncol.* **2021**, *11*, 622076. [CrossRef] [PubMed]
16. Mazur, M.; Eder, P.; Kolodziejczak, B.; Sobieska, M.; Krela-Kazmierczak, I.; Grzymislawski, M. Long-term prognostic utility of selected acute phase proteins in colorectal cancer. *Pol. Arch. Intern. Med.* **2019**, *129*, 292–294. [CrossRef]
17. Biswas, B.; Rastogi, S.; Khan, S.A.; Mohanti, B.K.; Sharma, D.N.; Sharma, M.C.; Mridha, A.R.; Bakhshi, S. Outcomes and prognostic factors for Ewing-family tumors of the extremities. *J. Bone Jt. Surg. Am.* **2014**, *96*, 841–849. [CrossRef]
18. Whelan, J.; Le Deley, M.C.; Dirksen, U.; Le Teuff, G.; Brennan, B.; Gaspar, N.; Hawkins, D.S.; Amler, S.; Bauer, S.; Bielack, S.; et al. High-Dose Chemotherapy and Blood Autologous Stem-Cell Rescue Compared with Standard Chemotherapy in Localized High-Risk Ewing Sarcoma: Results of Euro-E.W.I.N.G.99 and Ewing-2008. *J. Clin. Oncol.* **2018**, *36*, JCO2018782516. [CrossRef] [PubMed]
19. Aggerholm-Pedersen, N.; Maretty-Kongstad, K.; Keller, J.; Baerentzen, S.; Safwat, A. The Prognostic Value of Serum Biomarkers in Localized Bone Sarcoma. *Transl. Oncol.* **2016**, *9*, 322–328. [CrossRef] [PubMed]
20. Nakamura, T.; Grimer, R.J.; Gaston, C.L.; Watanuki, M.; Sudo, A.; Jeys, L. The prognostic value of the serum level of C-reactive protein for the survival of patients with a primary sarcoma of bone. *Bone Jt. J.* **2013**, *95-B*, 411–418. [CrossRef]
21. Li, Y.J.; Yang, X.; Zhang, W.B.; Yi, C.; Wang, F.; Li, P. Clinical implications of six inflammatory biomarkers as prognostic indicators in Ewing sarcoma. *Cancer Manag. Res.* **2017**, *9*, 443–451. [CrossRef]
22. Xu, K.; Lou, Y.; Sun, R.; Liu, Y.; Li, B.; Li, J.; Huang, Q.; Wan, W.; Xiao, J. Establishment of a Nomogram-Based Model for Predicting the Prognostic Value of Inflammatory Biomarkers and Preoperative D-Dimer Level in Spinal Ewing's Sarcoma Family Tumors: A Retrospective Study of 83 Patients. *World Neurosurg.* **2019**, *121*, e104–e112. [CrossRef]
23. Biswas, B.; Shukla, N.K.; Deo, S.V.; Agarwala, S.; Sharma, D.N.; Vishnubhatla, S.; Bakhshi, S. Evaluation of outcome and prognostic factors in extraosseous Ewing sarcoma. *Pediatr. Blood Cancer* **2014**, *61*, 1925–1931. [CrossRef] [PubMed]
24. Biswas, B.; Rastogi, S.; Khan, S.A.; Shukla, N.K.; Deo, S.V.; Agarwala, S.; Mohanti, B.K.; Sharma, M.C.; Vishnubhatla, S.; Bakhshi, S. Developing a prognostic model for localized Ewing sarcoma family of tumors: A single institutional experience of 224 cases treated with uniform chemotherapy protocol. *J. Surg. Oncol.* **2015**, *111*, 683–689. [CrossRef] [PubMed]
25. Biswas, B.; Thakar, A.; Mohanti, B.K.; Vishnubhatla, S.; Bakhshi, S. Prognostic factors in head and neck Ewing sarcoma family of tumors. *Laryngoscope* **2015**, *125*, E112–E117. [CrossRef] [PubMed]
26. Fujiwara, T.; Fukushi, J.; Yamamoto, S.; Matsumoto, Y.; Setsu, N.; Oda, Y.; Yamada, H.; Okada, S.; Watari, K.; Ono, M.; et al. Macrophage infiltration predicts a poor prognosis for human ewing sarcoma. *Am. J. Pathol.* **2011**, *179*, 1157–1170. [CrossRef] [PubMed]

27. Pahl, J.H.; Kwappenberg, K.M.; Varypataki, E.M.; Santos, S.J.; Kuijjer, M.L.; Mohamed, S.; Wijnen, J.T.; van Tol, M.J.; Cleton-Jansen, A.M.; Egeler, R.M.; et al. Macrophages inhibit human osteosarcoma cell growth after activation with the bacterial cell wall derivative liposomal muramyl tripeptide in combination with interferon-gamma. *J. Exp. Clin. Cancer Res.* **2014**, *33*, 27. [CrossRef] [PubMed]
28. Meyers, P.A. Muramyl Tripeptide-Phosphatidyl Ethanolamine Encapsulated in Liposomes (L-MTP-PE) in the Treatment of Osteosarcoma. *Adv. Exp. Med. Biol.* **2020**, *1257*, 133–139. [PubMed]
29. Kokkali, S.; Kotsantis, I.; Magou, E.; Sophia, T.; Kormas, T.; Diakoumis, G.; Spathas, N.; Psyrri, A.; Ardavanis, A. The addition of the immunomodulator mifamurtide to adjuvant chemotherapy for early osteosarcoma: A retrospective analysis. *Investig. New Drugs* **2022**, *40*, 668–675. [CrossRef]
30. Punzo, F.; Bellini, G.; Tortora, C.; Pinto, D.D.; Argenziano, M.; Pota, E.; Paola, A.D.; Martino, M.D.; Rossi, F. Mifamurtide and TAM-like macrophages: Effect on proliferation, migration and differentiation of osteosarcoma cells. *Oncotarget* **2020**, *11*, 687–698. [CrossRef]

cancers

MDPI

Article

Secondary Malignancies after Ewing Sarcoma—Epidemiological and Clinical Analysis of an International Trial Registry

Isabelle Kaiser [1,2,3], Katja Kauertz [1,2,3], Stefan K. Zöllner [1,2,3], Wolfgang Hartmann [4], Thorsten Langer [5], Heribert Jürgens [6], Andreas Ranft [1,2,3] and Uta Dirksen [1,2,3,*]

1 Pediatrics III, University Hospital Essen, 45147 Essen, Germany
2 West German Cancer Center (WTZ), University Hospital Essen, 45147 Essen, Germany
3 German Cancer Consortium (DKTK), Essen/Düsseldorf, University Hospital Essen, 45147 Essen, Germany
4 Gerhard Domagk Institute for Pathology, University Hospital Muenster, 48149 Muenster, Germany
5 Pediatric Hematology and Oncology, LESS Group, University Medical Center Schleswig-Holstein, 23538 Luebeck, Germany
6 Pediatric Hematology and Oncology, University Children's Hospital Münster, 48149 Münster, Germany
* Correspondence: uta.dirksen@uk-essen.de

Simple Summary: Ewing sarcoma (EwS) is a malignant bone and soft tissue cancer that requires intensive treatment with multiple chemotherapies and either surgery, irradiation, or both as local therapy. For most survivors of EwS, long-term sequelae such as secondary malignant neoplasms (SMNs) other than EwS are concerning. Few studies suggest that SMNs after EwS are a rare but serious event. Comprehensive data are lacking. We reviewed consecutive EwS trials from the Cooperative Ewing Sarcoma Study (CESS) group to evaluate the features of SMNs in EwS patients. Our analysis revealed 101 cases of SMNs in 96 EwS patients. Solid SMNs were detected more frequently than hematologic SMNs, in 55.2% versus 44.8%. The latency between EwS diagnosis and SMN occurrence was longer for solid SMNs (median: 8.4 years) than for hematologic SMNs (median: 2.4 years) ($p < 0.001$). The survival rate after SMNs was 0.49, with solid SMNs having a significantly better prognosis. Our results confirm the need for a structured follow-up system.

Abstract: Ewing sarcoma (EwS) represents highly aggressive bone and soft tissue tumors that require intensive treatment by multi-chemotherapy, surgery and/or radiotherapy. While therapeutic regimens have increased survival rates, EwS survivors face long-term sequelae that include secondary malignant neoplasms (SMNs). Consequently, more knowledge about EwS patients who develop SMNs is needed to identify high-risk patients and adjust follow-up strategies. We retrospectively analyzed data from 4518 EwS patients treated in five consecutive EwS trials from the Cooperative Ewing Sarcoma Study (CESS) group. Ninety-six patients developed SMNs after primary EwS, including 53 (55.2%) with solid tumors. The latency period between EwS and the first SMN was significantly longer for the development of solid SMNs (median: 8.4 years) than for hematologic SMNs (median: 2.4 years) ($p < 0.001$). The cumulative incidence (CI) of SMNs in general increased over time from 0.04 at 10 years to 0.14 at 30 years; notably, the specific CI for hematologic SMNs remained stable over the different decades, whereas for solid SMNs it gradually increased over time and was higher for metastatic patients than in localized EwS patients (20 years: 0.14 vs. 0.06; $p < 0.01$). The clinical characteristics of primary EwS did not differ between patients with or without SMNs. All EwS patients received multi-chemotherapy with adjuvant radiotherapy in 77 of 96 (80.2%) patients, and the use of radiation doses ≥ 60 Gy correlated with the occurrence of SMNs. The survival rate after SMNs was 0.49, with a significantly better outcome for solid SMNs compared with hematologic SMNs (3 years: 0.70 vs. 0.24, respectively; $p < 0.001$). The occurrence of SMNs after EwS remains a rare event but requires a structured follow-up system because it is associated with high morbidity and mortality.

Citation: Kaiser, I.; Kauertz, K.; Zöllner, S.K.; Hartmann, W.; Langer, T.; Jürgens, H.; Ranft, A.; Dirksen, U. Secondary Malignancies after Ewing Sarcoma—Epidemiological and Clinical Analysis of an International Trial Registry. *Cancers* **2022**, *14*, 5920. https://doi.org/10.3390/cancers14235920

Academic Editor: Franck Verrecchia

Received: 28 September 2022
Accepted: 22 November 2022
Published: 30 November 2022

Keywords: childhood cancer; cancer survivor; Ewing sarcoma; secondary malignant neoplasms; secondary malignancy; radiotherapy

1. Introduction

Ewing sarcoma (EwS) is a rare, highly aggressive malignancy of small blue, round cells that occurs in bone and soft tissue and predominantly affects children, adolescents and young adults [1]. An incidence of 1.5 cases per million is observed in people of European descent [1–3]. Twenty to 25% of patients have metastases at the time of initial diagnosis, which is the most important prognostic factor for EwS [4].

Although 85% of all cases have a balanced chromosomal translocation in which Ewing sarcoma breakpoint region 1 (EWSR1) protein fuses with the Friend leukemia integration 1 (FLI1) transcription factor, resulting in the *EWSR1-FLI1* fusion oncogene [5–7], current first-line treatment of EwS does not include targeted therapy. EwS requires a multimodal therapeutic approach consisting of multiple cycles of multiagent chemotherapy and local therapy consisting of surgery, radiotherapy, or both modalities. The decision on local treatment can be discussed in specialized tumor boards [8]. Rational combinatorial and dose-intensifying strategies of the basic chemotherapeutic regimen have improved outcome in all patients except those with disseminated disease [9–11].

As a consequence of improved survival, long-term surveillance for late effects has become increasingly important because the occurrence of secondary malignant neoplasms (SMNs) causes high morbidity and mortality [12,13]. SMNs are defined as malignancies that occur during or after cancer treatment and are not detected before the initial cancer treatment. Histologically, SMNs present as distinct from the primary tumor [14].

Few reports of SMNs after EwS have been published, mostly in association with trial reports. Nevertheless, important aspects such as the clinical characteristics of EwS patients who develop SMNs, the types of SMNs, and the outcomes for patients with SMNs after EwS have not yet been adequately studied.

Several risk factors have been associated with SMNs [12,14]. Previous studies showed a leukemogenic potential of epipodophyllotoxins, anthracyclines, and alkylating agents [15–21] and an increased risk of solid SMNs after high-dose radiotherapy ($\geq$60 Gy) [22–25].

In the present study, we retrospectively investigated the epidemiology and clinical features of 96 identified patients with SMNs in an international cohort of 4518 patients treated for EwS in five consecutive clinical trials. Specifically, we aimed to identify specific clinical features in EwS patients and associated treatments that might predispose to SMNs.

2. Materials and Methods

2.1. Patient Cohorts and Eligibility Criteria

This retrospective analysis included data from 4518 international patients treated for EwS between 1980 and 2019 and registered in the Ewing Sarcoma Study Group of the GPOH (German Society of Pediatric Hematology and Oncology) database. The corresponding phase III clinical trials were Cooperative Ewing's Sarcoma Studies 1981 (CESS 81, 184 patients), Cooperative Ewing's Sarcoma Studies 1986 (CESS 86, 490 patients), European Intergroup Cooperative Ewing's Sarcoma Study 92 (EICESS 92, 875 patients), EUROpean Ewing tumor Working Initiative of National Groups—Ewing Tumor Studies-1999 (Euro-E.W.I.N.G. 99, 1548 patients), and Ewing 2008 (1421 patients).

Phase III randomized clinical trials were multicenter and nationwide or international in scope. Informed consent for long-term follow-up and data analysis was obtained from patients, parents, or guardians. Ethics committee approval was obtained at baseline. All trials are summarized in Table 1 and briefly described below.

Table 1. Development of chemotherapy and radiotherapy for Ewing sarcoma (EwS) in different trials from the Cooperative Ewing Sarcoma Study (CESS) group.

	EwS Trial	CESS 81	CESS 86		EICESS 92				EURO E.W.I.N.G. 99								Ewing 2008					
	Number of Cycles	12	12			14			14		14	8			8		14		14	8	14	15
	Risk Strata	-	SR	HR	SR		HR		R1 = SR		R2 = HR				R3 = VHR		R1 = SR		R2 = HR		R3 = VHR	
	Regimen	VACA	VACA	VAIA	VAIA + VACA	VAIA	VAIA	EVAIA	VIDE + VAC ♀	VIDE + VAI ♂	VIDE + VAI	VIDE + VAI + BU/MEL	VIDE + VAI + ME/ME	VIDE + VAI + TREO/MEL	VIDE + VAI + BU/MEL	VIDE + VAI	VIDE + VAC ♀	VIDE + VAI ♂	VIDE + VAI	VIDE + VAI + BU/MEL	VIDE + VAC	VIDE + VAC + TREO/MEL
Chemotherapeutic Agent and Dose	V (mg/m²)	24	24	24			21			21		10.5			10.5			21		10.5	21	
	A (mg/m²)	6	6	6			10.5			12		1.5			1.5			12		1.5	12	
	C (g/m²)	14.4	14.4	-	12		-		10.5	-	-	-			-		12	-	-	-	12	
	I (g/m²)	-	-	72	24		84		60	102		60			60		54	102		60	54	
	D (mg/m²)	480	480	480		420				360					360					360		
	E (g/m²)	-	-	-		-		6.3		2.7			3.15		2.7					2.7		
	BU (mg/m²)	-	-	-		-				-		600	-	-	600	-	-	-	-	600	-	-
	MEL (mg/m²)	-	-	-		-				-		140		140		-	-	-	-	140	-	140
	TREO (g/m²)	-	-	-		-				-			-	36	-	-	-	-	-	-	-	36
Irradiation Dose (Gy)	Preoperative	-	-			45				54.4					54.4					54.4		
	Definitive	46–60	45–60 ∩			55				44.8–54.5 ∩					44.8–54.5 ∩					45–54 ∩		
	Postoperative	36	45			45				44.8–54.4					44.8–54.4					45–54		

SR—standard-risk, HR—high-risk, VHR—very high-risk. Please see section "Materials and Methods" for more detailed information on risk groups including randomization of R1/R2/R3. V—vincristine, A—actinomycin D, C—cyclophosphamide, I—ifosfamide, D—doxorubicin (adriamycin), E—etoposide, BU—busulfan, ME—melphalan, etoposide, MEL—melphalan, TREO—treosulfan. ∩ integrated boost to tumor target volume following predefined radiotherapeutic strategies within trial.

In the CESS 81 trial (patient enrollment from 1981 to 1985), four cycles (nine weeks each) of VACA chemotherapy were administered, which included vincristine, actinomycin D, cyclophosphamide, and adriamycin. Two cycles of VACA were followed by local therapy consisting of either surgery, surgery plus postoperative radiotherapy (36 Gy), or definitive radiotherapy (46 Gy or 60 Gy, 36 Gy to the entire bone) [22,26].

In the CESS 86 trial (patient enrollment from 1986 to 1991), patients were divided into two risk groups according to tumor volume, tumor location, and metastatic status. Standard-risk (SR) patients had small (<100 mL) extremity tumors and were treated with VACA. If an initial tumor volume ≥100 mL, central tumor localization, or metastatic disease was diagnosed, patients were classified as high-risk (HR) and received a VAIA regimen (vincristine, actinomycin D, ifosfamide, adriamycin). Local therapy was given after one cycle of VAIA (week 9/10) and included surgery, postoperative radiotherapy (44.8/45 Gy), or definitive radiotherapy (44.8/45 Gy with a local boost up to 60 Gy). Patients received a total of four chemotherapy cycles with twelve courses [22,26,27].

In the EICESS 92 trial (patient enrollment from 1992 to 1999), patients were divided into SR and HR groups according to tumor volume and metastatic status, with a threshold of 100 mL, and chemotherapy doses were changed. SR patients (tumor volume < 100 mL) were randomly assigned to receive four courses of VAIA followed by either ten courses of VAIA or VACA. HR patients (≥100 mL and/or metastatic disease) received VAIA or VAIA plus etoposide (EVAIA) over 14 courses. Preoperative radiotherapy (45–55 Gy depending on the anticipated extent of resection) was initiated at EICESS 92. It was applied after the sixth week of chemotherapy if the soft tissue component had decreased by less than 50% after the second chemotherapy or if the tumor was deemed unresectable.

Definitive radiotherapy (up to 55 Gy) was administered after the fourth course. Patients with wide resection but poor histological response (≥10% viable tumor) or marginal resection but good response (<10% viable tumor) received postoperative radiotherapy at a dose of 45 Gy. For intralesional resection or marginal resection and poor histological response, up to 55 Gy was recommended. In patients with pulmonary metastases, the entire lung was irradiated (15 Gy if < 14 years and 18 Gy if ≥ 14 years) [26,28].

In the Euro-E.W.I.N.G. 99 trial (patient enrollment from 1999 to 2011), patients were stratified into three risk groups. All patients received six induction courses of VIDE (vincristine, ifosfamide, doxorubicin, etoposide) and one course of VAI (vincristine, actinomycin D, ifosfamide). The R1 patients with localized disease and either a favorable response to neoadjuvant chemotherapy (<10% viable tumor) or a small initial tumor volume (<200 mL) were randomly assigned to either adjuvant therapy with vincristine, actinomycin D, and cyclophosphamide (VAC) or vincristine, actinomycin D, and ifosfamide (VAI) [29]. R2loc patients with localized disease and either poor histological response to neoadjuvant chemotherapy (≥10% viable tumor) or a large initial tumor volume (≥200 mL) and the R2pulm patients with isolated lung metastases were randomized to receive seven courses of VAI versus high-dose chemotherapy with busulfan-melphalan followed by retransfusion of autologous hematopoietic stem cells [30]. R2pulm patients randomized to VAI also received whole-lung irradiation [31]. R3 patients with disseminated disease were enrolled in a non-randomized arm and were scheduled to receive high-dose chemotherapy [11]. In the Euro-E.W.I.N.G. 99 trial, definitive surgical resection was recommended whenever possible and postoperative radiotherapy for large primary tumors, unfavorable histological response, and marginal resection; preoperative radiotherapy was optional. Definitive radiotherapy was considered for unresectable tumors. The following doses were recommended for radiotherapy: preoperative radiotherapy 54.4 Gy, definitive radiotherapy 44.8 Gy (with a boost of 54 Gy, but in individual cases with a maximum boost of 64 Gy depending on tumor location and patients age), and postoperative radiotherapy 44.8 Gy–54.4 Gy [32].

The Ewing 2008 trial (patient enrollment from 2008 to 2019) was the follow-up trial of Euro E.W.I.N.G. 99 and stratified patients accordingly but asked randomized questions in all risk groups. All patients received the VIDE induction regimen. Standard-risk patients (tumors < 200 mL and/or favorable histologic response to induction chemotherapy) received sex-specific maintenance therapy with VAC in women and VAI in men, and patients were randomized to the addition of zolendronic acid or not. The R2 arms were adopted from the Euro E.W.I.N.G. 99 trial [30,31]. Patients with disseminated disease were defined as very high-risk (R3) and received either eight cycles of VAC or eight cycles of VAC plus a course of high-dose treosulfan-melphalan chemotherapy (followed by autologous stem cell reinfusion) after induction chemotherapy. Local therapy was delivered as surgery, whenever possible, and/or radiotherapy [30,31].

2.2. Follow-Up and Statistical Analysis

Cases of SMNs (secondary malignant neoplasm defined as cancer of a histological type other than EwS that occurs during or after cancer treatment and was not detected before the initial cancer treatment) were reported by participating institutions and confirmed by pathology report. Survival after SMNs was estimated using the Kaplan–Meier method and defined as the interval between the diagnosis of SMNs and the date of death or last contact. Cumulative incidences of SMNs were estimated with XLSTAT using competing risk analysis. SAS and SPSS were used for exploratory data analysis.

2.3. Literature Search

A Medline search of the PubMed database was performed using the following terms: "second malignant neoplasm," "SMN," "secondary cancer," "second malignancy", and "childhood cancer survivor".

3. Results

3.1. Patient Characteristics and Clinical Features of Primary Ewing Sarcoma

Of the 4518 EwS patients treated between 1980 and 2019 according to EwS study protocols, 101 cases of subsequent malignant neoplasms were retrospectively assigned to 96 patients. Five patients presented with two different malignant neoplasms after the primary diagnosis of EwS. The epidemiological and clinical characteristics of EwS patients with SMNs are summarized in Table 2. The median time from diagnosis of SMNs to last follow-up was 1.16 years (range: 0–19). The median time from diagnosis of EwS to diagnosis of SMNs was 4.9 years (0.1–28.1). Four patients were lost to follow-up. Loss to follow-up resulted from relocation, change of oncologist, or refusal of further contact for follow-up. The clinical characteristics of the five EwS patients with two SMNs are summarized in Supplement Table S1. In the following, only the results of patients with a single SMN are listed: 51 of the 96 patients with SMNs were female, and 45 were male. The mean age of EwS patients at diagnosis was 14.4 years and ranged from 2.4 to 68.6 years. Thirty-one patients (32.3%) had metastases at the time of EwS diagnosis. In most patients, metastases were located to the lung (18.6%), bone marrow (7.2%), and bone (13.4%), but not at the site of the primary tumor. None of the patients with SMNs had a reported family history of cancer predisposition or tumor-associated syndrome.

Table 2. Patient characteristics and clinical features of 96 primary Ewing sarcoma (EwS) patients with secondary malignant neoplasms (SMNs) in the CESS 81, CESS 86, EICESS 92, Euro-E.W.I.N.G. 99, and Ewing 2008 trials.

Attributable Distribution of Primary EwS Patients at Diagnosis	Number of Patients with SMNs (n, %)	Median Observation Time from Primary EwS Diagnosis to SMNs (Years)
EwS trial (n = 96)		
CESS 81	2 (of 184), 1.1%	21.7
CESS 86	16 (of 490), 3.3%	11.9
EICESS 92	21 (of 875), 2.4%	6
EURO E.W.I.N.G. 99	36 (of 1548), 2.3%	4.9
Ewing 2008	21 (of 1421), 1.5%	2.3
Sex (%) (n = 96)		
Male	45 (46.9%)	
Female	51 (53.1%)	
Metastases (n = 96)		
Yes	31 (32.3%)	
No	65 (67.7%)	
Age (n = 96)		
median (range)	14.4 (2.4–68.6) years	
Localization (n = 96)		
Cranium	5 (5.2%)	
Hand/foot	6 (6.3%)	
Upper limb	9 (9.4%)	
Lower limb	21 (21.9%)	
Axial skeleton	29 (30.2%)	
Pelvis	26 (27%)	

Clinical data of five patients with >1 subsequent malignant neoplasm are summarized in Supplementary Table S1.

In our study cohort, all patients received chemotherapy. Table 1 provides an overview of the type and dosage of chemotherapy. In all studies, systemic treatment was supplemented by local therapy using surgery and/or radiotherapy. Radiotherapy was given either as definitive therapy or as a local adjuvant approach after surgery. 77 of 96 patients (80.2%) received radiotherapy. In the CESS 86 study, 11 of 16 SMNs patients were treated with high-dose radiotherapy (≥60 Gy). These patients suffered predominantly from solid tumors (n = 9) such as sarcomas including osteosarcomas (n = 5) or carcinomas (n = 4).

3.2. Epidemiology of Secondary Malignant Neoplasms

The types of SMNs after primary EwS are summarized in Table 3 according to the EwS study protocols. The corresponding data on EwS patients with more than one SMN are summarized in Supplementary Table S1. The highest percentage of patients with SMNs was found in the CESS 86 study (3.1%), and the lowest percentage of SMNs was documented in patients in the CESS 81 study (1.1%). Solid tumors were detected more frequently than hematologic neoplasms after primary EwS, in 55.2% versus 44.8%. Carcinomas formed the largest group (n = 23) of all solid tumors. Among sarcomas, osteosarcoma was the most common type of sarcoma, followed by rhabdomyosarcoma. The most common hematologic neoplasms were acute myeloid leukemia (AML) and myelodysplastic syndrome (MDS). In two cases, incomplete documentation prevented clear classification between MDS and AML. Other SMNs included astrocytoma (n = 1), blastoma (n = 1), melanoma (n = 2), neuroendocrine tumor (n = 1), and pancreatic tumor (n = 1). The median time between EwS and SMNs was 5.4 years. The latency period was longer for solid tumors (median: 8.4 years) than for hematologic neoplasms (median: 2.4 years) (d = 1.22; $p < 0.001$). The median time between EwS and tertiary malignant neoplasm was 9.4 years, and the median time between SMNs and tertiary malignant neoplasm was 1.4 years.

Table 3. Types of secondary malignant neoplasms (SMNs) of 96 primary Ewing sarcoma (EwS) patients in the CESS 81, CESS 86, EICESS 92, Euro-E.W.I.N.G. 99, and Ewing 2008 trials.

EwS Trial	Type of SMNs	Number of Patients with SMNs (n, %)
Across trials (n = 96)	Solid	53 (55.2%)
	Hematologic	43 (44.8%)
CESS 81 (n = 2)	Solid	2 (100%)
	Osteosarcoma	0
	Other sarcoma	0
	Carcinoma	2
	Other	0
	Hematologic	0 (0%)
	Leukemia, lymphoma	0
	Myelodysplastic syndrome	0
CESS 86 (n = 16)	Solid	11 (73.3%)
	Osteosarcoma	4
	Other sarcoma	3
	Carcinoma	4
	Other	0
	Hematologic	4 (26.7%)
	Leukemia, lymphoma	3
	Myelodysplastic syndrome	1
EICESS 92 (n = 21)	Solid	12 (54.5%)
	Osteosarcoma	4
	Other sarcoma	2
	Carcinoma	5
	Other	1
	Hematologic	10 (45.5%)
	Leukemia, lymphoma	5
	Myelodysplastic syndrome	5

Table 3. *Cont.*

EwS Trial	Type of SMNs	Number of Patients with SMNs (*n*, %)
Euro E.W.I.N.G. 99 (*n* = 36)	Solid	18 (50%)
	Osteosarcoma	7
	Other sarcoma	2
	Carcinoma	7
	Other	2
	Hematologic	18 (50%)
	Leukemia, lymphoma	9
	Myelodysplastic syndrome	9
Ewing 2008 (*n* = 21)	Solid	9 (42.9%)
	Osteosarcoma	0
	Other sarcoma	2
	Carcinoma	5
	Other	2
	Hematologic	12 (57.1%)
	Leukemia, lymphoma	6
	Myelodysplastic syndrome	6

3.3. Cumulative Incidences and Outcome of Secondary Malignant Neoplasms

The cumulative incidence (CI) of SMNs was 0.04 (SE < 0.01) at 10 years, 0.07 (SE = 0.01) at 20 years, and 0.14 (SE = 0.03) at 30 years. For solid tumors, the specific CI was 0.02 (SE < 0.01) at 10 years, 0.06 (SE = 0.01) at 20 years, and 0.12 (SE = 0.03) at 30 years. For hematologic neoplasms, the specific CI was 0.02 (SE < 0.01) after each of 10, 20 and 30 years. For hematologic neoplasms, the specific CI reached a plateau after 8 years. Female patients had a higher risk than male patients (20 years: 0.11 vs. 0.05; $p = 0.03$), similarly for metastatic patients compared to localized patients (20 years: 0.14 vs. 0.06; $p < 0.01$). Age (</≥15 years) had no effect on cumulative incidence. The type of local treatment did not affect the incidence of SMNs in general, but the use of radiation doses ≥ 60 Gy correlated with the incidence of SMNs in particular. The survival rate after SMNs was 0.49 (SE = 0.06). It differed significantly between solid and hematologic SMNs for EwS patients (3 years: 0.70 vs. 0.24; $p < 0.001$) (Figure 1).

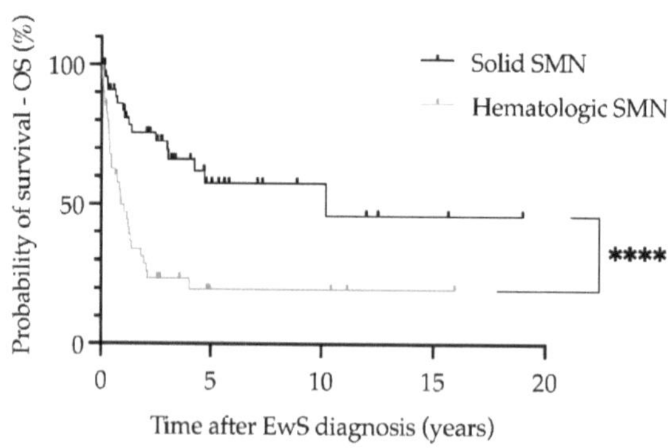

at risk

Solid SMN	48	12	5	2	0
Hematologic SMN	41	3	3	1	0

Figure 1. Survival of patients with primary Ewing sarcoma (EwS) and secondary malignant neoplasms (SMNs) as a function of secondary hematologic or solid malignancy in the CESS 81, CESS 86, EICESS 92, Euro-E.W.I.N.G. 99, and Ewing 2008 trials. Overall survival (OS) was calculated using the Kaplan–Meier method. **** $p < 0.001$. # at risk—number of patients at risk.

3.4. Literature Search

We identified the seven specific articles presented in Table 4 to compare and discuss risk-associated factors for SMNs after EwS. The articles included studies focused on the development of SMNs in patients with EwS or Ewing sarcoma family tumors. 55.9% of patients were female (4/7 articles) in a median cohort size of 381 EwS patients ($n = 7/7$). The median age at the time of EwS diagnosis was 14.7 years ($n = 5/7$) and the latency period was 8.2 years ($n = 6/7$). Cumulative incidences ranged from 4.7% at 10 years to 15% at 25 years. A solid tumor was identified as SMNs in 65.5% ($n = 6/7$). The tumors most frequently described in the articles were breast cancer, sarcoma including osteosarcoma, MDS, and leukemia including AML.

Table 4. Summary of data from secondary malignant neoplasms (SMNs) studies after primary tumors including Ewing sarcoma (EwS). AML—acute myeloid leukemia, BC—breast cancer, CI—cumulative incidence, CTX—chemotherapy, MDS—myelodysplastic syndrome, NA—not annotated; NMSC—Non-melanoma skin cancer, OS—osteosarcoma, RTX—radiotherapy, WLI—whole lung irradiation.

Study Details					EwS Characteristics at Diagnosis				Characteristics of Secondary Malignant Neoplasms						Comments
Author (Publication Year)	Reported Time Period (Time to Publication) (Years)	Cohort	Cohort Size (Patients)	Median Follow-Up (Range) (Years)	Median Age (Range) (Years)	Metastases (%)	Females (%)	Tumor Volume > 100 mL (%)	Number	Solid (%)	Predominant Type (%)	CI (%/Years)	Latency (Range) (Years)	Risk Factors	
Hawkins et al. (1996) [33]	1940–1983 (13)	EwS survivors	207	7.1	NA	NA	NA	NA	NA	NA	NA	5.4/20	NA	Sarcoma: CTX (alkylating agents, dose-dependent) Sarcoma: RTX ($^{4}/_{5}$ tumors in RTX field)	Selective description of secondary bone cancer after childhood cancer
Ginsberg et al. (2010) [34]	1970–1986 (24)	EwS	403	Alive: 23.0 (16–33) Deceased: 11.2 (5–28) *	13.5 (6–20)	NA	NA	NA	36	94.5	BC (36)	9/25	14.5 (4–32)	Solid: RTX ($p = 0.28$) BC: WLI	NMSC excluded
Kuttesch et al. (1996) [24]	1963–1990 (6)	EwS	266	9.5	14.2 (4.2–28; 90% < 21)	NA	56.25	NA	16	87.5	Sarcoma (62.5)	9.2/20	7.6 (3.5–25.7)	All SMN: RTX (>48 Gy) ($p = 0.043$) Sarcoma: RTX (100% in RTX field) ($p = 0.002$)	Combination of actinomycin D and RTX reduce risk
Dunst et al. (1998) [22]	1981–1991 (7)	EwS	674	5.1	13.25 (8–21)	25	87.5	50	8	37.5	AML (50)	4.7/15	6 (1.5–11.4)	Sarcoma: RTX (100% with RTX)	Selection bias for RTX

Table 4. *Cont.*

Study Details					EwS Characteristics at Diagnosis				Characteristics of Secondary Malignant Neoplasms						Comments
Author (Publication Year)	**Reported Time Period (Time to Publication) (Years)**	**Cohort**	**Cohort Size (Patients)**	**Median Follow-Up (Range) (Years)**	**Median Age (Range) (Years)**	**Metastases (%)**	**Females (%)**	**Tumor Volume > 100 mL (%)**	**Number**	**Solid (%)**	**Predominant Type (%)**	**CI (%/Years)**	**Latency (Range) (Years)**	**Risk Factors**	
Navid et al. (2008) [35]	1979–2004 (4)	EwS family of tumors	237		NA	8.3	50	NA	12	33.3	MDS/ leukemia (66.6)	4.7/10	3.3 (1.4–19.6)	Hematologic: CTX (alkylating agents, topoisomerase-II inhibitors, dose-dependent) All SMN: Localized stage ($p = 0.036$) Earlier treatment protocol ($p = 0.001$)	
Longhi et al. (2012) [36]	1983–2006 (6)	Localized EwS, <40 years	581	7.2	16.36 (6–39)	0	NA	NA	15	80	OS (40)	5/25	7 (1–21.1) Hematological: 3.1 Solid: 7.8	Female sex	
Friedmann et al. (2017) [37]	1974–2012 (5)	EwS, <40 years	300	7.8	MDS/AML: 17.4 (5–32) Solid: 14.6 (6–24)	30	30	30	15	60	MDS/AML (60)	15/25	10.9 (0.9–27.7) MDS/AML: 3.2 (0.9–4.6) Solid: 21.3 (10.5–27.7)	Hematologic: CTX (alkylating agents, topoisomerase-II inhibitors, dose-dependent)	NMSC and melanoma excluded

* Deceased patients within five years from EwS diagnosis excluded.

4. Discussion

A 2017 report from the German Childhood Cancer Registry showed that the 15-year survival rates for childhood cancer patients younger than 15 years have increased to 82% [38]. Refined treatment protocols for EwS aim to reduce therapy-related toxic effects, resulting in higher survival rates and longer follow-up [4,34]. However, one of the most devastating complications of primary cancer therapy is the development of a second malignancy.

The most recent report (2019) from the German Childhood Cancer Registry published a cumulative incidence of 6.8% for SMN in German cancer survivors (diagnosed < 18 years) within 30 years of diagnosis. Focusing on SMNs after EwS, the registry reported a cumulative incidence of 4.4% at 30 years (treated between 1981 and 2016) [39]. In contrast, a 2020 analysis of the Surveillance, Epidemiology and End Results Program (SEER) database by Friedman et al. published a cumulative incidence of 10.1% at 30 years (treated between 1970 and 1986) [40].

Our study reports comprehensive epidemiologic and clinical data on SMNs after primary EwS in a series of international EwS trials spanning 30 years of patient enrollment and standardized treatment, making it the largest and longest retrospective study of primary EwS patients with SMNs.

We determined cumulative incidences (CI) of 1.9%, 3.9%, 5%, 7.4% and 14% at 5, 10, 15, 20, and 30 years, respectively. The CI of SMNs in patients treated for EwS varies widely in the literature (Table 4). In a meta-analysis, Caruso et al. reported a CI ranging from less than 0.9% at 5 years to more than 20.5% at 30 years [12]. Friedman et al. (Memorial Sloan Kettering) described a CI of 15% at 25 years despite their more recent publication date, but this may be due to a long observation period of 38 years starting in 1974 [37]. Longhi et al. (Italian Sarcoma group) reported a CI of only 5% at 25 years but excluded patients with disseminated disease and thus those who received extensive treatment [36]. In the Children's Oncology Group's recent trial of the treatment of localized EwS (AEWS1031), patients were randomly assigned to two different regimens. In regimen A, patients received 17 cycles of compressed chemotherapy with a standard five-drug interval, while patients in regimen B received experimental therapy with five cycles of vincristine, topotecan, and cyclophosphamide within the 17 cycles. A numerical rate of 4.3% for SMNs was reported in a total of 626 patients. A cumulative incidence for comparison was not reported, nor was information on SMNs characteristics (e.g., tumor entity). Leavey et al. did not detect a significant difference in SMNs between the randomized study arms and hypothesized that vincristine, topotecan and cyclophosphamide had no effect on the risk of SMNs [41].

The median latency from primary EwS diagnosis to SMNs in our analysis was 5.4 years. Similar results were published by Kuttesch et al. (7.6 years) and are consistent with reported latencies of SMNs after other childhood tumors [24,37,42]. In our cohort, latency was related to the type of SMNs, with hematologic neoplasms (median 2.4 years) occurring earlier after primary EwS than solid tumors (median 8.4 years). The study by Friedman et al. reported a comparable median latency of 3.2 years for secondary AML/MDS. For solid tumors, the latency period in this study was 21.3 years, twice as long as our results. It remains unclear whether the above discrepancy is due to the limited number of only 300 patients observed by Friedman or to the non-Ewing round cell tumors included [37]. However, we suspect that the incidence of solid malignancies in our cohort may be underestimated because we included data from more recent trials with limited follow-up. In other studies, the incidence of hematologic neoplasms may be underestimated because the studies by Kuttesch et al. and Ginsberg et al. excluded patients who died within the first three and five years after diagnosis, respectively [24,34].

Our analysis shows that the risk of solid SMNs does not plateau but increases over time. Solid tumors commonly include carcinomas and osteosarcomas, and they can occur up to 28 years after the primary EwS diagnosis.

Several studies have found breast cancer to be the most common SMN in childhood cancer survivors [43–46]. A high incidence of breast cancer has been described especially after chest irradiation in Hodgkin's lymphoma and after total lung irradiation for EwS patients [43–45]. We observed a remarkably low number of breast cancer cases in our cohort (three cases exclusively in women). All patients received radiotherapy to the thoracic wall and developed tumors within the irradiation field. In the study by Schellong et al., 26 cases of breast cancer were detected in 590 Hodgkin's lymphoma patients, of which 25 tumors were located in the previously irradiated field, although the radiation doses for the supradiaphragmatic fields ranged only from 20 to 45 Gy [47]. In our patients, EwS was diagnosed at a median age of 15.6 years (range: 12–21 years), supporting the observations of Bhatia et al. who suggested an association between radiation exposure between the ages of 10 and 16 years and the development of breast cancer as SMNs. This could be caused by exposure of carcinogens in the growing breast tissue as shown by previous data [46]. We could not conclusively explain the low number of breast cancer cases detected in our cohort.

Several studies have confirmed initial status of EwS disease, young age at initial diagnosis (<10 years or <14 years), female gender, radiotherapy, alkylating agents as well as topoisomerase inhibitors, high-dose chemotherapy followed by autologous stem cell transplantation including granulocyte colony-stimulating factor (G-CSF), and tumor predisposition syndromes as risk factors for SMNs in EwS patients [12,24,48–50]. The median age at EwS diagnosis in our cohort was 16 years, which is consistent with the overall median age at EwS diagnosis (15 years) and previous reports of EwS with SMNs with a median age of 14.8 years (range: 0–40 years) (Table 4) [1]. While Navid et al. did not find an age-related risk, Kaatsch et al. showed that children younger than 10 years with EwS treated with radiotherapy were at higher risk for developing SMNs [35,51]. EwS patients with SMNs showed an inverse gender distribution compared with the typical distribution in EwS patients of 40% females and 60% males [3]. Female gender has also been associated with an increased risk of SMNs in other studies looking at the development of SMNs in specific childhood EwS and other cancer patients [27,43]. The long-term effects of radiation therapy on developing tissues remain uncertain, but tissue susceptibility to mutagenic effects may be higher in younger children [51]. Because there is a lack of studies on radiation therapy in children and young adults, the safety limits of radiation doses are unknown [52]. Radiation therapy is highly associated with the development of sarcomas of bone and soft tissue, especially osteosarcomas [53]. Among childhood cancers, EwS is treated with one of the highest doses of radiation [23]. Kuttesch et al. found that EwS patients who received radiation doses of > 48 Gy developed secondary sarcomas but not hematologic neoplasms. The highest risk was found in the group of patients irradiated with ≥ 60 Gy. All secondary sarcomas were located in or in close proximity to the primary irradiation field. The high percentage (92%) of patients treated with radiotherapy may also explain the high number of solid SMNs and the corresponding high cumulative incidences of SMNs [24]. In the study of Dunst et al., even no secondary sarcomas were detected in patients without prior radiotherapy [22]. In our study, patients who developed solid SMNs received higher doses of radiotherapy (mean 53.1 Gy) than hematologic SMNs (mean 49.1 Gy). Most EwS patients with SMNs in the CESS 86 trial were treated with high-dose radiotherapy (≥60 Gy). Solid SMNs were more likely to occur, of which > 50% were localized in the former irradiation area. Osteosarcomas as SMNs seemed to have reduced over time. We hypothesize that the risk decreased over time due to the amelioration in radiotherapy regimen (for example the improvement of irradiation plans as well as techniques). In general, the initiation of preoperative radiotherapy in the EICESS 92 may have also opened the option of surgery (instead of definitive radiotherapy) for tumors deemed as unresectable before and therefore reduced the applied dosage. Previous publications also assumed a risk association of radiotherapy with concomitant use of alkylating agents [24,25,33]. In previous studies of cancer survivors, alkylating agents, topoisomerase inhibitors (epipodophyllotoxins, e.g., etoposide), and the adjunctive use of anthracyclines [54] have been associated with the development of SMNs and may cause stem cell damage predisposing to secondary myelodysplastic syndrome and leukemia [15,17–21,48,55–57]. In addition to leukemic potential, exposure to anthracycline and alkylating agents has been reported to increase the risk of subsequent solid neoplasms [19,21,58,59]. Anthracyclines, topoisomerase inhibitors, and alkylating agents have been used extensively in EwS treatment (Table 1) [60]. While the number of cycles increased, the cumulative doses of most chemotherapeutic agents (vincristine, cyclophosphamide, doxorubicin, and etoposide) gradually decreased over the course of the various trials, with the exception of actinomycin D, whose dosage almost doubled (Table 1). The dose of ifosfamide was repeatedly readjusted in different risk arms [32]. Prognosis and survival in etoposide-induced leukemia remain extremely poor [56]. Etoposide was introduced in the high-risk arm of the EICESS 92 trial and was used to treat all risk arms with reduced doses starting in the EURO E.W.I.N.G. 99 trial [61]. After the introduction of etoposide in the EICESS 92 trial, more (45.5%) hematologic SMNs were detected than in the CESS 81 (0%) and CESS 86 (26.7%) trials without etoposide. However, the change in SMNs type after inclusion of etoposide may be due to several factors, including a longer follow-up period in the CESS 81/86 trials compared with subsequent trials. Finally, not only the cumulative dose but rather a combined use of chemotherapeutic classes such as anthracyclines with topoisomerase inhibitors or anthracyclines with alkylating agents might be crucial for the risk of leukemia [17,18]. Bhatia et al. confirmed that the risk of AML and MDS increased during treatments with a combination of anthracyclines (doxorubicin) and alkylating agents (cyclophosphamide, ifosfamide), as well as an increased cumulative dose of ifosfamide, cyclophosphamide, and doxorubicin [18]. Although there are several case reports of the association between chemotherapy and SMNs, few studies such as Bhatia et al.'s study have demonstrated a significant association between SMNs and the use of the latter chemotherapeutic agents [18,62]. It is also known that the use of

G-CSF in combination with etoposide increases the risk of secondary leukemia [57]. For EwS, G-CSF was first used in the Euro-E.W.I.N.G. 99 trial. Although we found an increase in hematologic SMNs after this time period, its efficacy as an adjuvant to chemotherapy remains uncertain.

With an improved survival, late effects and also SMNs come more and more into the focus of pediatric oncology research. It is uncertain whether SMNs are a matter of treatment or genetics. The role of tumor predisposition syndromes in both the development of EwS itself and the development of SMNs after EwS remains elusive; associations between tumor predisposition syndromes and EwS have not been described [63]. In our cohort, there was no case of documented cancer predisposition syndrome. However, it remains unclear whether patients with EwS are predisposed to SMNs and whether those at higher risk can be identified upfront.

5. Conclusions

In conclusion, SMNs in EwS after successful treatment of primary EwS are rare events, but their occurrence is highly associated with devastating morbidity and mortality and thus worse prognosis.

Because we found differences in the development of SMNs among different study protocols, the risk of SMNs may be influenced by the dose and type of systemic treatment and radiotherapy. The retrospective nature of the present study and the combinatorial use of similar chemotherapeutic agents in the different trial protocols make it difficult to clearly assess the risk of single or combinatorial administration of alkylating agents and anthracyclines. The small number of SMNs and retrospective analysis are important limitations. Furthermore, the small cohort sizes and the predefined population limit the comparability of the above studies [24,37]. Other limitations in our analysis resulted from the loss of follow-up, particularly due to the transition from pediatric to general oncologists. This gap in consistent medical care underscores the importance of continued oncologic care, such as in specialized facilities that focus on adolescents and young adults. Given the rarity of EwS, international collaboration is needed to develop surveillance strategies that both prevent loss to follow-up due to transitions in medical care from childhood to adolescence and extend the duration of follow-up in late adulthood to capture potential SMNs.

Supplementary Materials: The following supporting information can be downloaded at: https://www.mdpi.com/article/10.3390/cancers14235920/s1, Table S1: Patient characteristics and clinical features of 96 primary Ewing sarcoma (EwS) patients with >1 subsequent malignant neoplasm in the CESS 81, CESS 86, EICESS 92, Euro-E.W.I.N.G. 99, and Ewing 2008 trials.

Author Contributions: Conceptualization, I.K., S.K.Z. and U.D.; methodology, I.K. and A.R.; software, I.K., A.R. and S.K.Z.; validation, all authors; formal analysis, I.K. and A.R.; investigation, U.D.; data curation, I.K., K.K., W.H., T.L. and A.R.; writing—original draft preparation, I.K.; writing—review and editing, S.K.Z., A.R. and U.D.; visualization, I.K., A.R. and S.K.Z.; supervision: H.J. and U.D.; funding acquisition, U.D. All authors have read and agreed to the published version of the manuscript.

Funding: This research was funded by German Cancer Aid, grant number 102802, 7011218, 70113419; Gerd und Susanne Mayer Foundation.

Institutional Review Board Statement: The study was conducted in accordance with the Declaration of Helsinki and approved by the Institutional Review Board of Westfalische Wilhelms—Universitat Munster (protocol code EICESS92 and 8 May 1992).

Informed Consent Statement: Informed consent was obtained from all subjects involved in the study.

Data Availability Statement: The data presented in this study are available on request from the corresponding author.

Acknowledgments: The authors would like to express their sincere gratitude to S. Jabar for her administrative and statistical support and assistance.

Conflicts of Interest: The authors declare no conflict of interest.

References

1. Grunewald, T.G.P.; Cidre-Aranaz, F.; Surdez, D.; Tomazou, E.M.; de Alava, E.; Kovar, H.; Sorensen, P.H.; Delattre, O.; Dirksen, U. Ewing sarcoma. *Nat. Rev. Dis. Prim.* **2018**, *4*, 5. [CrossRef] [PubMed]
2. Worch, J.; Cyrus, J.; Goldsby, R.; Matthay, K.K.; Neuhaus, J.; DuBois, S.G. Racial differences in the incidence of mesenchymal tumors associated with EWSR1 translocation. *Cancer Epidemiol. Biomark. Prev.* **2011**, *20*, 449–453. [CrossRef] [PubMed]
3. Jawad, M.U.; Cheung, M.C.; Min, E.S.; Schneiderbauer, M.M.; Koniaris, L.G.; Scully, S.P. Ewing sarcoma demonstrates racial disparities in incidence-related and sex-related differences in outcome: An analysis of 1631 cases from the SEER database, 1973–2005. *Cancer* **2009**, *115*, 3526–3536. [CrossRef] [PubMed]
4. Gaspar, N.; Hawkins, D.S.; Dirksen, U.; Lewis, I.J.; Ferrari, S.; Le Deley, M.-C.; Kovar, H.; Grimer, R.; Whelan, J.; Claude, L.; et al. Ewing Sarcoma: Current Management and Future Approaches through Collaboration. *J. Clin. Oncol.* **2015**, *33*, 3036–3046. [CrossRef]
5. Delattre, O.; Zucman, J.; Plougastel, B.; Desmaze, C.; Melot, T.; Peter, M.; Kovar, H.; Joubert, I.; de Jong, P.; Rouleau, G.; et al. Gene fusion with an ETS DNA-binding domain caused by chromosome translocation in human tumours. *Nature* **1992**, *359*, 162–165. [CrossRef]
6. Zucman, J.; Melot, T.; Desmaze, C.; Ghysdael, J.; Plougastel, B.; Peter, M.; Zucker, J.M.; Triche, T.J.; Sheer, D.; Turc-Carel, C.; et al. Combinatorial generation of variable fusion proteins in the Ewing family of tumours. *EMBO J.* **1993**, *12*, 4481–4487. [CrossRef]
7. Aurias, A.; Rimbaut, C.; Buffe, D.; Zucker, J.M.; Mazabraud, A. Translocation involving chromosome 22 in Ewing's sarcoma. A cytogenetic study of four fresh tumors. *Cancer Genet. Cytogenet.* **1984**, *12*, 21–25. [CrossRef]
8. Kreyer, J.; Ranft, A.; Timmermann, B.; Juergens, H.; Jung, S.; Wiebe, K.; Boelling, T.; Schuck, A.; Vieth, V.; Streitbuerger, A.; et al. Impact of the Interdisciplinary Tumor Board of the Cooperative Ewing Sarcoma Study Group on local therapy and overall survival of Ewing sarcoma patients after induction therapy. *Pediatr. Blood Cancer* **2018**, *65*, e27384. [CrossRef]
9. Dirksen, U. Efficacy of add-on treosulfan and melphalan high-dose therapy in patients with high-risk metastatic Ewing sarcoma: Report from the International Ewing 2008R3 trial. *J. Clin. Oncol.* **2020**, *38*, 11501. [CrossRef]
10. Pappo, A.S.; Dirksen, U. Rhabdomyosarcoma, Ewing Sarcoma, and Other Round Cell Sarcomas. *J. Clin. Oncol.* **2017**, *36*, 168–179. [CrossRef]
11. Ladenstein, R.; Pötschger, U.; Le Deley, M.-C.; Whelan, J.; Paulussen, M.; Oberlin, O.; Van Den Berg, H.; Dirksen, U.; Hjorth, L.; Michon, J.; et al. Primary disseminated multifocal Ewing sarcoma: Results of the Euro-EWING 99 trial. *J. Clin. Oncol.* **2010**, *28*, 3284–3291. [CrossRef] [PubMed]
12. Caruso, J.; Shulman, D.S.; DuBois, S.G. Second malignancies in patients treated for Ewing sarcoma: A systematic review. *Pediatr. Blood Cancer* **2019**, *66*, e27938. [CrossRef] [PubMed]
13. Choi, D.K.; Helenowski, I.; Hijiya, N. Secondary malignancies in pediatric cancer survivors: Perspectives and review of the literature. *Int. J. Cancer* **2014**, *135*, 1764–1773. [CrossRef] [PubMed]
14. Bhatia, S.; Sklar, C. Second cancers in survivors of childhood cancer. *Nat. Rev. Cancer* **2002**, *2*, 124–132. [CrossRef]
15. Hijiya, N.; Ness, K.K.; Ribeiro, R.C.; Hudson, M.M. Acute leukemia as a secondary malignancy in children and adolescents: Current findings and issues. *Cancer* **2009**, *115*, 23–35. [CrossRef]
16. Pui, C.H.; Ribeiro, R.C.; Hancock, M.L.; Rivera, G.K.; Evans, W.E.; Raimondi, S.C.; Head, D.R.; Behm, F.G.; Mahmoud, M.H.; Sandlund, J.T.; et al. Acute myeloid leukemia in children treated with epipodophyllotoxins for acute lymphoblastic leukemia. *N. Engl. J. Med.* **1991**, *325*, 1682–1687. [CrossRef] [PubMed]
17. Heyn, R.; Khan, F.; Ensign, L.G.; Donaldson, S.S.; Ruymann, F.; Smith, M.A.; Vietti, T.; Maurer, H.M. Acute myeloid leukemia in patients treated for rhabdomyosarcoma with cyclophosphamide and low-dose etoposide on Intergroup Rhabdomyosarcoma Study III: An interim report. *Med. Pediatr. Oncol.* **1994**, *23*, 99–106. [CrossRef]
18. Bhatia, S.; Krailo, M.D.; Chen, Z.; Burden, L.; Askin, F.B.; Dickman, P.S.; Grier, H.E.; Link, M.P.; Meyers, P.; Perlman, E.; et al. Therapy-related myelodysplasia and acute myeloid leukemia after Ewing sarcoma and primitive neuroectodermal tumor of bone: A report from the Children's Oncology Group. *Blood* **2007**, *109*, 46–51. [CrossRef]
19. Travis, L.B.; Demark Wahnefried, W.; Allan, J.M.; Wood, M.E.; Ng, A.K. Aetiology, genetics and prevention of secondary neoplasms in adult cancer survivors. *Nat. Rev. Clin. Oncol.* **2013**, *10*, 289–301. [CrossRef]
20. Morton, L.M.; Onel, K.; Curtis, R.E.; Hungate, E.A.; Armstrong, G.T. The rising incidence of second cancers: Patterns of occurrence and identification of risk factors for children and adults. *Am. Soc. Clin. Oncol. Educ. Book* **2014**, *34*, e57–e67. [CrossRef]
21. Turcotte, L.M.; Neglia, J.P.; Reulen, R.C.; Ronckers, C.M.; van Leeuwen, F.E.; Morton, L.M.; Hodgson, D.C.; Yasui, Y.; Oeffinger, K.C.; Henderson, T.O. Risk, Risk Factors, and Surveillance of Subsequent Malignant Neoplasms in Survivors of Childhood Cancer: A Review. *J. Clin Oncol.* **2018**, *36*, 2145–2152. [CrossRef] [PubMed]
22. Dunst, J.; Ahrens, S.; Paulussen, M.; Rube, C.; Winkelmann, W.; Zoubek, A.; Harms, D.; Jurgens, H. Second malignancies after treatment for Ewing's sarcoma: A report of the CESS-studies. *Int. J. Radiat. Oncol. Biol. Phys.* **1998**, *42*, 379–384. [CrossRef] [PubMed]
23. Nguyen, F.; Rubino, C.; Guerin, S.; Diallo, I.; Samand, A.; Hawkins, M.; Oberlin, O.; Lefkopoulos, D.; De Vathaire, F. Risk of a second malignant neoplasm after cancer in childhood treated with radiotherapy: Correlation with the integral dose restricted to the irradiated fields. *Int. J. Radiat. Oncol. Biol. Phys.* **2008**, *70*, 908–915. [CrossRef] [PubMed]

24. Kuttesch, J.F.; Wexler, L.H.; Marcus, R.B.; Fairclough, D.; Weaver-McClure, L.; White, M.; Mao, L.; Delaney, T.F.; Pratt, C.B.; Horowitz, M.E.; et al. Second malignancies after Ewing's sarcoma: Radiation dose-dependency of secondary sarcomas. *J. Clin. Oncol.* **1996**, *14*, 2818–2825. [CrossRef] [PubMed]
25. Tucker, M.A.; D'Angio, G.J.; Boice, J.D., Jr.; Strong, L.C.; Li, F.P.; Stovall, M.; Stone, B.J.; Green, D.M.; Lombardi, F.; Newton, W.; et al. Bone sarcomas linked to radiotherapy and chemotherapy in children. *N. Engl. J. Med.* **1987**, *317*, 588–593. [CrossRef] [PubMed]
26. Schuck, A.; Ahrens, S.; Paulussen, M.; Kuhlen, M.; Könemann, S.; Rübe, C.; Winkelmann, W.; Kotz, R.; Dunst, J.; Willich, N.; et al. Local therapy in localized Ewing tumors: Results of 1058 patients treated in the CESS 81, CESS 86, and EICESS 92 trials. *Int. J. Radiat. Oncol. Biol. Phys.* **2003**, *55*, 168–177. [CrossRef]
27. Paulussen, M.; Ahrens, S.; Lehnert, M.; Taeger, D.; Hense, H.W.; Wagner, A.; Dunst, J.; Harms, D.; Reiter, A.; Henze, G.; et al. Second malignancies after ewing tumor treatment in 690 patients from a cooperative German/Austrian/Dutch study. *Ann. Oncol.* **2001**, *12*, 1619–1630. [CrossRef]
28. Paulussen, M.; Craft, A.W.; Lewis, I.; Hackshaw, A.; Douglas, C.; Dunst, J.; Schuck, A.; Winkelmann, W.; Köhler, G.; Poremba, C.; et al. Results of the EICESS-92 Study: Two randomized trials of Ewing's sarcoma treatment–cyclophosphamide compared with ifosfamide in standard-risk patients and assessment of benefit of etoposide added to standard treatment in high-risk patients. *J. Clin. Oncol.* **2008**, *26*, 4385–4393. [CrossRef]
29. Le Deley, M.-C.; Paulussen, M.; Lewis, I.; Brennan, B.; Ranft, A.; Whelan, J.; Le Teuff, G.; Michon, J.; Ladenstein, R.; Marec-Bérard, P.; et al. Cyclophosphamide compared with ifosfamide in consolidation treatment of standard-risk Ewing sarcoma: Results of the randomized noninferiority Euro-EWING99-R1 trial. *J. Clin. Oncol.* **2014**, *32*, 2440–2448. [CrossRef]
30. Whelan, J.; Le Deley, M.-C.; Dirksen, U.; Le Teuff, G.; Brennan, B.; Gaspar, N.; Hawkins, D.S.; Amler, S.; Bauer, S.; Bielack, S.; et al. High-Dose Chemotherapy and Blood Autologous Stem-Cell Rescue Compared with Standard Chemotherapy in Localized High-Risk Ewing Sarcoma: Results of Euro-E.W.I.N.G. 99 and Ewing-2008. *J. Clin. Oncol.* **2018**, *36*, 3110–3119. [CrossRef]
31. Dirksen, U.; Brennan, B.; Le Deley, M.-C.; Cozic, N.; Van Den Berg, H.; Bhadri, V.; Brichard, B.; Claude, L.; Craft, A.; Amler, S.; et al. High-Dose Chemotherapy Compared with Standard Chemotherapy and Lung Radiation in Ewing Sarcoma with Pulmonary Metastases: Results of the European Ewing Tumour Working Initiative of National Groups, 99 Trial and EWING 2008. *J. Clin. Oncol.* **2019**, *37*, 3192–3202. [CrossRef] [PubMed]
32. Juergens, C.; Weston, C.; Lewis, I.; Whelan, J.; Paulussen, M.; Oberlin, O.; Michon, J.; Zoubek, A.; Juergens, H.; Craft, A. Safety assessment of intensive induction with vincristine, ifosfamide, doxorubicin, and etoposide (VIDE) in the treatment of Ewing tumors in the EURO-E.W.I.N.G. 99 clinical trial. *Pediatr. Blood Cancer* **2006**, *47*, 22–29. [CrossRef] [PubMed]
33. Hawkins, M.M.; Wilson, L.M.; Burton, H.S.; Potok, M.H.; Winter, D.L.; Marsden, H.B.; Stovall, M.A. Radiotherapy, alkylating agents, and risk of bone cancer after childhood cancer. *J. Natl. Cancer Inst.* **1996**, *88*, 270–278. [CrossRef] [PubMed]
34. Ginsberg, J.P.; Goodman, P.; Leisenring, W.; Ness, K.K.; Meyers, P.A.; Wolden, S.L.; Smith, S.M.; Stovall, M.; Hammond, S.; Robison, L.L.; et al. Long-term survivors of childhood Ewing sarcoma: Report from the childhood cancer survivor study. *J. Natl. Cancer Inst.* **2010**, *102*, 1272–1283. [CrossRef] [PubMed]
35. Navid, F.; Billups, C.; Liu, T.; Krasin, M.J.; Rodriguez-Galindo, C. Second cancers in patients with the Ewing sarcoma family of tumours. *Eur. J. Cancer* **2008**, *44*, 983–991. [CrossRef]
36. Longhi, A.; Ferrari, S.; Tamburini, A.; Luksch, R.; Fagioli, F.; Bacci, G.; Ferrari, C. Late effects of chemotherapy and radiotherapy in osteosarcoma and Ewing sarcoma patients: The Italian Sarcoma Group Experience (1983–2006). *Cancer* **2012**, *118*, 5050–5059. [CrossRef] [PubMed]
37. Friedman, D.N.; Chastain, K.; Chou, J.F.; Moskowitz, C.S.; Adsuar, R.; Wexler, L.H.; Chou, A.J.; DeRosa, A.; Candela, J.; Magnan, H.; et al. Morbidity and mortality after treatment of Ewing sarcoma: A single-institution experience. *Pediatr. Blood Cancer* **2017**, *64*, e26562. [CrossRef]
38. Kaatsch, P.; Grabow, D.; Spix, C. *German Childhood Cancer Registry-Annual Report 2017 (1980–2016)*; Institute of Medical Biostatistics, Epidemiology and Informatics (IMBEI) at the University Medical Center of the Johannes Gutenberg University Mainz: Mainz, Germany, 2018; Volume 2018.
39. Erdmann, F.K.P.; Grabow, D.; Spix, C. *German Childhood Cancer Registry-Annual Report 2019 (1980–2018)*; Institute of Medical Biostatistics, Epidemiology and Informatics (IMBEI) at the University Medical Center of the Johannes Gutenberg University Mainz: Mainz, Germany, 2020; Volume 2019.
40. Friedman, D.L.; Whitton, J.; Leisenring, W.; Mertens, A.C.; Hammond, S.; Stovall, M.; Donaldson, S.S.; Meadows, A.T.; Robison, L.L.; Neglia, J.P. Subsequent neoplasms in 5-year survivors of childhood cancer: The Childhood Cancer Survivor Study. *J. Natl. Cancer Inst.* **2010**, *102*, 1083–1095. [CrossRef]
41. Leavey, P.J.; Laack, N.N.; Krailo, M.D.; Buxton, A.; Randall, R.L.; DuBois, S.G.; Reed, D.R.; Grier, H.E.; Hawkins, D.S.; Pawel, B.; et al. Phase III Trial Adding Vincristine-Topotecan-Cyclophosphamide to the Initial Treatment of Patients With Nonmetastatic Ewing Sarcoma: A Children's Oncology Group Report. *J. Clin. Oncol.* **2021**, *39*, 4029–4038. [CrossRef]
42. Sultan, I.; Rihani, R.; Hazin, R.; Rodriguez-Galindo, C. Second malignancies in patients with Ewing Sarcoma Family of Tumors: A population-based study. *Acta Oncol.* **2010**, *49*, 237–244. [CrossRef]
43. Neglia, J.P.; Friedman, D.L.; Yasui, Y.; Mertens, A.C.; Hammond, S.; Stovall, M.; Donaldson, S.S.; Meadows, A.T.; Robison, L.L. Second malignant neoplasms in five-year survivors of childhood cancer: Childhood cancer survivor study. *J. Natl. Cancer Inst.* **2001**, *93*, 618–629. [CrossRef] [PubMed]

44. Kenney, L.B.; Yasui, Y.; Inskip, P.D.; Hammond, S.; Neglia, J.P.; Mertens, A.C.; Meadows, A.T.; Friedman, D.; Robison, L.L.; Diller, L. Breast cancer after childhood cancer: A report from the Childhood Cancer Survivor Study. *Ann. Intern. Med.* **2004**, *141*, 590–597. [CrossRef] [PubMed]
45. Oeffinger, K.C.; Mertens, A.C.; Sklar, C.A.; Kawashima, T.; Hudson, M.M.; Meadows, A.T.; Friedman, D.L.; Marina, N.; Hobbie, W.; Kadan-Lottick, N.S.; et al. Chronic health conditions in adult survivors of childhood cancer. *N. Engl. J. Med.* **2006**, *355*, 1572–1582. [CrossRef] [PubMed]
46. Bhatia, S.; Robison, L.L.; Oberlin, O.; Greenberg, M.; Bunin, G.; Fossati-Bellani, F.; Meadows, A.T. Breast cancer and other second neoplasms after childhood Hodgkin's disease. *N. Engl. J. Med.* **1996**, *334*, 745–751. [CrossRef]
47. Schellong, G.; Riepenhausen, M.; Ehlert, K.; Bramswig, J.; Dorffel, W.; Schmutzler, R.K.; Rhiem, K.; Bick, U. Breast cancer in young women after treatment for Hodgkin's disease during childhood or adolescence–an observational study with up to 33-year follow-up. *Dtsch. Arztebl Int.* **2014**, *111*, 3–9.
48. Rodriguez-Galindo, C.; Poquette, C.A.; Marina, N.M.; Head, D.R.; Cain, A.; Meyer, W.H.; Santana, V.M.; Pappo, A.S. Hematologic abnormalities and acute myeloid leukemia in children and adolescents administered intensified chemotherapy for the Ewing sarcoma family of tumors. *J. Pediatr. Hematol. Oncol.* **2000**, *22*, 321–329. [CrossRef]
49. Paulides, M.; Kremers, A.; Stöhr, W.; Bielack, S.; Jürgens, H.; Treuner, J.; Beck, J.; Langer, T.; German Late Effects Working Group in the Society of Pediatric Oncology and Haematology (GPOH). Prospective longitudinal evaluation of doxorubicin-induced cardiomyopathy in sarcoma patients: A report of the late effects surveillance system (LESS). *Pediatr. Blood Cancer* **2006**, *46*, 489–495. [CrossRef]
50. Schiffman, J.D.; Wright, J. Ewing's Sarcoma and Second Malignancies. *Sarcoma* **2011**, *2011*, 736841. [CrossRef]
51. Kaatsch, P.; Reinisch, I.; Spix, C.; Berthold, F.; Janka-Schaub, G.; Mergenthaler, A.; Michaelis, J.; Blettner, M. Case-control study on the therapy of childhood cancer and the occurrence of second malignant neoplasms in Germany. *Cancer Causes Control* **2009**, *20*, 965–980. [CrossRef]
52. Constine, L.S.; Ronckers, C.M.; Hua, C.H.; Olch, A.; Kremer, L.C.M.; Jackson, A.; Bentzen, S.M. Pediatric Normal Tissue Effects in the Clinic (PENTEC): An International Collaboration to Analyse Normal Tissue Radiation Dose-Volume Response Relationships for Paediatric Cancer Patients. *Clin. Oncol.* **2019**, *31*, 199–207. [CrossRef]
53. Sheppard, D.G.; Libshitz, H.I. Post-radiation sarcomas: A review of the clinical and imaging features in 63 cases. *Clin. Radiol.* **2001**, *56*, 22–29. [CrossRef] [PubMed]
54. Granowetter, L.; Womer, R.; Devidas, M.; Krailo, M.; Wang, C.; Bernstein, M.; Marina, N.; Leavey, P.; Gebhardt, M.; Healey, J.; et al. Dose-intensified compared with standard chemotherapy for nonmetastatic Ewing sarcoma family of tumors: A Children's Oncology Group Study. *J. Clin. Oncol.* **2009**, *27*, 2536–2541. [CrossRef]
55. Mertens, A.C.; Liu, Q.; Neglia, J.P.; Wasilewski, K.; Leisenring, W.; Armstrong, G.T.; Robison, L.L.; Yasui, Y. Cause-specific late mortality among 5-year survivors of childhood cancer: The Childhood Cancer Survivor Study. *J. Natl. Cancer Inst.* **2008**, *100*, 1368–1379. [CrossRef] [PubMed]
56. Pui, C.H. Epipodophyllotoxin-related acute myeloid leukaemia. *Lancet* **1991**, *338*, 1468. [CrossRef] [PubMed]
57. Relling, M.V.; Boyett, J.M.; Blanco, J.G.; Raimondi, S.; Behm, F.G.; Sandlund, J.T.; Rivera, G.K.; Kun, L.E.; Evans, W.E.; Pui, C.H. Granulocyte colony-stimulating factor and the risk of secondary myeloid malignancy after etoposide treatment. *Blood* **2003**, *101*, 3862–3867. [CrossRef]
58. Teepen, J.C.; van Leeuwen, F.E.; Tissing, W.J.; van Dulmen-den Broeder, E.; van den Heuvel-Eibrink, M.M.; van der Pal, H.J.; Loonen, J.J.; Bresters, D.; Versluys, B.; Neggers, S.; et al. Long-Term Risk of Subsequent Malignant Neoplasms After Treatment of Childhood Cancer in the DCOG LATER Study Cohort: Role of Chemotherapy. *J. Clin. Oncol.* **2017**, *35*, 2288–2298. [CrossRef]
59. Henderson, T.O.; Rajaraman, P.; Stovall, M.; Constine, L.S.; Olive, A.; Smith, S.A.; Mertens, A.; Meadows, A.; Neglia, J.; Hammond, S.; et al. Risk factors associated with secondary sarcomas in childhood cancer survivors: A report from the childhood cancer survivor study. *Int. J. Radiat. Oncol. Biol. Phys.* **2012**, *84*, 224–230. [CrossRef]
60. Jurgens, H.; Dirksen, U. Ewing sarcoma treatment. *Eur. J. Cancer* **2011**, *47* (Suppl. S3), S366–S367. [CrossRef]
61. Whelan, J.; Hackshaw, A.; McTiernan, A.; Grimer, R.; Spooner, D.; Bate, J.; Ranft, A.; Paulussen, M.; Juergens, H.; Craft, A.; et al. Survival is influenced by approaches to local treatment of Ewing sarcoma within an international randomised controlled trial: Analysis of EICESS-92. *Clin. Sarcoma Res.* **2018**, *8*, 6. [CrossRef]
62. Li, X.; Li, W.; Mo, W.; Yang, Z. Acute lymphoblastic leukemia arising after treatment of Ewing sarcoma was misdiagnosed as bone marrow metastasis of Ewing sarcoma: A case report. *Medicine* **2018**, *97*, e9644. [CrossRef]
63. Gröbner, S.N.; Project, I.P.-S.; Worst, B.C.; Weischenfeldt, J.; Buchhalter, I.; Kleinheinz, K.; Rudneva, V.A.; Johann, P.D.; Balasubramanian, G.P.; Segura-Wang, M.; et al. Author Correction: The landscape of genomic alterations across childhood cancers. *Nature* **2018**, *559*, E10. [CrossRef] [PubMed]

cancers

Article

Ewing Sarcoma as Secondary Malignant Neoplasm—Epidemiological and Clinical Analysis of an International Trial Registry

Stefan K. Zöllner [1,2,3,†], Katja L. Kauertz [1,2,3,†], Isabelle Kaiser [1,2,3], Maximilian Kerkhoff [1,2,3], Christiane Schaefer [1,2,3], Madita Tassius [1,2,3], Susanne Jabar [1,2,3], Heribert Jürgens [4], Ruth Ladenstein [5], Thomas Kühne [6], Lianne M. Haveman [7], Michael Paulussen [8], Andreas Ranft [1,2,3,*] and Uta Dirksen [1,2,3,*]

1 Pediatrics III, University Hospital Essen, 45147 Essen, Germany
2 West German Cancer Center (WTZ), University Hospital Essen, 45147 Essen, Germany
3 German Cancer Consortium (DKTK), Essen/Düsseldorf, University Hospital Essen, 45147 Essen, Germany
4 Pediatric Hematology and Oncology, University Children's Hospital Münster, 48149 Münster, Germany
5 St. Anna Children's Cancer Research Institute and Medical University Vienna, 1090 Vienna, Austria
6 Department of Oncology and Haematology, University Children's Hospital Basel, 4056 Basel, Switzerland
7 Princess Máxima Center for Pediatric Oncology, 3584 Utrecht, The Netherlands
8 Pediatric Hematology and Oncology, Vestische Children's and Youth Hospital Datteln, University Witten/Herdecke, 45711 Datteln, Germany
* Correspondence: andreas.ranft@uk-essen.de (A.R.); uta.dirksen@uk-essen.de (U.D.)
† These authors contributed equally to this work.

Simple Summary: Ewing sarcoma (EwS) is a malignant bone and soft-tissue cancer that primarily affects adolescents and young adults. In rare cases, EwS develops as a secondary cancer; that is, after a previous cancer other than EwS. We collected information on all patients with EwS as a secondary cancer from three past international EwS studies to better understand affected patients and to identify potential at-risk patients. Forty-two patients with secondary EwS were identified, representing approximately 1.1% of all EwS cases. As primary cancers, patients suffered mainly from cancers of the blood-forming system, such as leukemia and lymphoma. We could not identify any risk factors for the development of EwS as a secondary cancer. Survival from a second cancer diagnosis with EwS is comparable to EwS as a first cancer diagnosis; therefore, patients with secondary EwS should also be offered complete therapy with the goal of cure, especially if the tumor is localized to only one site.

Abstract: Ewing sarcoma (EwS) is the second most common bone and soft tissue tumor, affecting primarily adolescents and young adults. Patients with secondary EwS are excluded from risk stratification in several studies and therefore do not benefit from new therapies. More knowledge about patients with EwS as secondary malignant neoplasms (SMN) is needed to identify at-risk patients and adapt follow-up strategies. Epidemiology, clinical characteristics, and survival analyses of EwS as SMN were analyzed in 3844 patients treated in the last three consecutive international EwS trials, EICESS 92, Euro-E.W.I.N.G. 99, and EWING 2008. Forty-two cases of EwS as SMN (approximately 1.1% of all patients) were reported, preceded by a heterogeneous group of malignancies, mainly acute lymphoblastic leukemias (n = 7) and lymphomas (n = 7). Three cases of EwS as SMN occurred in the presumed radiation field of the primary tumor. The median age at diagnosis of EwS as SMN was 19.4 years (range, 5.9–72) compared with 10.8 years (range, 0.9–51.2) for primary EwS. The median interval between first malignancy and EwS diagnosis was 7.4 years. The 3-year overall survival (OS)/event-free survival (EFS) was 0.69 (SE = 0.09)/0.53 (SE = 0.10) for localized patients and 0.36 (SE = 0.13)/0.29 (SE = 0.12) for metastatic patients (OS: p = 0.02; EFS: p = 0.03). Survival in patients with EwS as SMN did not differ between hematologic or solid primary malignancies. EwS as SMN is rare; however, survival is similar to that of primary EwS, and its risk-adjusted treatment should be curative, especially in localized patients.

Citation: Zöllner, S.K.; Kauertz, K.L.; Kaiser, I.; Kerkhoff, M.; Schaefer, C.; Tassius, M.; Jabar, S.; Jürgens, H.; Ladenstein, R.; Kühne, T.; et al. Ewing Sarcoma as Secondary Malignant Neoplasm—Epidemiological and Clinical Analysis of an International Trial Registry. *Cancers* **2022**, *14*, 5935. https://doi.org/10.3390/cancers14235935

Academic Editors: Torsten Kessler and Jaume Mora

Received: 29 September 2022
Accepted: 28 November 2022
Published: 30 November 2022

Keywords: childhood cancer; cancer survivor; Ewing sarcoma; secondary malignant neoplasms; secondary malignancy

1. Introduction

Overall survival rates for many pediatric cancers are improving decade by decade. In Europe, the 5-year survival rate for children diagnosed with malignant neoplasms reached 79.1% between 2000 and 2007, largely due to new advances in therapy and supportive care [1]. Currently, there are more than 500,000 survivors of childhood cancer in Europe [2]. As this trend continues, minimizing acute and long-term toxicity associated with treatment is of paramount importance, because most children with cancer continue to suffer from significant treatment-related toxicities [3]. Secondary malignant neoplasms (SMN) are recognized as late sequelae of cancer therapy in children [4]. SMN represent subsequent, distinct tumor entities in the same patient. They are histologically distinct from both the primary tumor and metastases from the primary tumor and usually occur within 30 years of the diagnosis of a first malignancy [5]. The marked differences in occurrence and the role of specific therapeutic exposures have led to the classification of SMN into two distinct categories: radiation-induced solid SMN and chemotherapy-induced myelodysplastic syndrome (MDS)/acute myeloid leukemia (AML) [6].

Ewing sarcoma (EwS) is a rare but highly aggressive bone and soft-tissue malignancy that usually affects adolescents and young adults [7]. EwS is one of the small blue round-cell sarcomas characterized by chromosomal translocations in which ETS transcription factors are fused to a member of the FET gene family [8]. The most common tumor-specific chimeric transcription factor, EWSR1-FLI1, consists of the Ewing sarcoma breakpoint region 1 protein (EWSR1) and an ETS family gene such as the Friend leukemia integration 1 transcription factor (FLI1) [7,9]. Therapies targeting EWSR1-FLI1 would provide a tumor-specific targeted approach but have yet to be established in routine clinical practice. Currently, the standard treatment for EwS patients is multimodal and includes intensive polychemotherapy, as well as surgery and/or radiotherapy for local therapy [10].

Previous studies have described a greater than ninefold increased risk of developing sarcoma in childhood cancer survivors compared with the general population [11]. While EwS survivors are at increased risk for SMN compared with the age- and sex-matched general population, EwS itself may also present as SMN [12,13]. EwS as SMN accounts for a minority of all EwS cases, for which limited data are available [13–16]. Despite its rare occurrence, patients with EwS as SMN appear to have clinical differences and a worse prognosis than patients with primary EwS [17].

To supplement data on this patient group, we retrospectively analyzed the epidemiology and clinical features of patients with EwS as SMN over a 27-year period who were enrolled in EwS trials and registries. The aim of this study was to retrospectively describe the incidence, epidemiology, and clinical characteristics, as well as the risk factors of patients with EwS as SMN. In one of the largest clinical studies on this topic, we have identified forty-two patients with EwS as SMN, representing approximately 1.1% of all reported EwS cases. The present study suggests that the outcome of patients with EwS as SMN is determined by the staging at diagnosis and not by the diagnosis of a secondary malignancy per se [4]. Survival in EwS as SMN does not differ between hematologic or solid primary malignancies, and risk factors for EwS as SMN appear to be similar to risk factors of other non-EwS SMN [18].

2. Materials and Methods

2.1. Patient Cohorts and Eligibility Criteria

From 1991 to 2019, three consecutive and multinational EwS trials, namely the European Intergroup Cooperative Ewing's Sarcoma Studies EICESS 92 (ClinicalTrials.gov identifier: NCT00002516), the European Ewing Tumor Working Initiative of National

Groups 1999 Euro-E.W.I.N.G. 99 (ClinicalTrials.gov identifier: NCT00020566), and the EWING 2008 (ClinicalTrials.gov identifier: NCT00987636), enrolled 3844 patients with primary EwS and EwS as SMN. All trial protocols were conducted in accordance with the ethical principles of the Declaration of Helsinki and the guidelines for Good Clinical Practice, reviewed by the appropriate institutional review boards, and approved by an independent ethics committee.

Several trials aimed to optimize treatment and outcome for EwS patients. Because primary malignancy was an exclusion criterion for randomization, patients with EwS as SMN were not eligible for randomization and were treated according to standard protocols. Patients with EwS as SMN were included as registry patients in the EWING registry of the GPOH (German Society of Pediatric Hematology and Oncology), which allowed analysis of data on patients not registered as trial patients. The GPOH database was last updated in June 2019.

The differences in risk stratification and treatment strategies across the different Cooperative Ewing Sarcoma Study (CESS)-led EwS trials is shown in Table 1. In the EICESS 92 protocol, patients with localized tumors and a tumor volume of <100 mL were classified into a standard-risk (SR) stratification, in contrast with high-risk (HR) patients who presented with metastasis or a tumor volume of ≥100 mL at diagnosis. Treatment consisted of vincristine, actinomycin D, ifosfamide, and adriamycin (VAIA) for standard-risk patients and VAIA plus etoposide (EVAIA) for high-risk patients. Local control included surgery, radiotherapy, or a combination of both. In the Euro-E.W.I.N.G. 99 trial, four different risk groups were defined, namely R1, R2pulm, R2loc, and R3. Standard-risk patients assigned to R1 either had a good histologic response to neoadjuvant treatment (<10% viable cells), a tumor with a baseline volume of <200 mL and after resection at diagnosis, or were treated with radiotherapy alone as local treatment. Patients with localized disease and either a poor histologic response (≥10% viable cells) to induction chemotherapy or with a tumor volume of ≥200 mL who were either unresected at diagnosis or initially resected were included in R2. Patients who had disseminated disease at diagnosis were assigned to R3. At the start of the Euro-E.W.I.N.G. 99 trial, treatment regimens were further stratified for patients with isolated pulmonary metastasis who were assigned to R2pulm. In induction chemotherapy, patients received 6 cycles of VIDE followed by treatment with vincristine, actinomycin D, Ifosfamide (VAI), and for local therapy, either surgery, definitive irradiation, or surgery and irradiation combined. In addition, R1 patients received 7 courses of vincristine, actinomycin D, cyclophosphamide (VAC) or 7 courses of vincristine, actinomycin D, ifosfamide, adriamycin (VAI) after induction chemotherapy. R2loc patients received 7 courses of VAI or 1 course of VAI and high-dose chemotherapy with busulfan and melphalan (Bu-Mel), whereas R2pulm patients received all 7 courses of VAI consolidation chemotherapy followed by lung irradiation or Bu-Mel. The EWING 2008 trial was a joint protocol of international EwS trial groups. Patients were stratified into R1 patients with good histologic response (<10% viable tumor cells) and localized disease, R2loc patients with poor histologic response (≥10% viable tumor cells) and localized disease, R2pulm patients with exclusive lung metastases, and R3 patients with disseminated disease at diagnosis, i.e., lung metastases and metastases to other sites (Table 1).

The GPOH database of these three studies was screened for EwS as SMN. Patients were identified based on the following criteria: First malignancy in all age groups, histologically confirmed diagnosis of EwS as SMN (non-EwS round cell sarcoma were excluded), treated analogously to the EICESS 92, Euro-E.W.I.N.G. 99, and EWING 2008 trial protocols. Patients with EwS as a third or fourth malignancy were not included in this study.

Table 1. Development of risk stratification and chemotherapy for Ewing sarcoma (EwS) in different trials from the Cooperative Ewing Sarcoma Study (CESS) group. SR—standard-risk, HR—high-risk, VHR—very high-risk. Please see "Materials and Methods" section for more detailed information on risk strata, including randomization of R1/R2/R3. Vol.—tumor volume at diagnosis, pulmonary mets—exclusive lung metastases at diagnosis. V—vincristine, A—actinomycin D, C—cyclophosphamide, I—ifosfamide, D—doxorubicin (adriamycin), E—etoposide, BU—busulfan, ME—melphalan, etoposide, MEL—melphalan, TREO—treosulfan.

EwS Trial	EICESS 92				EURO E.W.I.N.G. 99								EWING 2008					
Number of cycles	14				14		14	8	8				14		14	8	14	15
Risk strata	SR		HR		R1 = SR		R2 = HR		R3 = VHR				R1 = SR		R2 = HR		R3 = VHR	
Clinical criteria for risk strata	localized vol. < 100 mL		metastasized vol ≥ 100 mL		localized vol. ≥ 200 mL <10% viable cells *		locolazied/ pulmonary mets vol. ≥ 200 mL ≥10% viable cells *		metastasized				localized <10% viable cells *		localized/ pulmonary mets ≥10% viable cells *		metastasized	
Chemotherapeutic regimen	VAIA + VACA	VAIA	VAIA	EVAIA	VIDE + VAC ♀	VIDE + VAI ♂	VIDE + VAI	VIDE + VAI + BU/MEL	VIDE + VAI + ME/ME	VIDE + VAI + TREO/MEL	VIDE + VAI + BU/MEL	VIDE + VAI	VIDE + VAC ♀	VIDE + VAI ♂	VIDE + VAI	VIDE + VAI + BU/MEL	VIDE + VAC	VIDE + VAC + TREO/MEL

* viable tumor cells in histological tumor resection specimen.

2.2. Statistical Analysis

Descriptive statistics were calculated using the SPSS Statistics 29 (IBM Corporation, Armonk, NY), SAS 9.4 (SAS Institute, Cary, NC, USA), and MEDCALC software packages. Overall survival (OS) and event-free survival (EFS) were calculated using the Kaplan–Meier method. Univariate comparisons were estimated using the log-rank test. Fischer's exact test was used in the analysis of contingency tables.

3. Results

In this retrospective analysis of the EWING registry of the GPOH, a total of 3844 patients had been enrolled in the EICESS 92, the Euro-E.W.I.N.G. 99, and the EWING 2008 trials. We identified forty-two patients with EwS as SMN with a median follow-up of 2.6 years, ranging from 0.1 to 14.6 years. According to the different trials, 8 (0.9%), 16 (1.0%), and 18 (1.3%) patients in EICESS 92, Euro-E.W.I.N.G. 99, and EWING 2008, respectively, had EwS as SMN.

3.1. Epidemiology of Primary Malignancies

EwS as SMN was preceded by a heterogeneous group of primary malignancies. Sixteen patients were diagnosed with neoplasms of the hematopoietic and lymphoid tissue as first malignancy. Acute lymphoblastic leukemia (ALL) was present in almost half of these cases (n = 7). Seven cases of EwS as SMN occurred after Hodgkin lymphoma (n = 4) and non-Hodgkin lymphoma (n = 3). Four cases of breast cancer and two cases each of osteosarcoma, retinoblastoma, melanoma, and rhabdomyosarcoma were noted. One patient was initially diagnosed with Langerhans cell histiocytosis. One case each of nephroblastoma, hepatoblastoma, malignant hemangiopericytoma, testicular teratoma, renal cell carcinoma, cervical cancer, and ovarian germ cell tumor was reported. Only one patient was diagnosed with Li-Fraumeni syndrome, a hereditary cancer predisposition syndrome [19]. In this patient, the first malignancy was a mediastinal rhabdoid tumor and the patient developed EwS as SMN seven years later. Figure 1 shows the localization of the first malignancy in relation to the location of EwS as SMN.

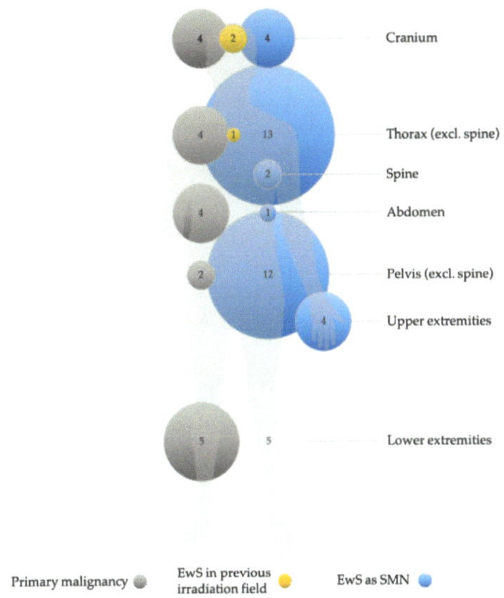

Figure 1. Localization and frequency of primary malignancies and Ewing sarcoma as secondary malignant neoplams (EwS as SMN), and occurrence of EwS as SMN in previously irradiated fields. Non-EwS malignancies are shown with gray circles, EwS cases with blue circles, and cases of EwS as SMN in previously irradiated fields with yellow circles. The numbers indicate the number of tumors per localization. The size of the circles is proportional to the number of cases. Missing circles mean no tumor case in this localization. Data for primary malignancies, EwS as SMN, and EwS as SMN in previously irradiated localizations (i.e., information on treatment modalities) were available in $n = 36$, $n = 41$, and $n = 27$, respectively. Nine patients were irradiated in the primary tumor setting.

3.2. Patient Characteristics and Clinical Features of Ewing Sarcoma as Secondary Malignant Neoplasms

The clinical features and patient characteristics of EwS as SMN cases in the different trials are summarized in both Tables 2 and 3. Table 2 presents the clinical characteristics of patients with EwS as SMN vs. EwS as primary malignancy. The gender distribution of EwS as SMN was balanced, with slightly more female (52.4%) than male (47.6%) patients. The majority of patients were older than 14 years (66.7%) at the time of EwS diagnosis. The median age at diagnosis of the first malignancy was 10.8 years (range 0.9 to 51.2), whereas EwS as SMN occurred at a median age of 19.4 years, with a range of 5.9 to 72 years. The median interval between the diagnosis of the first malignancy and the diagnosis of EwS as SMN was 7.4 years, whereas the first cases of EwS as SMN occurred as early as 1 year after the primary tumor. The longest interval between primary EwS and EwS as SMN was 41.4 years. There was no significant difference between the type of primary tumor of hematologic or solid origin with respect to the latency to EwS as SMN ($p = 0.55$; Figure 2).

Comparison between EwS as primary malignany and EwS as SMN did not reveal statistically significant differences in clinical features at diagnosis and response to treatment after induction therapy, except for tumor localization at diagnosis ($p = 0.03$; Table 2). We observed fewer EwS as SMN in the lower extremities and more thoracic and extraosseous EwS as SMN.

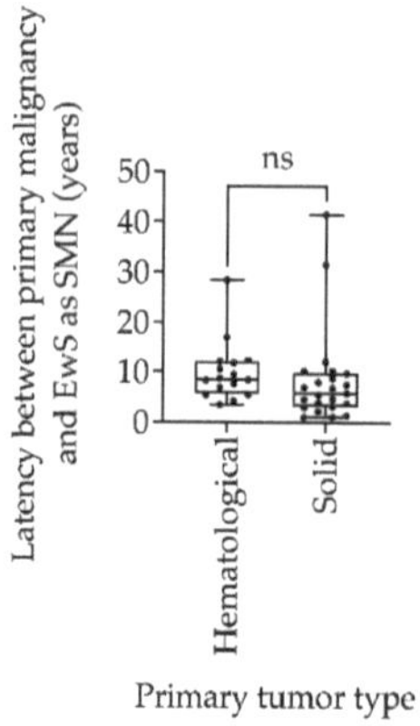

Figure 2. Box-and-whisker plot of latency between primary tumor and Ewing sarcoma as secondary malignant neoplasms (EwS as SMN) compared between hematologic and solid primary tumors in the EICESS 92, Euro-E.W.I.N.G. 99, and EWING 2008 trials. Statistical analysis by unpaired *t* test, ns = non-significant ($p = 0.55$).

EwS as SMN occurred most frequently in the thorax (26.8%), pelvis (22%), extraosseous sites including the abdomen (22%), and both the cranium and lower extremities (9.75% each) (Figure 1 and Table 2). Localized disease was present in 28 (67%) patients, 4/8 in the EICESS 92 trial, 10/16 in the Euro-E.W.I.N.G. 99 trial, and 14/18 in the EWING 2008 trial. Accordingly, 14 (33%) patients with EwS as SMN had metastatic disease at diagnosis. According to the Euro-E.W.I.N.G. 99 trial, risk stratification strategies distinguished between exclusive lung metastases and disseminated disease with lung metastases and metastases to other sites. In this context, six patients were diagnosed with disseminated EwS as SMN, two of whom had exclusive lung metastases in the Euro-E.W.I.N.G. 99 trial. In the EWING 2008 trial, one patient was diagnosed with disseminated disease without lung metastases, and three patients had isolated lung metastases (Table 2). Overall, disease stage did not differ statistically between primary tumor type and EwS as SMN patients ($p = 0.25$). Notably, six patients with disseminated EwS disease had a solid primary tumor, but only one patient with disseminated disease had ALL as an initial tumor. The tumor status of patients with EwS as SMN at the time of diagnosis is shown in Table 3. The majority of patients (60.5%) had a tumor volume < 200 mL at diagnosis (Table 2).

3.3. Treatment Management of Primary Malignancies

Despite the heterogeneity of the first malignancies, therapy of the primary diseases followed GPOH- and non-GPOH-standardized national trials, e.g., ALL/NHL-BFM-86 (ALL), ALL-BFM 95 (ALL), COSS-86 (osteosarcoma), MAKEI 96 (germ cell tumor), or CWS-86 (soft-tissue sarcoma; EwS excluded). Across trials, nine patients (33%) received radiotherapy for the primary tumor before developing EwS as SMN. In three of the nine irradiated patients, EwS as SMN occurred near the site where the primary malignancy had been localized and irradiated (Figure 1).

3.4. Treatment Management of Ewing Sarcoma as Secondary Malignant Neoplasms

Treatment of EwS followed the specific trial protocols. In the EICESS 92 trial, primary malignancies were an exclusion criterion for randomization, and patients with EwS as SMN received standard therapy. In contrast, patients included in the Euro-E.W.I.N.G. 99 protocol were not excluded from stratification and received treatment according to the standard arms of the trial. In the EWING 2008 trial, patients with EwS as SMN received similar treatment, with zoledronic acid introduced in R1 patients and high-dose treosulfan and melphalan (Treo-Mel) chemotherapy in R3 patients. Further explanation of the trial design and risk strata can be found in the "Materials and Methods" section.

Table 2. Clinical features at diagnosis and response to therapy in comparison between Ewing sarcomas (EwS) as primary malignancy and EwS as secondary malignant neoplasms (SMN) in the EICESS 92, Euro-E.W.I.N.G. 99, and EWING 2008 trials. Fisher's exact test was used in the analysis of contingency tables.

Clinical Features at Diagnosis (Avaiblabe Data for EwS as SMN)		EICESS 92, Euro-E.W.I.N.G. 99, EWING 2008		*p* Value
		EwS as Primary Malignancy (%)	EwS as SMN (%)	
Gender (n = 42)	Male	2238 (58.9)	20 (47.6)	0.16
	Female	1564 (41.1)	22 (52.4)	
Age (n = 42)	<14 years	1609 (42.3)	14 (33.3)	0.27
	≥14 years	2193 (57.7)	28 (66.7)	
Tumor stage (n = 42)	Localized	2496 (67.1)	28 (66.7)	1.00
	Metastasized	1223 (32.9)	14 (33.3)	
Tumor localization (n = 41) #	Cranium	152 (4.1)	4 (9.75)	0.03
	Thorax (excl. spine)	628 (16.9)	11 (26.8)	
	Spine	252 (6.8)	1 (2.4)	
	Extraosseous # (incl. abdomen)	552 (14.9)	9 (22.0)	
	Pelvis (excl. spine)	841 (22.6)	9 (22.0)	
	Upper extremities	262 (7.0)	3 (7.3)	
	Lower extremities	1030 (27.7)	4 (9.75)	
Tumor volume (n = 38)	<200 mL	2000 (57.7)	23 (60.5)	0.75
	≥200 mL	1468 (42.3)	15 (39.5)	
Response assessment				
Histological regression (Salzer-Kuntschik *) (n = 18)	<10%	1375 (71.9)	10 (55.6)	0.19
	≥10%	538 (28.1)	8 (44.4)	

* viable tumor cells in histological tumor resection specimen # for adequate statistical analysis and comparison between EwS as first and only tumor and EwS as second malignancy, tumor location categories were adjusted and all non-osseous tumors were combined.

3.5. *Outcome of Patients with Ewing Sarcoma as Secondary Malignant Neoplasms*

Data on follow-up of patients with EwS as SMN were available from 41 patients with a median follow-up of 2.6 years (range 0.1–14.6). During the 27-year follow-up period, the 3-year overall survival (OS) was 0.69 (SE = 0.09) for patients with localized disease and 0.36 (SE = 0.13) for patients with metastatic disease (p = 0.02; Figure 3). The 3-year event-free survival (EFS) was 0.53 (SE = 0.10) for patients with localized disease and 0.29 (SE = 0.12) for patients with metastatic disease (p = 0.03; Figure 3). There was no significant difference in OS or EFS of patients with EwS as SMN between primary hematologic or primary solid malignancy ($p > 0.05$; Figure 4). Statistically, due to the small number of cases, a comparison of EFS/OS between the different EwS studies is not meaningful, partly because localized patients are mixed with metastatic patients. A similar limitation would have been the analysis of other prognostic factors known for primary EwS, such as tumor volume and histopathologic response to therapy.

Table 3. Patient characteristics and clinical features of each patient with Ewing sarcoma as secondary malignant neoplasms in the EICESS 92, Euro-E.W.I.N.G. 99, and EWING 2008 trials. SR—standard-risk Please see "Materials and Methods" section for more detailed information on risk strata, including randomization of R1/R2/R3. ALL—acute lymphoblastic leukemia. Outcome: 1 dead, 0 alive.

Trial	Patient (n)	Primary Malignancy	Latency (Years)	Risk Strata	Follow-Up (Years)	Outcome
EICESS 92	1	Retinoblastoma	12	R3	2.4	1
	2	Lymphoma	7.3	R2p	4.1	1
	3	Melanoma	3.1	SR	10.9	0
	4	Testicular teratoma	6.8	R3	1.7	1
	5	Cervix carcinoma	21	SR	4.3	1
	6 [#]	Rhabdoid tumor	7.4	R2p	2.0	1
	7	Malignant hemangiopericytoma	1.3	SR	4.5	1
	8	Osteosarcoma	2.9	SR	14	0
Euro-E.W.I.N.G. 99	9	ALL	11.8	R2pulm	3.68	1
	10	ALL	12.3	R2loc	0.93	1
	11	ALL	6.8	R1	8.45	0
	12	Retinoblastoma	6.2	R3	1.92	1
	13	ALL	8.3	R1	10.15	0
	14	Non-Hodgkin lymphoma	8.75	R1	14.6	0
	15	Ovarian germ cell tumor	3.8	R1	10.6	0
	16	ALL	5.4	R1	6.1	0
	17	Nephroblastoma	4.5	R1	11.7	0
	18	Hepatoblastoma	6.9	R3	1.1	1
	19	ALL	8.7	R3	0.9	1
	20	Hodgkin lymphoma	17.0	R2pulm	0.3	1
	21	Hodgkin lymphoma	8.4	R1	3.5	1
	22	Renal cell carcinoma	1.0	R3	1.3	1
	23	Hodgkin lymphoma	28.4	R1	2.6	1
	24	Langerhans cell histiocytosis	12.0	R1	1.9	1
EWING 2008	25	Neuroblastoma	4.6	R2loc	2.33	1
	26	Osteosarcoma	7.9	R2loc	1.0	1
	27	Breast Cancer	5.4	R3	8.0	0
	28	Sweat gland carcinoma	10.1	R2pulm	6.9	0
	29	Synovial sarcoma	31.5	R2loc	1.0	1
	30	Breast cancer	3.8	R1	6.2	0
	31	Rhabdomyosarcoma	10.2	R1	6.3	1
	32	ALL	10.6	R1	6.1	0
	33	Seminoma	41.4	R2pulm	0.1	1
	34	Chronic myeloid leukemia	3.5	R1	0.9	1
	35	Rhabdomyosarcoma	5.8	R2loc	4.7	0
	36	Non-Hodgkin lymphoma	4.1	R2pulm	2.8	0
	37	Non-Hodgkin lymphoma	5.3	R1	2.4	0
	38	Hepatocellular carcinoma	9.7	NA	2.5	0
	39	Breast cancer	2.1	NA	2.4	0
	40	Hodgkin lymphoma	9.5	R1	1.5	0
	41	Liposarcoma	1.1	NA	0.8	1
	42	Breast cancer	NA	NA	NA	0

[#] patient with Li-Fraumeni syndrome.

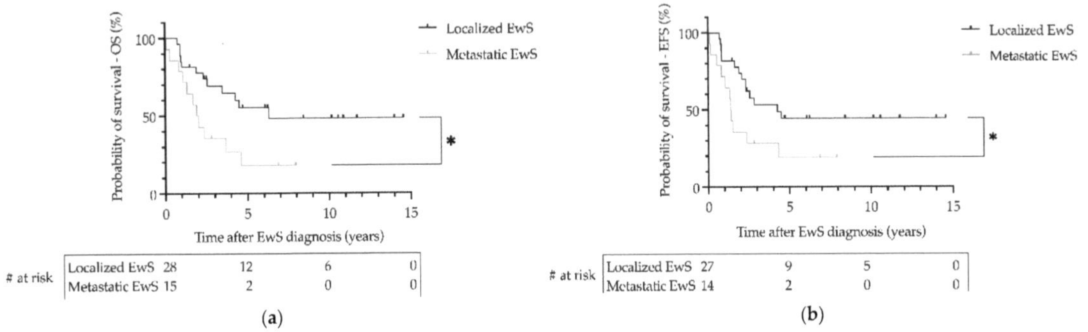

Figure 3. Survival of patients withEwing sarcoma (EwS) as secondary malignant neoplasms as a function of metastatic status at diagnosis in the EICESS 92, Euro-E.W.I.N.G. 99, and EWING 2008 trials. (**a**) Overall survival (OS) and (**b**) event-free survival (EFS) were calculated using the Kaplan–Meier method. * (**a**) $p = 0.02$, (**b**) $p = 0.03$. # at risk—numbers at risk.

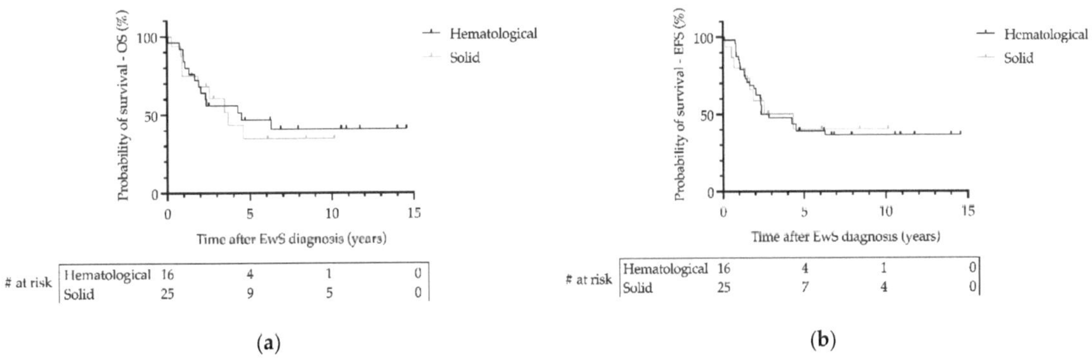

Figure 4. Survival of patients with Ewing sarcoma (EwS) as secondary malignant neoplasms as a function of type of primary hematological or primary solid malignancy in the EICESS 92, Euro-E.W.I.N.G. 99, and EWING 2008 trials. (**a**) Overall survival (OS) and (**b**) event-free survival (EFS) were calculated using the Kaplan–Meier method. Statistical analysis by Log-rank test. (**a**,**b**) $p > 0.05$. # at risk—numbers at risk.

4. Discussion

Cancer survivors are at high risk of developing long-term complications such as secondary malignant neoplasms (SMN) [20]. In childhood cancer survivors, the cumulative risk of developing a SMN 20 years after the primary diagnosis can be as high as 12% [21]. According to the German Childhood Cancer Registry, 6.8% (more than 1500 patients) of all German childhood cancer survivors diagnosed under the age of 15 were diagnosed with a SMN within 30 years of their first diagnosis between 2009 and 2018 [22].

EwS is a rare but highly aggressive bone and soft-tissue tumor [23]. Data on treatment management and survival of patients with EwS as SMN are conceivably limited [24]. In the present study, we analyzed the patient characteristics and survival data of patients with EwS as SMN in a heterogeneous group of patients with primary malignancies. EwS as SMN accounted for approximately 1.1% of all 3844 EwS cases treated between 1991 and 2019 in three consecutive and international EwS trials, i.e., EICESS 92, Euro-E.W.I.N.G. 99 and EWING 2008. The incidence of EwS as SMN is presumably lower than that of other solid bone and soft-tissue tumors such as rhabdomyosarcomas and osteosarcomas [25,26]. The median age of 19.4 years at diagnosis of EwS as SMN in our cohort is comparable to the general peak incidence age for primary EwS of 15 years [7]. While most SMN develop within 10 years, the incidence of SMN in survivors of childhood cancer increases with age, with a cumulative incidence of more than 20% 30 years after diagnosis of the primary cancer [4]. In one study sarcomas occurred a median of 11.8 years after the diagnosis for primary malignancy [11]. In comparison, the median time from primary malignancy to EwS as SMN in our study was 7.4 years, which is higher than latency times of 3.3 years to be calculated in other EwS-specific publications (6.9 years after primary hematologic malignancies, reviewed in [15]). There was no significant difference between the types of primary tumor of hematologic or solid origin regarding the timing of development of EwS as SMN (Figure 2). The short latency period compared with other secondary soft-tissue tumors supports close follow-up after completion of treatment for a primary malignancy [18,27]. However, given the low incidence of EwS as SMN, there is no reason to perform standardized radiologic screenings apart from routine follow-up. Alternative biomarkers such as liquid biopsies that can detect EwS at an early stage are not currently available.

In a previous study of 58 patients with EwS as SMN, affected patients showed poorer survival (34.3% vs. 52.2%), smaller tumor volumes (75.0% vs. 48.2% <8 cm), and more axial tumors (77.4% vs. 62.5%) compared with patients with primary EwS [17]. Our analysis showed that the initial tumor volume and the frequency and distribution of distant metastases at the time of diagnosis in EwS as SMN is similar to that in primary EwS (Table 2) [7]. The difference in tumor location in our study between primary and secondary EwS tumors must be critically questioned due to the low case number for EwS as SMN cases. Overall, survival rates in patients with both primary EwS and EwS as SMN are comparable, although this observation may be influenced again by the small number of cases in the present study. In a large previous study of EwS and PNET as SMN cases, these patients showed worse overall survival than primary EwS patients, and this was related to older age with comorbidity [17]. That we no longer saw this association may be due to the newer data and improved treatment options, including improved supportive care. Our prognostic analysis suggests that the outcome of patients with EwS as SMN is determined by the greatest risk factor of EwS, the spread of the disease at diagnosis, and not by the diagnosis of secondary malignancy per se; this is comparable to the results of previous studies [4,17]. As the most important unfavorable prognostic factor in both primary EwS and EwS as SMN is the presence of distant metastases at the time of diagnosis [28,29], staging remains crucial in the diagnosis of EwS as SMN. Although patients with EwS as SMN have similar outcomes compared with primary EwS patients, EwS as SMN are not eligible for randomization in either the current international Cooperative Ewing Sarcoma Study (CESS) registry or the upcoming CESS trial and are thus excluded from potential new therapeutics.

In our study, the primary malignancies that preceded EwS had a wide spectrum of diseases. Most patients had primary malignancies of the hematopoietic and lymphoid tissues. ALL was the most diagnosed primary malignancy. This is not surprising, as leukemia and lymphomas are the most common malignancies in children [30]. The 25-year cumulative incidence of a secondary malignancy in children surviving ALL was estimated to be 5.2% based on the Childhood Cancer Survivor Study (CCSS) [31]. In comparison, the same CCSS analysis showed a 20-year cumulative incidence of 1.7% for all secondary malignant neoplasms in survivors of childhood AML [32]. The frequency of hematologic malignancies preceding EwS as SMN in our study is at odds with previous findings that patients with a primary sarcoma were more likely to develop a secondary sarcoma [11]. The type of primary malignancy did not correlate with disease status or with the type of metastasis of EwS as SMN, and overall did not affect the prognostic outcome of patients with EwS as SMN (Table 3, Figure 4).

Given that patients with the most common primary malignancies in our cohort, ALL and lymphoma (11/24), were most likely to receive multichemotherapy with the alkylating agent cyclophosphamide and the topoisomerase inhibitor etoposide, and given previously published data on the risk of these agents to develop SMN [33–35], it is possible that these agents predispose for EwS as SMN among other SMN. Because of the heterogeneity of treatment protocols for primary malignancies, variable doses of chemotherapeutic agents, lack of clinical data on patients, and small EwS case series, the specific risk of these chemotherapeutic agents for EwS as SMN cannot be conclusively assessed. Controversy surrounds the effect of radiation: for ALL survivors, irradiation is thought to be responsible for most SMNs [36]. Two previous reports described an increased risk of secondary sarcomas in patients who received radiotherapy during treatment for their first malignancy [11,17]. In the present study, EwS as SMN occurred within the presumed former irradiation field in one-third of initially irradiated patients (Figure 1). Previous studies correlated EwS as SMN with the location of the previous radiation field in 12.1% of patients [17], while individual case reports may well show higher rates without accounting for publication bias [15]. While most EwS as SMN cases are likely due to established clinical risk factors for secondary malignancies, such as irradiation and/or chemotherapy, few cases may have a biological predisposition in the form of underlying mutations. There are several genetic syndromes with links to soft-tissue and bone sarcomas, whereas EwS does not appear to be an associated index tumor, unlike osteosarcoma and embryonal rhabdomyosarcoma [37]. In EwS, a balanced chromosomal translocation t(11;22)(q12;q24) leading to the specific fusion protein EWSR1–FLI1 appears to be genetically responsible for the tumor [38]. Apart from this translocation, other genetic alterations or mutations are rare [7]. Mutations of the p53 tumor suppressor gene are detected in 5–7% of EwS cases [39]. In our study, we identified only one case of EwS as SMN with an underlying Li-Fraumeni syndrome characterized by germline mutations of the tumor-suppressor gene p53 [40]. It must be taken into account that we did not have information on germline mutations in most patients with EwS as SMN. We tried to retrospectively examine the tumor samples available to us with regard to P53 expression, but the amount of available tumor material, as well as the interpretation, did not allow any clear conclusions, so this information was not explicitly included in the study. In conclusion, it remains unclear to what extent the type of treatment of the first malignancy or a genetic predisposition contributed to the development of EwS as SMN cases, or whether these cases were incidental findings.

5. Conclusions

In recent years, therapy management has been modified to achieve a better balance between acute therapy-related toxicity, long-term sequelae, and efficacy to prolong survival. The successes in treating children with cancer should not be overshadowed by the occurrence of secondary malignancies, but patients and healthcare providers need to be aware of risk factors for secondary malignancies, including sarcomas, so that surveillance is targeted, and early prevention strategies are implemented. EwS as SMN are rare and occur within 10 years of primary diagnosis. Overall, clinical surveillance is complicated by the lack of clear clinical features in primary disease or treatment that precede the development of EwS as SMN.

The survival rate of EwS as SMN is comparable to that of primary EwS, and patients with EwS as SMN should be treated with curative intent.

Author Contributions: Conceptualization, U.D., K.L.K., A.R., H.J., M.P. and S.K.Z.; methodology, A.R.; software, A.R., S.J. and S.K.Z.; formal analysis, A.R.; data curation, K.L.K., I.K., S.J., C.S., L.M.H., T.K., R.L. and M.T.; writing—original draft preparation, K.L.K. and S.K.Z.; writing—review and editing, I.K., M.K., C.S., M.T. and U.D.; visualization, K.L.K., A.R. and S.K.Z.; supervision, U.D., H.J. and M.P. All authors have read and agreed to the published version of the manuscript.

Funding: This work was supported by Deutsche Krebshilfe DKH M43/92/Jü2, 70-2551 Jü3, and 102802 for the EICESS 92 trial, the EURO-E.W.I.N.G. 99 trial, and the Ewing 2008 trial, respectively. Additional funding was granted through Deutsche Krebshilfe (DKH 70112018) for staff funding, which helped with the evaluation.

Institutional Review Board Statement: All studies investigated in this study were based on a positive ethics vote by the Ethics Committee of the Westphalia-Lippe Medical Association (Ärztekammer Westfalen-Lippe) of the Westphalian Wilhelms University (WWU) of Münster. For the EICESS 92 trial, the EURO-E.W.I.N.G. 99 trial, and the Ewing 2008 trial, the positive ethics vote was received on 8 May 1992, 3 March 1999 and 16 December 2008, respectively.

Informed Consent Statement: Informed consent was obtained from all subjects involved in the study.

Data Availability Statement: The data presented in this study are available on request from the corresponding author.

Conflicts of Interest: The authors declare no conflict of interest.

References

1. Gatta, G.; Botta, L.; Rossi, S.; Aareleid, T.; Bielska-Lasota, M.; Clavel, J.; Dimitrova, N.; Jakab, Z.; Kaatsch, P.; Lacour, B.; et al. Childhood cancer survival in Europe 1999-2007: Results of EUROCARE-5–a population-based study. *Lancet Oncol.* **2014**, *15*, 35–47. [CrossRef] [PubMed]
2. CCI Europe. Available online: https://ccieurope.eu/ (accessed on 19 July 2022).
3. Bhakta, N.; Liu, Q.; Ness, K.K.; Baassiri, M.; Eissa, H.; Yeo, F.; Chemaitilly, W.; Ehrhardt, M.J.; Bass, J.; Bishop, M.W.; et al. The cumulative burden of surviving childhood cancer: An initial report from the St Jude Lifetime Cohort Study (SJLIFE). *Lancet* **2017**, *390*, 2569–2582. [CrossRef]
4. Bhatia, S.; Sklar, C. Second cancers in survivors of childhood cancer. *Nat. Rev. Cancer* **2002**, *2*, 124–132. [CrossRef]
5. Curtis, R.E.; Freedman, D.M.; Ron, E.; Ries, L.A.G.; Hacker, D.G.; Edwards, B.K.; Tucker, M.A.; Fraumeni, J.F., Jr. (Eds.) *New Malignancies Among Cancer Survivors: SEER Cancer Registries, 1973-2000*; NIH Publ. No. 05-5302; National Cancer Institute NIH: Bethesda, MD, USA, 2006.
6. Ng, A.K.; Kenney, L.B.; Gilbert, E.S.; Travis, L.B. Secondary malignancies across the age spectrum. *Semin. Radiat. Oncol.* **2010**, *20*, 67–78. [CrossRef] [PubMed]
7. Grunewald, T.G.P.; Cidre-Aranaz, F.; Surdez, D.; Tomazou, E.M.; de Alava, E.; Kovar, H.; Sorensen, P.H.; Delattre, O.; Dirksen, U. Ewing sarcoma. *Nat. Rev. Dis. Prim.* **2018**, *4*, 5. [CrossRef] [PubMed]
8. Sand, L.G.; Szuhai, K.; Hogendoorn, P.C. Sequencing Overview of Ewing Sarcoma: A Journey across Genomic, Epigenomic and Transcriptomic Landscapes. *Int J. Mol. Sci.* **2015**, *16*, 16176–16215. [CrossRef]
9. Delattre, O.; Zucman, J.; Plougastel, B.; Desmaze, C.; Melot, T.; Peter, M.; Kovar, H.; Joubert, I.; de Jong, P.; Rouleau, G.; et al. Gene fusion with an ETS DNA-binding domain caused by chromosome translocation in human tumours. *Nature* **1992**, *359*, 162–165. [CrossRef]
10. Gaspar, N.; Hawkins, D.S.; Dirksen, U.; Lewis, I.J.; Ferrari, S.; Le Deley, M.C.; Kovar, H.; Grimer, R.; Whelan, J.; Claude, L.; et al. Ewing Sarcoma: Current Management and Future Approaches Through Collaboration. *J. Clin. Oncol.* **2015**, *33*, 3036–3046. [CrossRef]
11. Henderson, T.O.; Rajaraman, P.; Stovall, M.; Constine, L.S.; Olive, A.; Smith, S.A.; Mertens, A.; Meadows, A.; Neglia, J.P.; Hammond, S.; et al. Risk factors associated with secondary sarcomas in childhood cancer survivors: A report from the childhood cancer survivor study. *Int. J. Radiat. Oncol. Biol. Phys.* **2012**, *84*, 224–230. [CrossRef]
12. Schiffman, J.D.; Wright, J. Ewing's Sarcoma and Second Malignancies. *Sarcoma* **2011**, *2011*, 736841. [CrossRef]
13. Wolpert, F.; Grotzer, M.A.; Niggli, F.; Zimmermann, D.; Rushing, E.; Bode-Lesniewska, B. Ewing's Sarcoma as a Second Malignancy in Long-Term Survivors of Childhood Hematologic Malignancies. *Sarcoma* **2016**, *2016*, 5043640. [CrossRef] [PubMed]
14. Applebaum, M.A.; Worch, J.; Matthay, K.K.; Goldsby, R.; Neuhaus, J.; West, D.C.; Dubois, S.G. Clinical features and outcomes in patients with extraskeletal Ewing sarcoma. *Cancer* **2011**, *117*, 3027–3032. [CrossRef] [PubMed]
15. Hiramoto, N.; Kobayashi, Y.; Nomoto, J.; Maruyama, D.; Watanabe, T.; Tochigi, N.; Furuta, K.; Takeda, K.; Chuman, H.; Yagyu, S.; et al. Ewing sarcoma arising after treatment of diffuse large B-cell lymphoma. *Jpn. J. Clin. Oncol.* **2013**, *43*, 417–421. [CrossRef] [PubMed]
16. Vanhapiha, N.; Knuutila, S.; Vettenranta, K.; Lohi, O. Burkitt lymphoma and Ewing sarcoma in a child with Williams syndrome. *Pediatr. Blood Cancer* **2014**, *61*, 1877–1879. [CrossRef]
17. Applebaum, M.A.; Goldsby, R.; Neuhaus, J.; DuBois, S.G. Clinical features and outcomes in patients with secondary Ewing sarcoma. *Pediatr. Blood Cancer* **2013**, *60*, 611–615. [CrossRef]
18. Choi, D.K.; Helenowski, I.; Hijiya, N. Secondary malignancies in pediatric cancer survivors: Perspectives and review of the literature. *Int. J. Cancer* **2014**, *135*, 1764–1773. [CrossRef]
19. Srivastava, S.; Wang, S.; Tong, Y.A.; Pirollo, K.; Chang, E.H. Several mutant p53 proteins detected in cancer-prone families with Li-Fraumeni syndrome exhibit transdominant effects on the biochemical properties of the wild-type p53. *Oncogene* **1993**, *8*, 2449–2456.
20. Armstrong, G.T.; Liu, W.; Leisenring, W.; Yasui, Y.; Hammond, S.; Bhatia, S.; Neglia, J.P.; Stovall, M.; Srivastava, D.; Robison, L.L. Occurrence of multiple subsequent neoplasms in long-term survivors of childhood cancer: A report from the childhood cancer survivor study. *J. Clin. Oncol.* **2011**, *29*, 3056–3064. [CrossRef]
21. Robison, L.L.; Mertens, A. Second tumors after treatment of childhood malignancies. *Hematol. Oncol. Clin. N. Am.* **1993**, *7*, 401–415. [CrossRef]

22. Erdmann, F.; Kaatsch, P.; Grabow, D.; Spix, C. *German Childhood Cancer Registry—Annual Report 2019 (1980–2018)*; Institute of Medical Biostatistics, Epidemiology and Informatics (IMBEI) at the University Medical Center of the Johannes Gutenberg University Mainz: Mainz, Germany, 2020.
23. Zöllner, S.K.; Amatruda, J.F.; Bauer, S.; Collaud, S.; de Alava, E.; DuBois, S.G.; Hardes, J.; Hartmann, W.; Kovar, H.; Metzler, M.; et al. Ewing Sarcoma-Diagnosis, Treatment, Clinical Challenges and Future Perspectives. *J. Clin. Med.* **2021**, *10*, 1685. [CrossRef]
24. Spunt, S.L.; Rodriguez-Galindo, C.; Fuller, C.E.; Harper, J.; Krasin, M.J.; Billups, C.A.; Khoury, J.D. Ewing sarcoma-family tumors that arise after treatment of primary childhood cancer. *Cancer* **2006**, *107*, 201–206. [CrossRef] [PubMed]
25. Mirabello, L.; Troisi, R.J.; Savage, S.A. Osteosarcoma incidence and survival rates from 1973 to 2004: Data from the Surveillance, Epidemiology, and End Results Program. *Cancer* **2009**, *115*, 1531–1543. [CrossRef] [PubMed]
26. Henderson, T.O.; Whitton, J.; Stovall, M.; Mertens, A.C.; Mitby, P.; Friedman, D.; Strong, L.C.; Hammond, S.; Neglia, J.P.; Meadows, A.T.; et al. Secondary sarcomas in childhood cancer survivors: A report from the Childhood Cancer Survivor Study. *J. Natl. Cancer Inst.* **2007**, *99*, 300–308. [CrossRef] [PubMed]
27. Zichova, A.; Eckschlager, T.; Ganevova, M.; Malinova, B.; Luks, A.; Kruseova, J. Subsequent neoplasms in childhood cancer survivors. *Cancer Epidemiol.* **2020**, *68*, 101779. [CrossRef] [PubMed]
28. Cotterill, S.J.; Ahrens, S.; Paulussen, M.; Jurgens, H.F.; Voute, P.A.; Gadner, H.; Craft, A.W. Prognostic factors in Ewing's tumor of bone: Analysis of 975 patients from the European Intergroup Cooperative Ewing's Sarcoma Study Group. *J. Clin. Oncol.* **2000**, *18*, 3108–3114. [CrossRef]
29. Ladenstein, R.; Potschger, U.; Le Deley, M.C.; Whelan, J.; Paulussen, M.; Oberlin, O.; van den Berg, H.; Dirksen, U.; Hjorth, L.; Michon, J.; et al. Primary disseminated multifocal Ewing sarcoma: Results of the Euro-EWING 99 trial. *J. Clin. Oncol.* **2010**, *28*, 3284–3291. [CrossRef]
30. Kremeike, K.; Juergens, C.; Alz, H.; Reinhardt, D. Patients' Adherence in the Maintenance Therapy of Children and Adolescents with Acute Lymphoblastic Leukemia. *Klin. Padiatr.* **2015**, *227*, 329–334. [CrossRef]
31. Mody, R.; Li, S.; Dover, D.C.; Sallan, S.; Leisenring, W.; Oeffinger, K.C.; Yasui, Y.; Robison, L.L.; Neglia, J.P. Twenty-five-year follow-up among survivors of childhood acute lymphoblastic leukemia: A report from the Childhood Cancer Survivor Study. *Blood* **2008**, *111*, 5515–5523. [CrossRef]
32. Mulrooney, D.A.; Dover, D.C.; Li, S.; Yasui, Y.; Ness, K.K.; Mertens, A.C.; Neglia, J.P.; Sklar, C.A.; Robison, L.L.; Davies, S.M.; et al. Twenty years of follow-up among survivors of childhood and young adult acute myeloid leukemia: A report from the Childhood Cancer Survivor Study. *Cancer* **2008**, *112*, 2071–2079. [CrossRef]
33. Cooper, S.L.; Brown, P.A. Treatment of pediatric acute lymphoblastic leukemia. *Pediatr. Clin. N. Am.* **2015**, *62*, 61–73. [CrossRef]
34. Moore, M.J. Clinical pharmacokinetics of cyclophosphamide. *Clin. Pharm.* **1991**, *20*, 194–208. [CrossRef] [PubMed]
35. Flinn, I.W.; Neuberg, D.S.; Grever, M.R.; Dewald, G.W.; Bennett, J.M.; Paietta, E.M.; Hussein, M.A.; Appelbaum, F.R.; Larson, R.A.; Moore, D.F., Jr.; et al. Phase III trial of fludarabine plus cyclophosphamide compared with fludarabine for patients with previously untreated chronic lymphocytic leukemia: US Intergroup Trial E2997. *J. Clin. Oncol.* **2007**, *25*, 793–798. [CrossRef] [PubMed]
36. Neglia, J.P.; Meadows, A.T.; Robison, L.L.; Kim, T.H.; Newton, W.A.; Ruymann, F.B.; Sather, H.N.; Hammond, G.D. Second neoplasms after acute lymphoblastic leukemia in childhood. *N. Engl. J. Med.* **1991**, *325*, 1330–1336. [CrossRef] [PubMed]
37. Eulo, V.; Lesmana, H.; Doyle, L.A.; Nichols, K.E.; Hirbe, A.C. Secondary Sarcomas: Biology, Presentation, and Clinical Care. *Am. Soc. Clin. Oncol. Educ. Book* **2020**, *40*, 1–12. [CrossRef]
38. Grunewald, T.G.; Bernard, V.; Gilardi-Hebenstreit, P.; Raynal, V.; Surdez, D.; Aynaud, M.M.; Mirabeau, O.; Cidre-Aranaz, F.; Tirode, F.; Zaidi, S.; et al. Chimeric EWSR1-FLI1 regulates the Ewing sarcoma susceptibility gene EGR2 via a GGAA microsatellite. *Nat. Genet.* **2015**, *47*, 1073–1078. [CrossRef]
39. Tirode, F.; Surdez, D.; Ma, X.; Parker, M.; Le Deley, M.C.; Bahrami, A.; Zhang, Z.; Lapouble, E.; Grossetete-Lalami, S.; Rusch, M.; et al. Genomic landscape of Ewing sarcoma defines an aggressive subtype with co-association of STAG2 and TP53 mutations. *Cancer Discov.* **2014**, *4*, 1342–1353. [CrossRef]
40. Randall, R.L.; Lessnick, S.L.; Jones, K.B.; Gouw, L.G.; Cummings, J.E.; Cannon-Albright, L.; Schiffman, J.D. Is There a Predisposition Gene for Ewing's Sarcoma? *J. Oncol.* **2010**, *2010*, 397632. [CrossRef]

cancers

MDPI

Article

Correlation of Transcriptomics and FDG-PET SUVmax Indicates Reciprocal Expression of Stemness-Related Transcription Factor and Neuropeptide Signaling Pathways in Glucose Metabolism of Ewing Sarcoma

Carolin Prexler [1], Marie Sophie Knape [1], Janina Erlewein-Schweizer [2], Wolfgang Roll [3], Katja Specht [2], Klaus Woertler [4], Wilko Weichert [2,5], Irene von Luettichau [1,6], Claudia Rossig [7,8], Julia Hauer [1], Guenther H. S. Richter [9,10], Wolfgang Weber [5,11] and Stefan Burdach [1,2,5,12,*]

1 Department of Pediatrics and Children's Cancer Research Center, Kinderklinik München Schwabing, Klinikum Rechts der Isar, Fakultät für Medizin, Technische Universität München, 80804 Munich, Germany
2 Institute of Pathology, Technische Universität München, 81675 Munich, Germany
3 Department of Nuclear Medicine, University Hospital Münster, Albert-Schweitzer-Campus 1 A1, 48149 Munster, Germany
4 Musculoskeletal Radiology Section, Klinikum Rechts der Isar, Technische Universität München, 81675 Munich, Germany
5 German Cancer Consortium (DKTK), Partner Site Munich, 81675 Munich, Germany
6 ERN PaedCan, 1090 Vienna, Austria
7 Department of Pediatric Hematology and Oncology, University Children's Hospital Muenster, 48149 Muenster, Germany
8 Cells-in-Motion Cluster of Excellence (EXC 1003-CiM), University of Muenster, 48149 Muenster, Germany
9 Department of Pediatrics, Division of Oncology and Hematology, Charite–Universitätsmedizin Berlin, Augustenburger Platz 1, 13353 Berlin, Germany
10 German Cancer Consortium (DKTK), Partner Site Berlin, 13353 Berlin, Germany
11 Department of Nuclear Medicine, Klinikum Rechts der Isar, Technische Universität München, 81675 Munich, Germany
12 Academy of Translational Medicine and Department of Molecular Oncology–British Columbia Cancer Research Centre, University of British Columbia, 675 West 10th Avenue, Vancouver, BC V5Z 1L3, Canada
* Correspondence: stefan.burdach@tum.de

Citation: Prexler, C.; Knape, M.S.; Erlewein-Schweizer, J.; Roll, W.; Specht, K.; Woertler, K.; Weichert, W.; von Luettichau, I.; Rossig, C.; Hauer, J.; et al. Correlation of Transcriptomics and FDG-PET SUVmax Indicates Reciprocal Expression of Stemness-Related Transcription Factor and Neuropeptide Signaling Pathways in Glucose Metabolism of Ewing Sarcoma. *Cancers* **2022**, *14*, 5999. https://doi.org/10.3390/cancers14235999

Academic Editor: Franck Verrecchia

Received: 18 October 2022
Accepted: 28 November 2022
Published: 5 December 2022

Simple Summary: Survival rates for metastatic or early recurring Ewing sarcoma (EwS) are dismal, and therapies have severe side and long-term effects. Both aspects require new therapeutic options and targets for treatment. Risk stratification enables individualized treatment and may reduce the burden of side effects. Our radiogenomics study provides novel candidates at the gene expression level to explain the mechanisms of malignancy. We retrospectively analyzed 19 EwS samples (17 patients) and integrated functional genomics assessed by gene expression and functional imaging assessed by FDG-PET, which has the potential to better characterize these highly malignant tumors. The identified genes and pathways can serve as a starting point for prospective experimental and clinical studies of new therapeutic interventions. Thus, this study opens new opportunities for future studies to improve the outcome of patients with poor survival rates.

Abstract: Background: In Ewing sarcoma (EwS), long-term treatment effects and poor survival rates for relapsed or metastatic cases require individualization of therapy and the discovery of new treatment methods. Tumor glucose metabolic activity varies significantly between patients, and FDG-PET signals have been proposed as prognostic factors. However, the biological basis for the generally elevated but variable glucose metabolism in EwS is not well understood. Methods: We retrospectively included 19 EwS samples (17 patients). Affymetrix gene expression was correlated with maximal standardized uptake value (SUVmax) using machine learning, linear regression modelling, and gene set enrichment analyses for functional annotation. Results: Expression of five genes correlated (*MYBL2*, *ELOVL2*, *NETO2*) or anticorrelated (*FAXDC2*, *PLSCR4*) significantly with SUVmax (adjusted p-value $\leq$ 0.05). Additionally, we identified 23 genes with large SUVmax effect size, which were significantly enriched for "neuropeptide Y receptor activity (GO:0004983)" (adjusted p-value = 0.0007).

The expression of the members of this signaling pathway (*NPY, NPY1R, NPY5R*) anticorrelated with SUVmax. In contrast, three transcription factors associated with maintaining stemness displayed enrichment of their target genes with higher SUVmax: *RNF2*, E2F family, and *TCF3*. Conclusion: Our large-scale analysis examined comprehensively the correlations between transcriptomics and tumor glucose utilization. Based on our findings, we hypothesize that stemness may be associated with increased glucose uptake, whereas neuroectodermal differentiation may anticorrelate with glucose uptake.

Keywords: Ewing sarcoma; 18-FDG-PET; SUVmax; radiogenomics; transcriptomics

1. Introduction

Ewing sarcoma (EwS) is a malignancy of bone or soft tissue. Incidence and prognosis decrease with age [1,2]. Survival rates have increased from 10% in the 1970s to 70–80% in 2000 for localized disease [1,3], due to multimodal approaches combining chemotherapy, surgery, and radiation. However, the 5-year survival rate for patients with advanced stages is still <30% [1,3]. Little therapeutic progress has been made in the last two decades [4–6]. Stratification of therapeutic intensity is critical in children because of long-term effects [1]. Toxicity and poor survival rates require individualization of therapy with new approaches [7]. Better mechanistic understanding may help to address these issues.

Resistance to targeted therapies is related to heterogeneity and plasticity of tumors [8]. Biopsy of a single lesion is the diagnostic gold standard. However, the complexity of a systemic disease cannot be mapped with such a locally restricted procedure. Whole-body molecular imaging might aid in overcoming these limitations. Radiomics offers new opportunities for analyzing these large imaging datasets. The combination of advanced image analysis with tissue-based genomic data—called radiogenomics—allows an in-depth characterization of the disease.

EwS cells show a stem cell-like phenotype with poor differentiation [9,10]. The exact cell of origin is still under debate [9,11–15]. Genetically, EwS is uniformly characterized [1,2,10,16–18]. Epigenetics and transcriptional regulation are thought to account for variability [1,2,9,16,19]. However, it is unclear how they relate to imaging.

18-F-fluorodeoxyglucose positron emission tomography (18-F-FDG PET) is a functional imaging method that quantitatively measures glucose uptake. A common parameter is the maximum standardized uptake value (SUVmax), the highest rate of FDG uptake in the tumor indicating metabolic tumor activity. Quantitative FDG-PET parameters, and in particular, SUVmax, have been shown to be potentially prognostic in several cancers, both in primary and recurrence, pre-treatment and post neoadjuvant therapy, correlating with tumor growth, worse survival, poor prognosis, advanced stage, and worse course of disease [20–35]. SUVmax correlates with worse outcome and stage of the disease in primary EwS [36–41] while studied less in recurrence. However, primary and relapsed cases show similar SUVmax [42]. To further investigate the role of SUVmax in EwS, we use it as a quantitative phenotype in our study.

The aim of the present study was to characterize variable glucose uptake in EwS by identifying correlations of SUVmax and transcriptomics. These correlations indicate which genes and pathways may be more or less active in tumors with regard to SUVmax. Given the potential prognostic value of SUVmax suggested by the literature, we hypothesize that such genes or pathways may help improve our mechanistic understanding and qualify for experimental validation.

2. Materials and Methods

2.1. Dataset: The Munich Cohort (n = 19)

We included all patients suffering from EwS in the Children's Hospital Schwabing and in the Klinikum rechts der Isar of the Technical University of Munich in the years 2011 to 2019 who fulfilled our inclusion criteria. The criteria were: all patients suffering from primary or relapsed EwS, aged up to 40 years, with image data and tissue sampled from the same lesion. Patients with pre-therapeutic primary disease or pre-therapeutic relapse were classified as "untreated", otherwise "under therapy", at the time of PET imaging or tissue sampling. The time interval between tissue sampling and imaging had to be maximum 6 weeks for untreated patients, and maximum 2 weeks for patients under therapy, thus ensuring that tissue sample and PET reflected the same biological characteristics of the tumor. This yielded 19 samples from 17 patients (Table 1, Supplementary Table S1). From 2 patients, 2 samples each were included, representing different lesions at different time points in the disease course.

Table 1. Patients' characteristics of our EwS cohort.

		Number	**Fraction**
Total		**19**	**1**
Sex	Female Male	11 8	0.58 0.42
Disease state	Primary disease Relapse	5 14	0.26 0.74
Sample type	Tumor Metastasis	5 14	0.26 0.74
Therapy	Untreated Under therapy	12 7	0.63 0.37
Age at PET (all)	Number Range Median Mean	19 3–31 14 14.8	1
Age at PET (≤15 years)	Number Range Median Mean	11 3–15 10 9.2	0.58
Age at PET (>15 years)	Number Range Median Mean	8 17–31 21 22.5	0.42
Imaging modality	PET-CT PET-MR	15 4	0.79 0.21
SUVmax (all)	Number Range Median Mean	19 1.898–21.269 5.387 6.819	1
SUVmax low	Number Range Median Mean	9 1.898–5.084 2.756 3.207	0.47
SUVmax high	Number Range Median Mean	10 5.387–21.269 9.035 10.07	0.53

The data were analyzed retrospectively. The registry study was approved by the local ethics committee (reference 223/16S) and complies with the Declaration of Helsinki. All patients gave their written consent.

2.2. Gene Expression Data

2.2.1. Tissue

Frozen tumor samples for expression analysis were obtained from biopsies or resection specimens and passed the quality control of experienced pathologists. Sample preparation followed the Affymetrix protocol and was previously described [12]. Gene expression was measured using [HuGene-1_0-st] Affymetrix Human Gene 1.0 ST Array.

2.2.2. Preprocessing

The microarray data was processed using Robust Multichip Average (RMA) preprocessing [43–45] (R package oligo [46]), including background subtraction, quantile normalization, and median-polish summarization of probe sets to genes (log2 expression values). For summarization, Brainarray (version 24) [47–49] was used, yielding 20,722 genes.

2.2.3. Filtering

We applied two filtering steps on our gene expression data. In the first step, we excluded genes with low expression (average expression below 10), as microarray chips are not precise at low expression levels [50]. A total of 20,524/20,722 genes remained.

For the second filtering step, we wanted to focus on genes that might be related to survival in EwS. To obtain such "potential survival genes", we applied a machine learning approach to external datasets. We collected all EwS datasets from the Gene Expression Omnibus (GEO) database [51] (as of March 2018) of gene expression plus survival data, which yielded 3 datasets: GSE63155 [52], GSE17618 [53], GSE63156 [52]. The survival data were used to split the patients into long-term survival (overall survival, OS $\geq$ 5 years) or short-term survival (OS < 5 years). Ambiguous patients (i.e., patients lost to follow-up within 5 years) were excluded, as it is unknown whether they died before or after the 5-year cutoff. This yielded n = 31 for GSE63155 (12 short OS, 19 long OS), n = 40 for GSE17618 (22 short OS, 18 long OS), and n = 31 for GSE63156 (9 short OS, 22 long OS). Each dataset was analyzed separately to avoid bias of different data sources. These external expression data were preprocessed in the same way as our expression data described above. To reduce dimensionality, 50% of the genes with low coefficient of variation were removed, as these were supposed to be not useful for the discrimination task. Random forest classification was applied to each dataset to predict binary OS: short OS vs. long OS (R package caret [54], method "rf" using 1000 trees). Due to class imbalance in all datasets, downsampling to equal class sizes was performed to avoid bias for one class. Then, random forest models were generated for each dataset in terms of repeated 10-fold cross-validation. Each model randomly chose genes and from this subset selected the gene that best discriminated between samples with long and short OS. This gene was incorporated into a decision tree. This way, each model built 1000 decision trees, whose ensemble vote would be used for classifying a new, unseen sample as "long OS" or "short OS". The genes, which were used in the decision trees, contributed information to the discrimination task. This way, the model identified a set of genes that were informative for predicting survival time in each dataset. We called the overlap of the 3 gene sets "potential survival genes" (1491 genes). Functional annotation of these genes was obtained using DAVID Bioinformatics Resources (v6.8) [55,56], category UP_KEYWORDS. Adjusted p-values < 0.01 (Benjamini) were considered significant. The "potential survival genes" were used to filter the expression data in our cohort (1376/1491 genes, Supplementary Table S2).

2.3. Imaging Data

We analyzed PET computed tomography (PET-CT)/PET magnetic resonance (PET-MR) to obtain SUVmax for those lesions that were used for gene expression measurements. All patients underwent 18F-FDG PET imaging for diagnostic or staging purposes in our institution.

The PETs were controlled for quality using OsiriX DICOM viewer [57], and tumors were delineated using cuboids by experienced nuclear medicine physicians. Exact delineations were obtained based on a standardized uptake value (SUV) threshold of 40% of SUVmax for each lesion. Image features were calculated according to the image biomarker standardization initiative (IBSI) [58] using PyRadiomics [59] v3.0.1 with standard settings (Python version 3.8 as of October 2020). Spatial resampling to 4 × 4 × 4 mm was applied. For image discretization, a fixed bin size of 0.5 was used. SUVmax was obtained from the original image, i.e., no filters were applied (PyRadiomics feature "original_firstorder_Maximum").

2.4. Statistical Analysis of SUVmax and Clinical Variables

2.4.1. Distribution of SUVmax Regarding Clinical Variables

We tested for equal distribution of SUVmax values regarding the clinical variables: sex (male, female), disease state (primary disease, relapse), sample type (tumor, metastasis), therapy (under therapy, untreated), and age ($\leq$15 years, >15 years). A two-sided Welch two-sample t-test was applied (R function stats::t.test). *p*-values < 0.05 were considered significant

2.4.2. Survival Analysis

We applied univariate Kaplan–Meier analyses for OS with log-rank tests (R package survival [60,61], functions survfit and ggsurvplot). We tested sex, disease state, age, and SUV categories (samples split by median SUVmax into low SUV vs. high SUV).

We built multivariate Cox proportional hazards models (R package survival [60,61], function coxph), including continuous SUVmax and disease state. A forest plot was generated using R package survminer [62], function ggforest. *p*-values < 0.05 were considered significant.

2.5. Correlation Analysis

2.5.1. Linear Regression

We used linear regression ("least squares") to correlate gene expression with SUVmax. Moderated t-statistics were calculated based on an empirical Bayes method in the R package limma [63]. We applied the limma-trend method to fit a trend to the prior variances. Limma was run on all genes with average expression above 10 (20,524 genes). Afterwards, the model results were filtered for "potential survival genes" (1376 genes). Adjusted *p*-values < 0.05 (Benjamini–Hochberg) were considered significant.

In general, when comparing expression between two groups, a doubling of gene expression is usually considered to be of interest. When using fold changes on the log2 expression values (logFC), a doubling of gene expression corresponds to a logFC cutoff of 1. For the linear regression used in this analysis, we defined relations of high effect size by transferring the standard abs(logFC) cutoff of 1 to regression modeling, which corresponded to abs(slope) >0.146. A total of 23 genes fulfilled this criterion.

The heatmap depicting Z-scaled gene expression in all samples with clinical data in the side bars was generated using R function GMD::heatmap.3 [64]. The dendrograms are based on Euclidean distance and average linkage.

2.5.2. Enrichment Analyses

The results of linear regression were further tested in enrichment analyses for functional annotation by applying 2 tools: Enrichr [65–67] and gene set enrichment analysis (GSEA v4.1.0) [68,69]. The set of 23 genes of high effect size was analyzed using Enrichr (applied in September 2021). We focused on pathways in "Reactome 2016" and the Gene Ontology (GO) knowledgebase, including "GO biological process 2021", "GO molecular function 2021" and "GO cellular component 2021". Adjusted *p*-values < 0.01 were considered significant.

Additionally, we performed GSEA based on all 1376 genes ranked by their correlation with SUVmax (decreasing slope). For each gene set, a normalized enrichment score (NES) is calculated, which indicates the extent to which this gene set is enriched at the top or bottom of the ranked gene list. *p*-values are corrected for multiple testing (false discovery rate, FDR, q-value). If GSEA is used for hypothesis generation, the developers suggest a less stringent cutoff for significance. Thus, FDR < 0.1 was considered significant.

3. Results

An overview of all analysis steps is provided in Figure 1. In general, we used R Statistical Software (v4.0.2) [70] for data analysis, statistical testing, and plot generation.

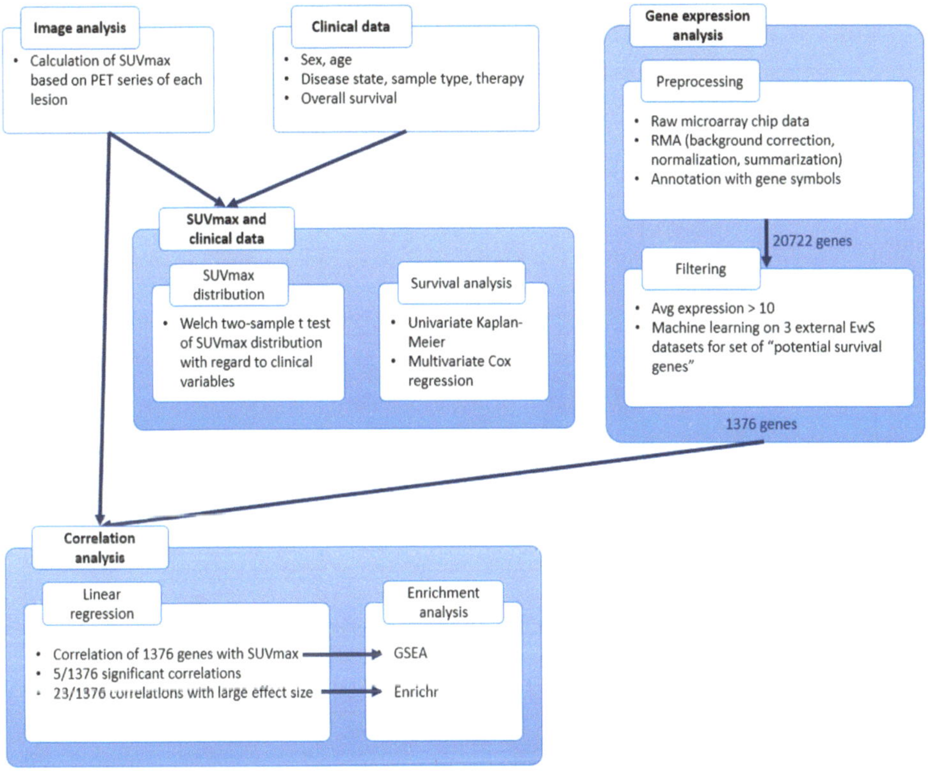

Figure 1. Workflow scheme. Overview of analysis steps. Positron emission tomography (PET) images are used to calculate maximal standardized uptake values (SUVmax). SUVmax distribution is analyzed with regard to clinical data, and a survival analysis is performed. Gene expression data is preprocessed and filtered, and afterwards correlated with SUVmax. The results of the correlation analysis are further annotated using enrichment analyses.

3.1. The Munich Cohort (n = 19)

Out of 75 patients referred to our institution during the duration of the study, only 17 (Table 1, Supplementary Table S1) met all our quality standards. We included 19 samples (primary disease n = 5, recurrence n = 14) from 17 EwS patients (female n = 10, male n = 7), aged three to thirty-one years. Tissue samples for expression analysis were obtained from one lesion each (tumor n = 5, metastasis n = 14) before (n = 12) or during (n = 7) treatment. From this tissue, expression of all genes was assessed by Affymetrix Gene Chip analysis. Of the same lesion, we also analyzed PET-CT (n = 15)/PET-MR (n = 4) to obtain SUVmax. The glucose uptake varied in our cohort (Figure 2a). SUVmax showed a spectrum of 1.9 to 21.3, with a median of 5.4. Based on the median, the samples were split into two groups: lesions with low SUV (n = 9, SUVmax 1.9 to 5.1, median 2.8) and lesions with high SUV (n = 10, SUVmax 5.4 to 21.3, median 9.0).

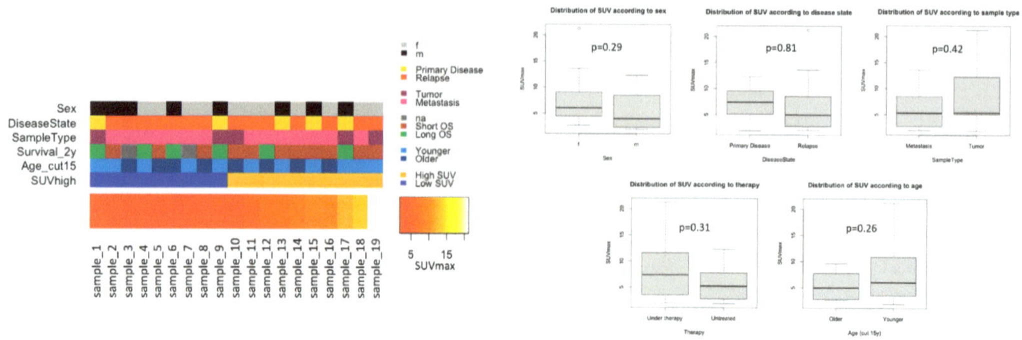

(a) (b)

Figure 2. Clinical data. (**a**) 19 EwS samples ordered by increasing SUVmax having a range of 1.9 to 21.3 (bottom). On top, additional clinical information is provided about sex (female or male), disease state (primary disease or relapse), sample type (tumor or metastasis), 2-year overall survival (OS), age (≤15 years or >15 years), and categorical partitioning of the samples into high SUV or low SUV (split by median SUVmax). (**b**) Boxplots showing distribution of SUVmax values with regard to clinical variables: sex (female or male), disease state (primary disease or relapse), sample type (tumor or metastasis), therapy (under therapy or untreated), and age (≤15 years or >15 years). *p*-Values from the Welch two-sample *t* test.

3.2. Statistical Analysis of SUVmax and Clinical Variables

3.2.1. SUVmax Distribution with Regard to Clinical Variables

We explored the correlation of SUVmax with clinical variables in our cohort. Potential correlations would render the clinical variables confounding factors and introduce bias in subsequent analyses, in which we correlated gene expression with SUVmax. We found no significant difference in SUVmax distribution with respect to sex, disease state, sample type, therapy, and age (Figure 2b).

3.2.2. Survival Analysis

Univariate Kaplan–Meier analysis (Supplementary Figure S1a) indicated no significant correlation of clinical variables (sex, disease state, and age) or SUVmax with OS (all p-values > 0.05). To investigate continuous SUVmax and incorporate more variables, a multivariate Cox proportional hazards model was built in the next step. We used SUVmax and added disease state, as it has shown a trend in Kaplan–Meier analysis (p-value = 0.09). In combination, both variables showed significant correlation with survival (overall p-value = 0.02 in log-rank test, Supplementary Figure S1b): higher SUVmax (p = 0.02, hazard ratio, HR, [95% confidence interval (CI)] = 1.2 [1.0; 1.3]), and relapse (p = 0.05, HR [95% CI] = 5.0 [1.0; 24.9]) were associated with increased risk of death.

3.3. Gene Expression Analysis

We analyzed the expression of 20,722 genes passing preprocessing and quality control (Figure 1, right panel). To overcome the problem of high dimensionality in such large-scale analyses, we performed two filtering steps. First, we excluded genes with low expression. In the second filtering step, we focused on genes that are potentially linked to survival in EwS. To obtain such genes, we used a machine learning approach on three external, public EwS datasets: GSE63155, GSE17618, GSE63156 (Figure 3a). The model identified 1491 genes that were predictive in all three external datasets, denoted as "potential survival genes". Functional annotation of these genes yielded phosphoprotein, alternative splicing, polymorphism, acetylation, cytoplasm, cell division, cell cycle, Golgi apparatus, DNA replication, disease mutation, mitosis, cell junction, and endoplasmic reticulum (Figure 3b). We focused on the "potential survival genes" for the correlation analysis with SUVmax.

After both filtering steps, we obtained a total of 1376 genes, whose log2 expression values were used for further analysis.

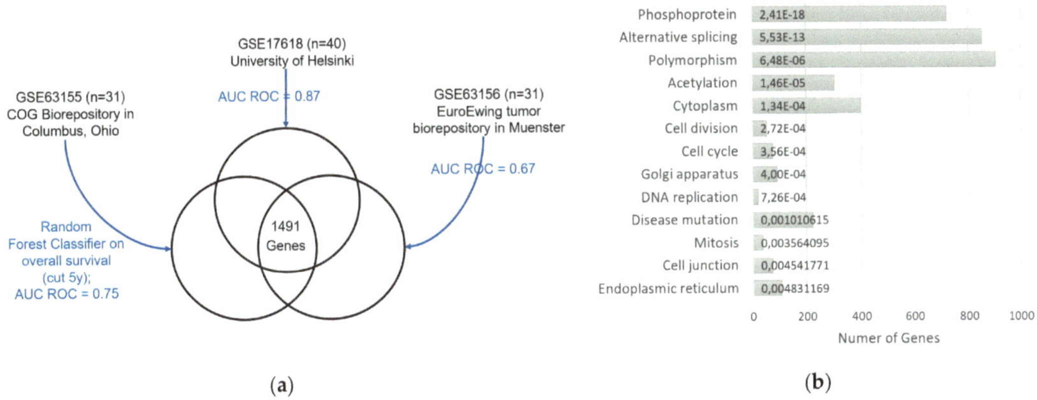

Figure 3. Diagram of machine learning analysis (second filtering step in gene expression analysis) in order to obtain "potential survival genes" in Ewing sarcoma (EwS). (**a**) Random forest classifiers are applied to 3 public datasets (GSE63155, GSE17618, GSE63156) in 10-fold cross-validation, obtaining an area under the receiver operating characteristic curve (ROC AUC) of 0.67 to 0.87. These models yield genes that are predictive for survival for each dataset. The intersection of these 3 gene sets contains 1491 genes, which we consider as "potential survival genes" in EwS. (**b**) DAVID functional annotation of "potential survival genes". Thirteen terms in category UP-KEYWORDS obtained significant p-values (Benjamini adjusted $p < 0.01$). For each term, the number of annotated genes among the "potential survival genes" and the adjusted p-value is given.

3.4. Correlation Analysis of SUVmax and Gene Expression

3.4.1. Linear Regression

To investigate correlations of gene expression and SUVmax (Figure 1, lower panel), we applied linear regression modeling for all 1376 genes that remained after the preprocessing and filtering steps. The volcano plot (Figure 4a) illustrates the effect size (i.e., the slope of the regression line) and r^2. For 645/1376 genes, the regression slope was positive, for 731/1376 genes, the regression slope was negative. A volcano plot depicting slopes and adjusted p-values is supplied in the supplement (Supplementary Figure S2).

3.4.2. Significant Association of SUVmax and Five Genes

A total of 5/1376 genes showed a significant correlation with SUVmax (adjusted $p < 0.05$). Of these, 3/5 genes were positively associated with SUVmax (Figure 4b, top row): *MYBL2* showed a slope of the regression line of 0.149 (95% CI [0.088; 0.211]), indicating that the gene expression doubled over 6.69 SUV units ($r^2 = 0.58$). *ELOVL2* showed a slope of 0.148 (95% CI [0.084; 0.212]), indicating that the expression doubled over 6.76 SUV units ($r^2 = 0.54$). *NETO2* showed a slope of 0.157 (95% CI [0.086; 0.228]), indicating that the expression doubled over 6.38 SUV units ($r^2 = 0.51$). In contrast, 2/5 genes were negatively associated with SUVmax (Figure 4b, bottom row): *FAXDC2* showed a slope of −0.176 (95% CI [−0.246; −0.107]), indicating that the expression halved over 5.67 SUV units ($r^2 = 0.59$). *PLSCR4* showed a slope of −0.223 (95% CI [−0.321; −0.125]), indicating that the expression halved over 4.49 SUV units ($r^2 = 0.51$).

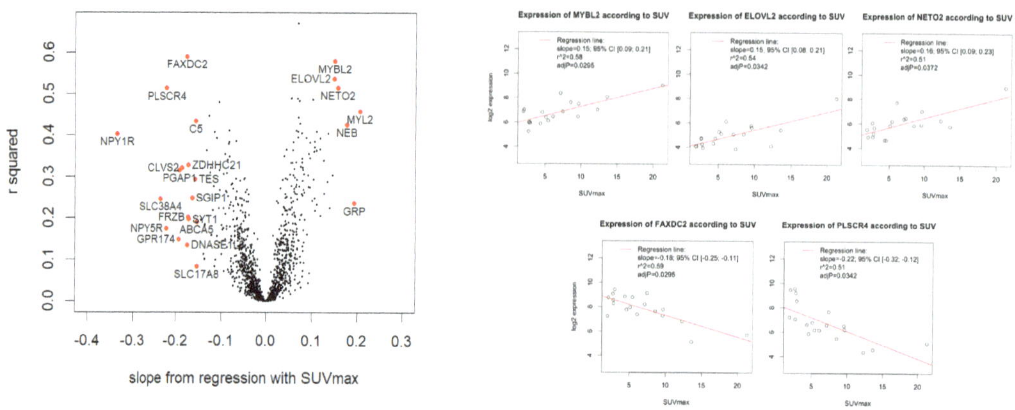

(**a**) (**b**)

Figure 4. Results from linear regression modeling of SUVmax and gene expression. (**a**) Volcano plot. For each gene tested, the slope of the regression line and r^2 (Pearson correlation) are given. Five genes are significantly correlated with SUVmax with adjusted *p*-value < 0.05 (FAXDC2, MYBL2, PLSCR4, ELOVL2, and NETO2). In addition, 23 genes show a high effect size of abs(slope) >0.146 (labelled genes). (**b**) Scatterplots of significant correlations of SUVmax and log2 gene expression. Expression of MYBL2, ELOVL2, and NETO2 is positively associated with SUVmax (top row), whereas FAXDC2 and PLSCR4 are negatively associated (bottom row).

3.4.3. 23 Genes with High Effect Size

Next, we examined genes correlating with SUVmax with high effect size. Effect size is important, i.e., how much gene expression changes in relation to SUVmax. Genes with a high difference in expression levels in relation to SUVmax are more likely to have an impact of biological relevance. To define genes with high effect size, we chose a cutoff for the slope: abs(slope) > 0.146. Thereby, we obtained 23 genes with high effect size (Figure 4a labelled genes, Supplementary Table S3). The majority of genes was negatively correlated with SUVmax (17/23), 6/23 positively. The 23 genes included the five significant correlations, so the previously described five genes had a high effect size by our definition.

The expression of the 23 genes in the 19 samples is illustrated in a heatmap (Figure 5a), together with clinical variables and a stratification into two groups of low SUV or high SUV according to median SUVmax. Visual inspection of the hierarchical clustering of the samples based on expression showed that the two tumors with the highest glucose uptake in our cohort (SUVmax 13.6 and 21.3) and two tumors with low SUVmax (2.7 and 4.3) clustered apart from the other samples. The remaining samples split into two groups: a cluster with lower SUVmax (5/6 in low SUV group), and a cluster with higher SUVmax (7/9 in high SUV group). As this sample clustering was based on genes whose expression changed strongly in relation to SUVmax, it reflected the samples' spectrum of SUVmax values. This suggested that there was an expression signature for metabolic activity.

The hierarchical clustering of the genes displayed two clusters. In the smaller cluster, 5/6 genes correlated positively with SUVmax. In the larger cluster, 16/17 genes correlated negatively with SUVmax. Thus, the clustering reflected the two directions of association with SUVmax.

3.5. Enrichment Analyses

After analyzing single correlations of gene expression and SUVmax, we looked for shared pathways and processes that correlated with glucose uptake. The aim was to summarize and generalize the results of the correlation analysis on a functional level. We investigated whether there were pathways and annotated gene sets distinguishing tumors with varying glucose uptake by applying different approaches of enrichment analysis. Enrichment analyses are robust to false positive findings because many genes are considered at once. This is an advantage, especially with a small sample size.

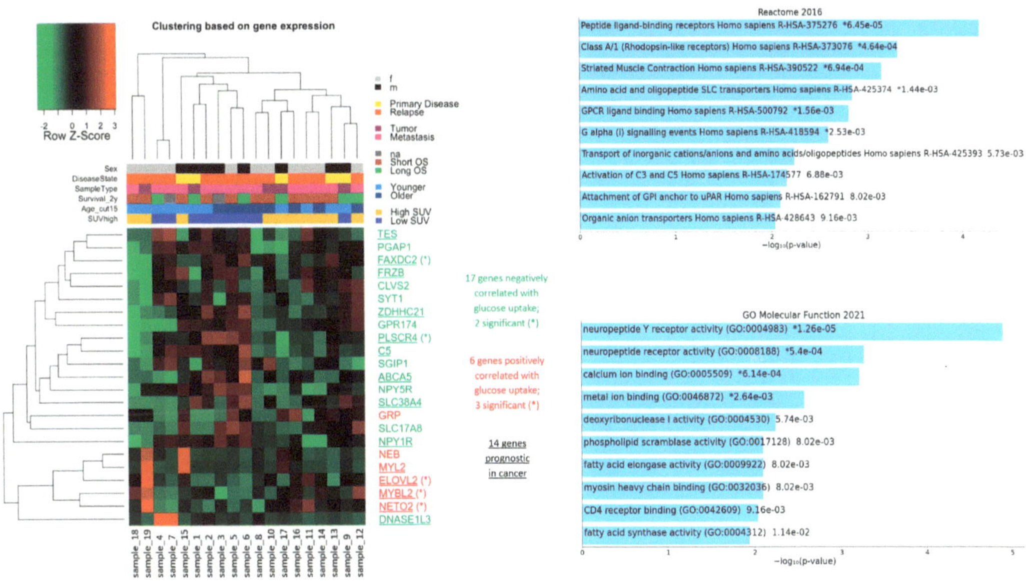

(a) (b)

Figure 5. Twenty-three genes with high effect (abs(slope) > 0.146) in linear regression with SUVmax. (**a**) Heatmap depicting their expression in our cohort. On top, clinical data is provided about sex (male or female), disease state (primary disease or relapse), sample type (tumor or metastasis), 2-year OS, age (≤15 years or >15 years), and categorical partitioning of the samples into high SUV or low SUV (split by median SUVmax). (**b**) Enrichments among the 23 genes with high effect size found by Enrichr (raw *p*-values are provided, * indicates adjusted *p*-value < 0.05). Top: enrichments of the "Reactome Database". Considering terms with adjusted *p*-value < 0.01 as significant, one term obtains a significant adjusted *p*-value (q = 0.0043): "Peptide ligand-binding receptors Homo sapiens R-HAS-375276". Bottom: enrichments of "GO molecular function". One term obtains a significant adjusted *p*-value (q = 0.0007): "neuropeptide Y receptor activity (GO:0004983)".

3.5.1. Enrichments among the 23 Genes with High Effect Size

First, we examined the functional annotation of genes with high effect size. For the 23 genes that correlated with SUVmax with high effect size, we scanned their annotation for their prognostic value in cancer entities (Supplementary Table S3), and then performed enrichment analysis for shared pathways and function.

According to the Human Protein Atlas [71], 14/23 genes are related with survival in several cancer types. High expression of *ABCA5*, *C5*, *DNASE1L3*, *ELOVL2*, *FAXDC2*, *NPY1R*, *SLC38A4*, *TES*, and *ZDHHC21* predicts a favorable outcome in breast, liver, pancreatic, renal, and urothelial cancer. In contrast, high expression of *FRZB*, *MYBL2*, *MYL2*, *NETO2*, *PLSCR4*, and *TES* predicts an unfavorable outcome in endometrial, head and neck, liver, pancreatic, and renal cancer.

We tested for enrichment of pathways and functions systematically using the tool Enrichr [65–67]. The set of 23 genes with high effect size showed significant enrichment for the Reactome pathway "Peptide ligand-binding receptors Homo sapiens R-HAS-375276" with q = 0.0043 (Figure 5b, top), which included a subset of the rhodopsin-like G protein-coupled receptor (GPCR) family. This enrichment was due to four genes: *C5*, *NPY1R*,

NPY5R, and *GRP*. There were no significant enrichments in GO biological process and GO cellular component. However, the 23 genes were significantly enriched for the GO molecular function "neuropeptide Y receptor activity (GO:0004983)" with q-value = 0.0007 (Figure 5b, bottom). NPY receptors are also rhodopsin-like receptors. This enrichment was based on the two genes, *NPY1R* and *NPY5R*.

As the NPY receptors contributed to both enrichments, we further investigated the role of the NPY pathway in our dataset. The expression of the genes in the NPY signaling axis was decreased with increasing SUVmax (Supplementary Figure S3). The signaling molecule *NPY* showed a slope of the regression line of −0.136 (95% CI [−0.263; −0.009]), implying expression halved per 7.37 SUV units (r^2 = 0.18). The NPY receptor *NPY1R* showed a slope of −0.334 (95% CI [−0.514; −0.155]), indicating expression halved per 2.99 SUV units (r^2 = 0.40). *NPY5R*, another receptor, showed a slope of −0.222 (95% CI [−0.433; −0.010]), that is, expression halved per 4.51 SUV units (r^2 = 0.17). In contrast, two other NPY receptors (*NPY2R* and pseudogene *NPY6R*) were expressed constantly, regardless of SUVmax. *NPY2R* showed a slope of 0.00007 (95% CI [−0.043; 0.043]; r^2 = 0.00), and *NPY6R* showed a slope of 0.009 (95% CI [−0.030; 0.048]; r^2 = 0.01). In addition, there were two paralogs of *NPY*, namely, *PYY* and *PPY*. Their expression was independent of SUVmax as well: *PYY* showed a slope of −0.005 (95% CI [−0.056;0.046]; r^2 = 0.00), and *PPY* showed a slope of 0.015 (95% CI [−0.025; 0.054]; r^2 = 0.03).

All in all, of the 23 genes that were strongly associated with SUVmax, most are associated with survival in several cancer entities, and therefore play a role in distinguishing subgroups in entities other than EwS. Our findings, namely, that these genes have different expression levels in EwS tumors with high or low glucose uptake potentially indicating a different prognosis, suggest that they may also distinguish subgroups in EwS. Looking for functional similarities of the 23 genes, we found that the NPY signaling axis seems to correlate negatively with glucose uptake.

3.5.2. GSEA Based on Regression Results of All 1376 Genes

In addition to enrichments in the set of 23 genes with high effect size, we investigated enrichments across the entire results obtained from linear regression. To this end, we used GSEA, which has several advantages compared to enrichment methods working on a set of genes of interest. First, no cutoff has to be chosen to determine the gene set for enrichment testing, which makes the whole analysis less arbitrary. Second, GSEA utilizes much more information because the entire linear regression results serve as input, namely, a list of genes ranked by their correlation with SUVmax. Using this ranking of genes, the direction of the association is considered: whether enrichment for a gene set occurs at the top of the list (among genes that correlate positively with SUVmax) or at the bottom (among genes that correlate negatively with SUVmax). Thus, GSEA provides a broader view of which functional gene sets are related to glucose uptake, compared to the analysis limited to the 23 genes with high effect size.

With GSEA, we tested different categories of gene sets. First, we used "hallmark gene sets" (H), which provided initial insight and a general overview of all categories. We then focused on more specific aspects. To investigate pathways, we used "curated gene sets: canonical pathway" (C2cp), which summarized pathways from five databases (BIOCARTA, KEGG, PID, REACTOME, and WikiPathways). We also scanned for transcription factors (TFs) whose targets were positively or negatively correlated with SUVmax. Thus, we used "regulatory target gene sets: TF targets" (C3tft), which contained gene sets that shared TF binding sites or motifs. Finally, we tested "ontology gene sets" (C5) containing terms from GO and the Human Phenotype Ontology (HPO).

The results of GSEA—the number of enriched terms found for each category—is listed in Supplementary Table S4. Across all categories, we observed more significant enrichments among genes with positive regression slope, i.e., genes upregulated in tumors with higher SUVmax, than enrichments among genes with negative regression slope, i.e., genes upregulated in tumors with lower SUVmax. In H, there were three terms enriched among genes with positive slope, in C2cp there were 16 terms, and in C5 there were 71 terms—all representing basic functions such as cell cycle, DNA replication and repair, transcription, cytoskeleton, actin–myosin interaction, muscle, and muscle development. These processes seem to be upregulated in EwS cells relative to increased glucose uptake and reflect raised metabolic activity.

However, we also found enrichments that cover more specific aspects. Looking at category C2cp for canonical pathways, two pathways were significantly enriched among genes with negative slope (Figure 6a): "REACTOME_PEPTIDE_LIGAND_BINDING_RECEPTORS" (NES −1.90; q = 0.042) and "WP_GPCRS_CLASS_A_RHODOPSINLIKE" (NES −1.86; q = 0.036), which both involve rhodopsin-like GPCRs. For both terms, the enrichments were predominantly due to genes of the NPY signaling axis (so-called core enrichment, Supplementary Table S5a,b). These findings mirrored the results of the previous enrichment analysis of Enrichr on the 23 genes with high effect size.

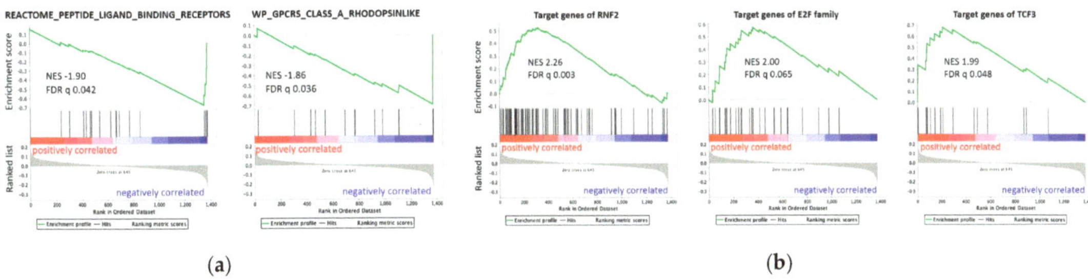

Figure 6. Enrichment plots of gene set enrichment analysis (GSEA) analysis on 1376 genes ordered by positive to negative correlation with SUVmax. (**a**) For canonical pathways (C2cp), 2 gene sets of rhodopsin-like receptors are enriched among genes that are negatively associated with SUVmax: "REACTOME_PEPTIDE_LIGAND_BINDING_RECEPTORS" and "WP_GPCRS_CLASS_A_RHODOPSINLIKE". (**b**) Three transcription factors (in C3tft) show enrichment of their target genes among genes that are positively associated with SUVmax: RNF2, E2F family, TCF3.

Furthermore, in category C3tft of TF targets, three terms showed significant enrichment. The targets of *RNF2* (NES 2.26; q = 0.003), of the E2F family of TFs (NES 2.00; q = 0.065), and of *TCF3* (NES 1.99; q = 0.048) all showed enrichment among genes positively associated with SUVmax (Figure 6b, Supplementary Table S5c–e). This may indicate that the activity of *RNF2*, the E2F family, and *TCF3* was related to the glucose uptake by EwS cells.

In summary, we identified pathways and TFs that characterized tumors in terms of glucose uptake. Tumors with higher glucose uptake had more terms than tumors with lower glucose uptake, especially terms referring to increased turnover, such as cell cycle, replication, and transcription. This may also reflect the increased glucose uptake. Furthermore, the activity of the three TFs *RNF2*, the E2F family, and *TCF3* might be positively associated with glucose uptake, whereas rhodopsin-like receptor pathways might be negatively associated with glucose uptake in EwS tumors.

4. Discussion

Our large-scale analysis comprehensively examined the correlations between gene expression and variable glucose uptake from PET. EwS has long been known for its predilection for glucose utilization and FDG-PET sensitivity [72], and inhibiting glycolysis was promising [73]. We found genes, signaling pathways, and TFs characterizing EwS tumors in terms of SUVmax, i.e., glucose uptake and consecutive glycolysis.

Most studies investigating the potential prognostic value of PET have in common that high SUV values indicate a more aggressive stage of disease and worse survival, although the exact cutoff values vary. Nevertheless, it is established that higher glucose uptake indicates malignancy, in general associated with the Warburg effect [74,75]. As PET signals such as SUVmax are supposed to be prognostic in EwS, we anticipate that our results provide novel insight to explain the mechanisms of malignancy, and potentially reveal new therapeutic options in the future.

4.1. Limitations

Due to the low prevalence, studies of EwS have a small sample size [2]. As the radiogenomics approach is not established for pediatric sarcoma yet, public datasets of imaging and expression data are not available. However, we are presently validating our findings on an external dataset.

Another limitation is the heterogeneity of clinical variables. There are multiple prognostic factors even in EwS, such as age, ethnic background, localized or metastatic disease, primary site, tumor volume, response to therapy, and primary disease or recurrence, in which the time to recurrence also has an influence [76]. With this complex interplay of risk factors, it is futile to stratify patients just by one risk factor. Limited sample sizes do not allow to create subgroups with respect to multiple risk factors. To compensate for this, we checked for SUVmax distribution with respect to risk factors that were available for our dataset. Among the clinical variables, we found no significant differences in SUVmax distribution, suggesting that single risk factors do not introduce an obvious bias into our data.

A third limitation is the skewed SUVmax distribution in our cohort. The sample with a very high SUVmax of 21.3 (sample_19) may influence the linear regression more than one of the other samples with SUVmax between 1.9 to 13.6. To evaluate the sensitivity of our findings towards sample_19, we compared the results from regression analysis with and without sample_19. We conclude that the sample with high SUVmax has an impact on the correlation results in some cases. The significant results rely on the presence of sample_19, while the findings with high effect size are mostly independent of it. However, there is no reason to consider that the results based on the 18 samples are more correct than the results based on the 19 samples. For completeness, we have pointed out the problem that might be caused by the skewed SUVmax distribution in our cohort. To dispel doubts and to validate the results, we propose to repeat this analysis on a larger dataset with a better-covered range of SUVmax values.

As a consequence of the small, heterogeneous cohort, we focused on the results of robust methods, such as functional enrichment analyses. Based on our findings, we hypothesized patterns of gene expression associated with metabolic activity.

4.2. Negative Correlation of NPY Axis and SUVmax in Enrichment Analysis

In our enrichment analyses, the expression of genes in the NPY signaling axis and rhodopsin-like GPCRs were found to be decreased as glucose uptake indicating glycolysis increased.

NPY signaling utilizes rhodopsin-like receptors, thereby promoting inflammation [77,78] and differentiation. The role of *NPY* and its receptors in cancer is not completely understood. NPY receptors are overexpressed in different cancer entities [79]—yet studies are sometimes contradictory and suggest a very context-specific role of NPY signaling in cancers other than EwS [80–82].

NPY pathway expression and function in EwS has been studied as well [14,83–87]. *NPY* and its receptors *NPY1R* and *NPY5R* are targets of *EWS-FLI1*, and therefore upregulated in EwS [86,88]. NPY signaling was shown to foster bone metastasis in vivo. The pro-metastatic and proliferation signaling of NPY was conveyed by the receptors *NPY2R* and *NPY5R* [87,88]. However, Tilan et al. [86] showed in vitro that NPY signaling via the receptors *NPY1R* and *NPY5R* promoted cell death. A survival analysis of a publicly available EwS dataset on the R2 platform (Savola dataset, n = 44) showed significantly longer overall and event-free survival for tumors with high *NPY1R* or *NPY5R* expression. In our cohort, the expression of *NPY, NPY1R,* and *NPY5R* was negatively associated with SUVmax. We infer from this negative association that there is no uniform upregulation of these genes in all EwS tumors. Instead, the expression seems to be related to the glucose uptake of the tumor. We hypothesize that there is more *NPY* signaling promoting cell death in EwS tumors with low glucose uptake, possibly associated with neuroectodermal differentiation.

4.3. Spectrum of Stemness to Differentiation

Strikingly, many of our findings had a link to stemness or differentiation.

A total of 4/5 genes that were significantly associated with SUVmax in our EwS cohort have functions regarding stemness or differentiation: *FAXDC2* plays a role in megakaryocyte differentiation; *NETO2* activates tumorigenic, stemness-related signaling pathways [89,90]; *ELOVL2* is upregulated in glioma stem cells in glioblastoma, which is mediated by stem cell enhancers like *SOX2* [91]; *MYBL2* maintains an undifferentiated state of cells [92].

Furthermore, the three TFs whose targets were enriched among the genes that were positively correlated with SUVmax are associated with maintaining stemness: *RNF2* [12,93], the E2F family [94], and *TCF3* [95].

Because of these functional links described in the literature, we consider the possibility that the genes and TFs associated with stemness in other tumors or tissues are also involved in stemness in EwS, and may imply increased stemness in tumors with higher glucose uptake. This hypothesis demands further experimental validation.

Additionally, based on the findings of the NPY axis, we speculate about potential neuroectodermal and endothelial differentiation of cells at lower SUVmax. Similar findings were reported in studies on esophageal cancer [29] and lung carcinoma [96], where the authors found a correlation of high SUVmax and poorly differentiated tumors.

A stem cell-like phenotype is a basic characteristic of EwS, and maintaining stemness plays an important role [9,10,12]. Our group [12] and Sheffield et al. [19] found that EwS tumors exhibit a spectrum of stemness varying from stem-like towards mesenchymal, neuroectodermal, or endothelial differentiation. However, correlation analyses with outcome were not performed. An analysis of stemness and prognosis was conducted by Stahl et al. [11], who found a trend of high *HIF1A* expression—reflecting stemness—and shortened overall and event-free survival in EwS. This is in concordance with the general observation that stemness in cancer is predominantly associated with poor prognosis [97].

Given our findings described above, we hypothesize that the spectrum of stemness to differentiation, which was described for EwS tumors by Sheffield et al. [19], could be reflected in the SUVmax of the tumor indicating glycolytic activity. Note that we observe correlations on our study where cause and effect are unknown.

5. Conclusions

In our study, we characterized EwS tumors in terms of their variable glucose uptake measured as SUVmax in PET. With this, we aimed to identify novel mechanistic candidates, as SUVmax can be prognostic based on the current literature. Since EwS tumors are quite uniform at the genomic level, we assessed gene expression. Due to the low incidence of EwS, we focused on the results of enrichment analyses, as these are more robust to single false positive findings. Thus, we identified correlations between SUVmax and neuroectodermal signaling pathways or pathways downstream of stemness-related TFs. We hypothesize that stemness may be associated with increased glucose uptake. Furthermore, increased differentiation may correlate with low glucose uptake. These mechanistic candidates warrant further validation in an external cohort, as well as in experimental and clinical settings. They may eventually lead to new therapeutic options.

Our study tested the potential of a radiogenomic approach to discover novel candidates to explain the mechanisms of malignancy in EwS. Prospective data collection and validation is needed in clinical practice in the future. Thus, if we are to reap the benefits of prospective large-scale analyses, future clinical practice must be adapted accordingly.

Supplementary Materials: The following supporting information can be downloaded at: https://www.mdpi.com/article/10.3390/cancers14235999/s1, Supplementary Figure S1: Survival analysis. Supplementary Figure S2: Volcano plot showing results from linear regression modeling of SUVmax and gene expression. Supplementary. Figure S3: Correlation of SUVmax and the NPY axis. Supplementary Table S1: Patients' metadata including clinical data, genetic data, tissue data, survival data, and image data. Supplementary Table S2: List of 1491 "potential survival genes" from external datasets, of which 1376 are present in our cohort. Supplementary Table S3: Results from linear regression analysis correlating gene expression and SUVmax. Supplementary Table S4: Results from GSEA analysis on 1376 genes ranked by their correlation with SUVmax. Supplementary Table S5: List of genes in GSEA core enrichments, which contribute most to the enrichments found.

Author Contributions: Conceptualization, I.v.L., G.H.S.R., W.W. (Wolfgang Weber) and S.B.; Data curation, C.P. and M.S.K.; Formal analysis, C.P. and J.E.-S.; Funding acquisition, S.B.; Investigation, C.P., M.S.K. and J.E.-S.; Methodology, C.P.; Project administration, I.v.L., W.W. (Wolfgang Weber) and S.B.; Resources, W.R., K.S., K.W., W.W. (Wilko Weichert), C.R. and J.H.; Supervision, S.B.; Validation, C.P. and J.E.-S.; Visualization, C.P.; Writing—original draft, C.P. and S.B.; Writing—review and editing, C.P., M.S.K., J.E.-S., W.R., I.v.L., C.R., J.H., G.H.S.R., W.W. (Wolfgang Weber) and S.B. All authors have read and agreed to the published version of the manuscript.

Funding: This research was funded by Wilhelm Sander-Stiftung (#2015.028.1) and Cura Placida Children's Cancer Research Foundation.

Institutional Review Board Statement: The registry study was conducted according to the guidelines of the Declaration of Helsinki and approved by the Institutional Ethics Committee of TUM School of Medicine (reference 223/16S, 05-31-2016).

Informed Consent Statement: Informed consent was obtained from all subjects involved in the study.

Data Availability Statement: The gene expression data generated for this study have been deposited in NCBI's Gene Expression Omnibus and are accessible through GEO Series accession number GSE218758 (https://www.ncbi.nlm.nih.gov/geo/query/acc.cgi?acc=GSE218758) accessed on 17 October 2022. Raw PET data were generated at the Department of Nuclear Medicine, Klinikum rechts der Isar, Technische Universität München, Munich, Germany. Raw PETs are not publicly available due to patient privacy requirements. The derived data are available within the article. In addition, publicly available datasets were analyzed in this study. These data can be found in the Gene Expression Omnibus database through GEO Series accession numbers GSE63155, GSE17618, and GSE63156.

Acknowledgments: We thank all colleagues who participated in this project for their help and support, fruitful discussions, and professional contribution: Wiebke Seemann, Sebastian Kleiner, Oxana Schmidt, Katja Gall, Denise Lasher-Epple, Hans Rechl, Carolin Knebel, Mona Mustafa, Markus Schwaiger, Uta Dirksen, and Hans-Werner Mewes. Special thanks go to Nadine Kliese for her administrative support, to Julia Hauer for her general support and advice, and to all employees of the Tissue Bank of the Klinikum rechts der Isar and the Technical University of Munich, especially Melanie Boxberg, Julia Slotta-Huspenina, Karl-Friedrich Becker, and Katja Steiger. Last but not least, we thank all patients and their families for providing their data for research.

Conflicts of Interest: Stefan Burdach has an ownership interest in PDL BioPharma and served as a consultant to EOS Biotechnology Inc., Bayer AG, and Swedish Orphan Biovitrum AB. Stefan Burdach held US and EU intellectual properties in gene-expression analysis. The other authors declare no conflict of interest.

References

1. Grünewald, T.G.P.; Cidre-Aranaz, F.; Surdez, D.; Tomazou, E.M.; de Álava, E.; Kovar, H.; Sorensen, P.H.; Delattre, O.; Dirksen, U. Ewing sarcoma. *Nat. Rev. Dis. Prim.* **2018**, *4*, 5. [CrossRef]
2. Tirode, F.; Surdez, D.; Ma, X.; Parker, M.; Le Deley, M.C.; Bahrami, A.; Zhang, Z.; Lapouble, E.; Grossetête-Lalami, S.; Rusch, M.; et al. Genomic Landscape of Ewing Sarcoma Defines an Aggressive Subtype with Co-Association of STAG2 and TP53 Mutations. *Cancer Discov.* **2014**, *4*, 1342–1353. [CrossRef]
3. Schmidkonz, C.; Krumbholz, M.; Atzinger, A.; Cordes, M.; Goetz, T.I.; Prante, O.; Ritt, P.; Schaefer, C.; Agaimy, A.; Hartmann, W.; et al. Assessment of treatment responses in children and adolescents with Ewing sarcoma with metabolic tumor parameters derived from 18F-FDG-PET/CT and circulating tumor DNA. *Eur. J. Nucl. Med. Mol. Imaging* **2020**, *47*, 1564–1575. [CrossRef]

4. Burdach, S.; Jürgens, H.; Peters, C.; Nürnberger, W.; Mauz-Körholz, C.; Körholz, D.; Paulussen, M.; Pape, H.; Dilloo, D.; Koscielniak, E. Myeloablative radiochemotherapy and hematopoietic stem-cell rescue in poor-prognosis Ewing's sarcoma. *J. Clin. Oncol.* **1993**, *11*, 1482–1488. [CrossRef]
5. Koch, R.; Gelderblom, H.; Haveman, L.; Brichard, B.; Jürgens, H.; Cyprova, S.; Berg, H.V.D.; Hassenpflug, W.; Raciborska, A.; Ek, T.; et al. High-Dose Treosulfan and Melphalan as Consolidation Therapy Versus Standard Therapy for High-Risk (Metastatic) Ewing Sarcoma. *J. Clin. Oncol.* **2022**, *40*, 2307–2320. [CrossRef]
6. Haveman, L.M.; van Ewijk, R.; van Dalen, E.C.; Breunis, W.B.; Kremer, L.C.; Berg, H.V.D.; Dirksen, U.; Merks, J.H. High-dose chemotherapy followed by autologous haematopoietic cell transplantation for children, adolescents, and young adults with primary metastatic Ewing sarcoma. *Cochrane Database Syst. Rev.* **2021**, *9*, CD011405. [CrossRef] [PubMed]
7. Burdach, S.E.G.; Westhoff, M.-A.; Steinhauser, M.F.; Debatin, K.-M. Precision medicine in pediatric oncology. *Mol. Cell. Pediatr.* **2018**, *5*, 6. [CrossRef]
8. Burdach, S. Molecular Precision Chemotherapy: Overcoming Resistance to Targeted Therapies? *Clin. Cancer Res.* **2014**, *20*, 1064–1066. [CrossRef]
9. Sand, L.G.L.; Szuhai, K.; Hogendoorn, P.C.W. Sequencing Overview of Ewing Sarcoma: A Journey across Genomic, Epigenomic and Transcriptomic Landscapes. *Int. J. Mol. Sci.* **2015**, *16*, 16176–16215. [CrossRef]
10. Agelopoulos, K.; Richter, G.H.; Schmidt, E.; Dirksen, U.; von Heyking, K.; Moser, B.; Klein, H.-U.; Kontny, U.; Dugas, M.; Poos, K.; et al. Deep Sequencing in Conjunction with Expression and Functional Analyses Reveals Activation of FGFR1 in Ewing Sarcoma. *Clin. Cancer Res.* **2015**, *21*, 4935–4946. [CrossRef]
11. Stahl, D.; Gentles, A.J.; Thiele, R.; Gütgemann, I. Prognostic profiling of the immune cell microenvironment in Ewing's Sarcoma Family of Tumors. *OncoImmunology* **2019**, *8*, e1674113. [CrossRef]
12. Richter, G.H.; Plehm, S.; Fasan, A.; Rössler, S.; Unland, R.; Bennani-Baiti, I.M.; Hotfilder, M.; Löwel, D.; von Luettichau, I.; Mossbrugger, I.; et al. EZH2 is a mediator of EWS/FLI1 driven tumor growth and metastasis blocking endothelial and neuro-ectodermal differentiation. *Proc. Natl. Acad. Sci. USA* **2009**, *106*, 5324–5329. [CrossRef]
13. Ewing, J. Diffuse endothelioma of bone. *CA Cancer J. Clin.* **1972**, *22*, 95–98. [CrossRef]
14. Staege, M.S.; Hutter, C.; Neumann, I.; Foja, S.; Hattenhorst, U.E.; Hansen, G.; Afar, D.; Burdach, S.E.G. DNA Microarrays Reveal Relationship of Ewing Family Tumors to Both Endothelial and Fetal Neural Crest-Derived Cells and Define Novel Targets. *Cancer Res.* **2004**, *64*, 8213–8221. [CrossRef]
15. Schmidt, D.; Harms, D.; Burdach, S. Malignant peripheral neuroectodermal tumours of childhood and adolescence. *Virchows Arch.* **1985**, *406*, 351–365. [CrossRef] [PubMed]
16. Crompton, B.D.; Stewart, C.; Taylor-Weiner, A.; Alexe, G.; Kurek, K.C.; Calicchio, M.L.; Kiezun, A.; Carter, S.L.; Shukla, S.A.; Mehta, S.S.; et al. The Genomic Landscape of Pediatric Ewing Sarcoma. *Cancer Discov.* **2014**, *4*, 1326–1341. [CrossRef]
17. Delattre, O.; Zucman, J.; Plougastel, B.; Desmaze, C.; Melot, T.; Peter, M.; Kovar, H.; Joubert, I.; De Jong, P.; Rouleau, G.; et al. Gene fusion with an ETS DNA-binding domain caused by chromosome translocation in human tumours. *Nature* **1992**, *359*, 162–165. [CrossRef]
18. Burchill, S.A. Molecular abnormalities in Ewing's sarcoma. *Expert Rev. Anticancer Ther.* **2008**, *8*, 1675–1687. [CrossRef]
19. Sheffield, N.C.; Pierron, G.; Klughammer, J.; Datlinger, P.; Schönegger, A.; Schuster, M.; Hadler, J.; Surdez, D.; Guillemot, D.; Lapouble, E.; et al. DNA methylation heterogeneity defines a disease spectrum in Ewing sarcoma. *Nat. Med.* **2017**, *23*, 386–395. [CrossRef]
20. Singh, I.; Bikas, A.; Garcia, C.A.; Desale, S.; Wartofsky, L.; Burman, K.D. 18F-FDG-PET Suv as a Prognostic Marker of Increasing Size in Thyroid Cancer Tumors. *Endocr. Pract. Off. J. Am. Coll. Endocrinol. Am. Assoc. Clin. Endocrinol.* **2017**, *23*, 182–189. [CrossRef]
21. Floberg, J.; Zhang, J.; DeWees, T.; Markovina, S.; Grigsby, P.; Schwarz, J. Pre-Treatment [F-18]FDG-PET SUVmax as a Prognostic and Radiogenomic Marker in Cervical Cancer. *Int. J. Radiat. Oncol.* **2018**, *102*, S82–S83. [CrossRef]
22. Diao, W.; Tian, F.; Jia, Z. The prognostic value of SUVmax measuring on primary lesion and ALN by 18F-FDG PET or PET/CT in patients with breast cancer. *Eur. J. Radiol.* **2018**, *105*, 1–7. [CrossRef]
23. Zhang, Q.; Gao, X.; Wei, G.; Qiu, C.; Qu, H.; Zhou, X. Prognostic Value of MTV, SUVmax and the T/N Ratio of PET/CT in Patients with Glioma: A Systematic Review and Meta-Analysis. *J. Cancer* **2019**, *10*, 1707–1716. [CrossRef]
24. Perrone, A.M.; Dondi, G.; Lima, G.M.; Castellucci, P.; Tesei, M.; Coluccelli, S.; Gasparre, G.; Porcelli, A.M.; Nanni, C.; Fanti, S.; et al. Potential Prognostic Role of 18F-FDG PET/CT in Invasive Epithelial Ovarian Cancer Relapse. A Preliminary Study. *Cancers* **2019**, *11*, 713. [CrossRef]
25. Kim, C.-Y.; Jeong, S.Y.; Chong, G.O.; Son, S.H.; Jung, J.-H.; Kim, D.-H.; Lee, S.-W.; Ahn, B.-C.; Lee, J. Quantitative metabolic parameters measured on F-18 FDG PET/CT predict survival after relapse in patients with relapsed epithelial ovarian cancer. *Gynecol. Oncol.* **2015**, *136*, 498–504. [CrossRef]
26. Torizuka, T.; Tanizaki, Y.; Kanno, T.; Futatsubashi, M.; Naitou, K.; Ueda, Y.; Ouchi, Y. Prognostic Value of ^{18}F-FDG PET in Patients with Head and Neck Squamous Cell Cancer. *Am. J. Roentgenol.* **2009**, *192*, W156–W160. [CrossRef]
27. Pankowska, V.; Malkowski, B.; Wedrowski, M.; Wedrowska, E.; Roszkowski, K. FDG PET/CT as a survival prognostic factor in patients with advanced renal cell carcinoma. *Clin. Exp. Med.* **2019**, *19*, 143–148. [CrossRef]

28. Berghmans, T.; Dusart, M.; Paesmans, M.; Hossein-Foucher, C.; Buvat, I.; Castaigne, C.; Scherpereel, A.; Mascaux, C.; Moreau, M.; Roelandts, M.; et al. Primary Tumor Standardized Uptake Value (SUVmax) Measured on Fluorodeoxyglucose Positron Emission Tomography (FDG-PET) is of Prognostic Value for Survival in Non-small Cell Lung Cancer (NSCLC): A Systematic Review and Meta-Analysis (MA) by the European Lung Cancer Working Party for the IASLC Lung Cancer Staging Project. *J. Thorac. Oncol.* **2008**, *3*, 6–12. [CrossRef]
29. Cerfolio, R.J.; Bryant, A.S. Maximum Standardized Uptake Values on Positron Emission Tomography of Esophageal Cancer Predicts Stage, Tumor Biology, and Survival. *Ann. Thorac. Surg.* **2006**, *82*, 391–394; discussion 394–395. [CrossRef] [PubMed]
30. Pan, L.; Gu, P.; Huang, G.; Xue, H.; Wu, S. Prognostic significance of SUV on PET/CT in patients with esophageal cancer: A systematic review and meta-analysis. *Eur. J. Gastroenterol. Hepatol.* **2009**, *21*, 1008–1015. [CrossRef]
31. Lee, W.; Oh, M.; Kim, J.S.; Park, Y.; Kwon, J.W.; Jun, E.; Song, K.B.; Lee, J.H.; Hwang, D.W.; Yoo, C.; et al. Metabolic activity by FDG-PET/CT after neoadjuvant chemotherapy in borderline resectable and locally advanced pancreatic cancer and association with survival. *Br. J. Surg.* **2021**, *109*, 61–70. [CrossRef] [PubMed]
32. Hellwig, D.; Gröschel, A.; Graeter, T.P.; Hellwig, A.P.; Nestle, U.; Schäfers, H.-J.; Sybrecht, G.W.; Kirsch, C.-M. Diagnostic performance and prognostic impact of FDG-PET in suspected recurrence of surgically treated non-small cell lung cancer. *Eur. J. Nucl. Med.* **2006**, *33*, 13–21. [CrossRef] [PubMed]
33. Jiang, Y.; Hou, G.; Wu, F.; Zhu, Z.; Zhang, W.; Cheng, W. The maximum standardized uptake value and extent of peritoneal involvement may predict the prognosis of patients with recurrent ovarian cancer after primary treatment: A retrospective clinical study. *Medicine* **2020**, *99*, e19228. [CrossRef] [PubMed]
34. Sala, E.; Kataoka, M.; Pandit-Taskar, N.; Ishill, N.; Mironov, S.; Moskowitz, C.S.; Mironov, O.; Collins, M.A.; Chi, D.S.; Larson, S.; et al. Recurrent Ovarian Cancer: Use of Contrast-enhanced CT and PET/CT to Accurately Localize Tumor Recurrence and to Predict Patients' Survival. *Radiology* **2010**, *257*, 125–134. [CrossRef] [PubMed]
35. Franzius, C.; Schulte, M.; Hillmann, A.; Winkelmann, W.; Jürgens, H.; Bockisch, A.; Schober, O. Clinical value of positron emission tomography (PET) in the diagnosis of bone and soft tissue tumors. 3rd Interdisciplinary Consensus Conference "PET in Oncology": Results of the Bone and Soft Tissue Study Group. *Chirurgie* **2001**, *72*, 1071–1077. [CrossRef] [PubMed]
36. Hwang, J.P.; Lim, I.; Kong, C.-B.; Jeon, D.G.; Byun, B.H.; Kim, B.I.; Choi, C.W.; Lim, S.M. Prognostic Value of SUVmax Measured by Pretreatment Fluorine-18 Fluorodeoxyglucose Positron Emission Tomography/Computed Tomography in Patients with Ewing Sarcoma. *PLoS ONE* **2016**, *11*, e0153281. [CrossRef] [PubMed]
37. Jamet, B.; Carlier, T.; Campion, L.; Bompas, E.; Girault, S.; Borrely, F.; Ferrer, L.; Rousseau, M.; Venel, Y.; Kraeber-Bodéré, F.; et al. Initial FDG-PET/CT predicts survival in adults Ewing sarcoma family of tumors. *Oncotarget* **2017**, *8*, 77050–77060. [CrossRef]
38. Raciborska, A.; Bilska, K.; Drabko, K.; Michalak, E.; Chaber, R.; Pogorzala, M.; Polczynska, K.; Sobol, G.; Wieczorek, M.; Muszynskaroslan, K.; et al. Response to chemotherapy estimates by FDG PET is an important prognostic factor in patients with Ewing sarcoma. *Clin. Transl. Oncol.* **2016**, *18*, 189–195. [CrossRef]
39. Hawkins, D.S.; Schuetze, S.M.; Butrynski, J.E.; Rajendran, J.G.; Vernon, C.B.; Conrad, E.U.; Eary, J.F. [^{18}F]Fluorodeoxyglucose Positron Emission Tomography Predicts Outcome for Ewing Sarcoma Family of Tumors. *J. Clin. Oncol.* **2005**, *23*, 8828–8834. [CrossRef]
40. Salem, U.; Amini, B.; Chuang, H.H.; Daw, N.C.; Wei, W.; Haygood, T.M.; Madewell, J.E.; Costelloe, C.M. ^{18}F-FDG PET/CT as an Indicator of Survival in Ewing Sarcoma of Bone. *J. Cancer* **2017**, *8*, 2892–2898. [CrossRef]
41. Annovazzi, A.; Ferraresi, V.; Anelli, V.; Covello, R.; Vari, S.; Zoccali, C.; Biagini, R.; Sciuto, R. [18F]FDG PET/CT quantitative parameters for the prediction of histological response to induction chemotherapy and clinical outcome in patients with localised bone and soft-tissue Ewing sarcoma. *Eur. Radiol.* **2021**, *31*, 7012–7021. [CrossRef]
42. Macpherson, R.E.; Pratap, S.; Tyrrell, H.; Khonsari, M.; Wilson, S.; Gibbons, M.; Whitwell, D.; Giele, H.; Critchley, P.; Cogswell, L.; et al. Retrospective audit of 957 consecutive 18F-FDG PET–CT scans compared to CT and MRI in 493 patients with different histological subtypes of bone and soft tissue sarcoma. *Clin. Sarcoma Res.* **2018**, *8*, 9. [CrossRef] [PubMed]
43. Bolstad, B.M.; Irizarry, R.A.; Astrand, M.; Speed, T.P. A comparison of normalization methods for high density oligonucleotide array data based on variance and bias. *Bioinformatics* **2003**, *19*, 185–193. [CrossRef] [PubMed]
44. Irizarry, R.A.; Bolstad, B.M.; Collin, F.; Cope, L.M.; Hobbs, B.; Speed, T.P. Summaries of Affymetrix GeneChip probe level data. *Nucleic Acids Res.* **2003**, *31*, e15. [CrossRef] [PubMed]
45. Irizarry, R.A.; Hobbs, B.; Collin, F.; Beazer-Barclay, Y.D.; Antonellis, K.J.; Scherf, U.; Speed, T. Exploration, normalization, and summaries of high density oligonucleotide array probe level data. *Biostatistics* **2003**, *4*, 249–264. [CrossRef] [PubMed]
46. Carvalho, B.S.; Irizarry, R.A. A framework for oligonucleotide microarray preprocessing. *Bioinformatics* **2010**, *26*, 2363–2367. [CrossRef] [PubMed]
47. Dai, M.; Wang, P.; Boyd, A.D.; Kostov, G.; Athey, B.; Jones, E.G.; Bunney, W.E.; Myers, R.M.; Speed, T.P.; Akil, H.; et al. Evolving gene/transcript definitions significantly alter the interpretation of GeneChip data. *Nucleic Acids Res.* **2005**, *33*, e175. [CrossRef]
48. Lu, X.; Zhang, X. The effect of GeneChip gene definitions on the microarray study of cancers. *BioEssays* **2006**, *28*, 739–746. [CrossRef]
49. Sandberg, R.; Larsson, O. Improved precision and accuracy for microarrays using updated probe set definitions. *BMC Bioinform.* **2007**, *8*, 48. [CrossRef]
50. Draghici, S.; Khatri, P.; Eklund, A.C.; Szallasi, Z. Reliability and reproducibility issues in DNA microarray measurements. *Trends Genet.* **2006**, *22*, 101–109. [CrossRef]

51. Edgar, R.; Domrachev, M.; Lash, A.E. Gene Expression Omnibus: NCBI gene expression and hybridization array data repository. *Nucleic Acids Res.* **2002**, *30*, 207–210. [CrossRef] [PubMed]
52. Volchenboum, S.L.; Andrade, J.; Huang, L.; Barkauskas, D.A.; Krailo, M.; Womer, R.B.; Ranft, A.; Potratz, J.; Dirksen, U.; Triche, T.J.; et al. Gene expression profiling of Ewing sarcoma tumours reveals the prognostic importance of tumour-stromal interactions: A report from the Children's Oncology Group. *J. Pathol. Clin. Res.* **2015**, *1*, 83–94. [CrossRef] [PubMed]
53. Savola, S.; Klami, A.; Myllykangas, S.; Manara, C.; Scotlandi, K.; Picci, P.; Knuutila, S.; Vakkila, J. High Expression of Complement Component 5 (*C5*) at Tumor Site Associates with Superior Survival in Ewing's Sarcoma Family of Tumour Patients. *ISRN Oncol.* **2011**, *2011*, 168712. [CrossRef] [PubMed]
54. Kuhn, M. Caret: Classification and Regression Training, R package version 6.0-90. Available online: https://CRAN.R-project.org/package=caret:2021 (accessed on 1 November 2021).
55. Huang, D.W.; Sherman, B.T.; Lempicki, R.A. Systematic and integrative analysis of large gene lists using DAVID bioinformatics resources. *Nat. Protoc.* **2009**, *4*, 44–57. [CrossRef] [PubMed]
56. Sherman, B.T.; Hao, M.; Qiu, J.; Jiao, X.; Baseler, M.W.; Lane, H.C.; Imamichi, T.; Chang, W. DAVID: A web server for functional enrichment analysis and functional annotation of gene lists (2021 update). *Nucleic Acids Res.* **2022**, *50*, W216–W221. [CrossRef] [PubMed]
57. Rosset, A.; Spadola, L.; Ratib, O. OsiriX: An Open-Source Software for Navigating in Multidimensional DICOM Images. *J. Digit. Imaging* **2004**, *17*, 205–216. [CrossRef]
58. Zwanenburg, A.; Vallières, M.; Abdalah, M.A.; Aerts, H.J.W.L.; Andrearczyk, V.; Apte, A.; Ashrafinia, S.; Bakas, S.; Beukinga, R.J.; Boellaard, R.; et al. The Image Biomarker Standardization Initiative: Standardized Quantitative Radiomics for High-Throughput Image-based Phenotyping. *Radiology* **2020**, *295*, 328–338. [CrossRef]
59. van Griethuysen, J.J.M.; Fedorov, A.; Parmar, C.; Hosny, A.; Aucoin, N.; Narayan, V.; Beets-Tan, R.G.H.; Fillion-Robin, J.-C.; Pieper, S.; Aerts, H.J.W.L. Computational Radiomics System to Decode the Radiographic Phenotype. *Cancer Res.* **2017**, *77*, e104–e107. [CrossRef]
60. Therneau, T.M. A Package for Survival Analysis in R, R package version 3.2-13. Available online: https://CRAN.R-project.org/package=survival:2021 (accessed on 1 September 2021).
61. Therneau, T.; Grambsch, P. *Modeling Survival Data: Extending the Cox Model*; Springer: New York, NY, USA, 2000; ISBN 0-387-98784-3.
62. Kassambara, A.; Kosinski, M.; Biecek, P. Survminer: Drawing Survival Curves using 'ggplot2', R package version 0.4.9. 2021. Available online: https://CRAN.R-project.org/package=survminer (accessed on 17 October 2022).
63. Ritchie, M.E.; Belinda, P.; Wu, D.; Hu, Y.; Law, C.W.; Shi, W.; Smyth, G.K. limma powers differential expression analyses for RNA-sequencing and microarray studies. *Nucleic Acids Res.* **2015**, *43*, e47. [CrossRef]
64. Heatmap.3. Available online: https://raw.githubusercontent.com/obigriffith/biostar-tutorials/master/Heatmaps/heatmap.3.R (accessed on 1 January 2018).
65. Chen, E.Y.; Tan, C.M.; Kou, Y.; Duan, Q.; Wang, Z.; Meirelles, G.V.; Clark, N.R.; Ma'Ayan, A. Enrichr: Interactive and collaborative HTML5 gene list enrichment analysis tool. *BMC Bioinform.* **2013**, *14*, 128. [CrossRef]
66. Kuleshov, M.V.; Jones, M.R.; Rouillard, A.D.; Fernandez, N.F.; Duan, Q.; Wang, Z.; Koplev, S.; Jenkins, S.L.; Jagodnik, K.M.; Lachmann, A.; et al. Enrichr: A comprehensive gene set enrichment analysis web server 2016 update. *Nucleic Acids Res.* **2016**, *44*, W90–W97. [CrossRef] [PubMed]
67. Xie, Z.; Bailey, A.; Kuleshov, M.V.; Clarke, D.J.B.; Evangelista, J.E.; Jenkins, S.L.; Lachmann, A.; Wojciechowicz, M.L.; Kropiwnicki, E.; Jagodnik, K.M.; et al. Gene Set Knowledge Discovery with Enrichr. *Curr. Protoc.* **2021**, *1*, e90. [CrossRef]
68. Subramanian, A.; Tamayo, P.; Mootha, V.K.; Mukherjee, S.; Ebert, B.L.; Gillette, M.A.; Paulovich, A.; Pomeroy, S.L.; Golub, T.R.; Lander, E.S.; et al. Gene set enrichment analysis: A knowledge-based approach for interpreting genome-wide expression profiles. *Proc. Natl. Acad. Sci. USA* **2005**, *102*, 15545–15550. [CrossRef] [PubMed]
69. Mootha, V.K.; Lindgren, C.M.; Eriksson, K.-F.; Subramanian, A.; Sihag, S.; Lehar, J.; Puigserver, P.; Carlsson, E.; Ridderstråle, M.; Laurila, E.; et al. PGC-1α-responsive genes involved in oxidative phosphorylation are coordinately downregulated in human diabetes. *Nat. Genet.* **2003**, *34*, 267–273. [CrossRef] [PubMed]
70. R Core Team. *R: A language and Environment for Statistical Computing*; R Foundation for Statistical Computing: Vienna, Austria, 2020; Available online: https://www.R-project.org/ (accessed on 1 July 2020).
71. Human Protein Atlas. Available online: www.proteinatlas.org (accessed on 7 December 2021).
72. Burdach, S.E.G.; Combs, S.E.; Dürr, H.R.; Eisenhart-Rothe, R.v.; Lindner, L.H. Tumorzentrum München an den Medizinischen Fakultäten der Ludwig-Maximilians-Universität und der Technischen Universität München. Klinik und multidisziplinäre Therapie der Ewing-Sarkome, MANUAL Knochentumoren und Weichteilsarkome. Zuckschwerdt Verlag: München, Germany, 2017.
73. Yeung, C.; Gibson, A.E.; Issaq, S.H.; Oshima, N.; Baumgart, J.T.; Edessa, L.D.; Rai, G.; Urban, D.J.; Johnson, M.S.; Benavides, G.A.; et al. Targeting Glycolysis through Inhibition of Lactate Dehydrogenase Impairs Tumor Growth in Preclinical Models of Ewing Sarcoma. *Cancer Res.* **2019**, *79*, 5060–5073. [CrossRef] [PubMed]
74. Warburg, O. On Respiratory Impairment in Cancer Cells. *Science* **1956**, *124*, 269–270. [CrossRef] [PubMed]
75. Sipol, A.; Hameister, E.; Xue, B.; Hofstetter, J.; Barenboim, M.; Öllinger, R.; Jain, G.; Prexler, C.; Rubio, R.A.; Baldauf, M.C.; et al. MondoA Drives B-ALL Malignancy through Enhanced Adaptation to Metabolic Stress. *Blood* **2022**, *139*, 1184–1197. [CrossRef] [PubMed]

76. Stahl, M.; Ranft, A.; Paulussen, M.; Bölling, T.; Vieth, V.; Bielack, S.; Görtitz, I.; Braun-Munzinger, G.; Hardes, J.; Jürgens, H.; et al. Risk of recurrence and survival after relapse in patients with Ewing sarcoma. *Pediatr. Blood Cancer* **2011**, *57*, 549–553. [CrossRef]
77. Groneberg, D.A.; Folkerts, G.; Peiser, C.; Chung, K.F.; Fischer, A. Neuropeptide Y (NPY). *Pulm. Pharmacol. Ther.* **2004**, *17*, 173–180. [CrossRef]
78. Wheway, J.; Herzog, H.; Mackay, F. NPY and Receptors in Immune and Inflammatory Diseases. *Curr. Top. Med. Chem.* **2007**, *7*, 1743–1752. [CrossRef]
79. Li, J.; Tian, Y.; Wu, A. Neuropeptide Y receptors: A promising target for cancer imaging and therapy. *Regen. Biomater.* **2015**, *2*, 215–219. [CrossRef] [PubMed]
80. Dietrich, P.; Wormser, L.; Fritz, V.; Seitz, T.; De Maria, M.; Schambony, A.; Kremer, A.E.; Günther, C.; Itzel, T.; Thasler, W.E.; et al. Molecular crosstalk between Y5 receptor and neuropeptide Y drives liver cancer. *J. Clin. Investig.* **2020**, *130*, 2509–2526. [CrossRef] [PubMed]
81. Lv, X.; Zhao, F.; Huo, X.; Tang, W.; Hu, B.; Gong, X.; Yang, J.; Shen, Q.; Qin, W. Neuropeptide Y1 receptor inhibits cell growth through inactivating mitogen-activated protein kinase signal pathway in human hepatocellular carcinoma. *Med. Oncol.* **2016**, *33*, 70. [CrossRef] [PubMed]
82. Liu, J.; Wang, X.; Sun, J.; Chen, Y.; Li, J.; Huang, J.; Du, H.; Gan, L.; Qiu, Z.; Li, H.; et al. The Novel Methylation Biomarker NPY5R Sensitizes Breast Cancer Cells to Chemotherapy. *Front. Cell Dev. Biol.* **2022**, *9*, 798221. [CrossRef] [PubMed]
83. Körner, M.; Waser, B.; Reubi, J.C. High Expression of Neuropeptide Y1 Receptors in Ewing Sarcoma Tumors. *Clin. Cancer Res.* **2008**, *14*, 5043–5049. [CrossRef]
84. van Valen, F.; Keck, E.; Jurgens, H. Neuropeptide Y inhibits vasoactive intestinal peptide- and dopamine-induced cyclic AMP formation in human Ewing's sarcoma WE-68 cells. *FEBS Lett.* **1989**, *249*, 271–274. [CrossRef]
85. Kitlinska, J. Neuropeptide Y in neural crest-derived tumors: Effect on growth and vascularization. *Cancer Lett.* **2007**, *245*, 293–302. [CrossRef]
86. Tilan, J.U.; Lu, C.; Galli, S.; Izycka-Swieszewska, E.; Earnest, J.P.; Shabbir, A.; Everhart, L.M.; Wang, S.; Martin, S.; Horton, M.; et al. Hypoxia shifts activity of neuropeptide Y in Ewing sarcoma from growth-inhibitory to growth-promoting effects. *Oncotarget* **2013**, *4*, 2487–2501. [CrossRef]
87. Tilan, J.; Kitlinska, J. Neuropeptide Y (NPY) in tumor growth and progression: Lessons learned from pediatric oncology. *Neuropeptides* **2016**, *55*, 55–66. [CrossRef]
88. Lu, C.; Mahajan, A.; Hong, S.-H.; Galli, S.; Zhu, S.; Tilan, J.U.; Abualsaud, N.; Adnani, M.; Chung, S.; Elmansy, N.; et al. Hypoxia-activated neuropeptide Y/Y5 receptor/RhoA pathway triggers chromosomal instability and bone metastasis in Ewing sarcoma. *Nat. Commun.* **2022**, *13*, 2323. [CrossRef]
89. Fedorova, M.S.; Snezhkina, A.V.; Lipatova, A.V.; Pavlov, V.S.; Kobelyatskaya, A.A.; Guvatova, Z.G.; Pudova, E.A.; Savvateeva, M.V.; Ishina, I.A.; Demidova, T.B.; et al. NETO2 Is Deregulated in Breast, Prostate, and Colorectal Cancer and Participates in Cellular Signaling. *Front. Genet.* **2020**, *11*, 594933. [CrossRef] [PubMed]
90. Wang, X.; Bian, Z.; Hou, C.; Li, M.; Jiang, W.; Zhu, L. Neuropilin and tolloid-like 2 regulates the progression of osteosarcoma. *Gene* **2021**, *768*, 145292. [CrossRef]
91. Gimple, R.C.; Kidwell, R.L.; Kim, L.J.Y.; Sun, T.; Gromovsky, A.D.; Wu, Q.; Wolf, M.; Lv, D.; Bhargava, S.; Jiang, L.; et al. Glioma Stem Cell–Specific Superenhancer Promotes Polyunsaturated Fatty-Acid Synthesis to Support EGFR Signaling. *Cancer Discov.* **2019**, *9*, 1248–1267. [CrossRef] [PubMed]
92. Musa, J.; Aynaud, M.-M.; Mirabeau, O.; Delattre, O.; Grünewald, T.G. MYBL2 (B-Myb): A central regulator of cell proliferation, cell survival and differentiation involved in tumorigenesis. *Cell Death Dis.* **2017**, *8*, e2895. [CrossRef] [PubMed]
93. Sauvageau, M.; Sauvageau, G. Polycomb Group Genes: Keeping Stem Cell Activity in Balance. *PLOS Biol.* **2008**, *6*, e113. [CrossRef]
94. Kent, L.N.; Leone, G. The broken cycle: E2F dysfunction in cancer. *Nat. Rev. Cancer* **2019**, *19*, 326–338. [CrossRef]
95. Wu, C.-I.; Hoffman, J.; Shy, B.R.; Ford, E.M.; Fuchs, E.; Nguyen, H.; Merrill, B.J. Function of Wnt/β-catenin in counteracting Tcf3 repression through the Tcf3–β-catenin interaction. *Development* **2012**, *139*, 2118–2129. [CrossRef]
96. Duan, X.-Y.; Wang, W.; Li, M.; Li, Y.; Guo, Y.-M. Predictive significance of standardized uptake value parameters of FDG-PET in patients with non-small cell lung carcinoma. *Braz. J. Med. Biol. Res.* **2015**, *48*, 267–272. [CrossRef]
97. Ge, Y.; Gomez, N.C.; Adam, R.C.; Nikolova, M.; Yang, H.; Verma, A.; Lu, C.P.-J.; Polak, L.; Yuan, S.; Elemento, O.; et al. Stem Cell Lineage Infidelity Drives Wound Repair and Cancer. *Cell* **2017**, *169*, 636–650.e614. [CrossRef]

cancers

Review

Current State of Immunotherapy and Mechanisms of Immune Evasion in Ewing Sarcoma and Osteosarcoma

Valentina Evdokimova [1,†], Hendrik Gassmann [2,†], Laszlo Radvanyi [1,*] and Stefan E. G. Burdach [2,3,4,*]

1 Ontario Institute for Cancer Research, Toronto, ON M5G 0A3, Canada
2 Department of Pediatrics, Children's Cancer Research Center, Kinderklinik München Schwabing, TUM School of Medicine, Technical University of Munich, 80804 Munich, Germany
3 Translational Pediatric Cancer Research, Institute of Pathology, Technical University of Munich, 81675 Munich, Germany
4 German Cancer Consortium (DKTK), German Cancer Research Center (DKFZ), Partner Site Munich, 81675 Munich, Germany
* Correspondence: laszlo.radvanyi@oicr.on.ca (L.R.); stefan.burdach@tum.de (S.E.G.B.)
† These authors contributed equally to this work.

Simple Summary: Pediatric sarcomas, including Ewing sarcoma and osteosarcoma, were first to be treated with an anticancer vaccine 100 years ago. This review moves on from a historical perspective to the current progress and challenges of immunotherapy in these immunologically cold bone and soft tissue sarcomas. We discuss mechanisms of immune escape and immunosuppression employed by these tumors, and the potential novel directions of the research and therapy. The intention of this review is to stimulate alternative concepts and treatment strategies.

Abstract: We argue here that in many ways, Ewing sarcoma (EwS) is a unique tumor entity and yet, it shares many commonalities with other immunologically cold solid malignancies. From the historical perspective, EwS, osteosarcoma (OS) and other bone and soft-tissue sarcomas were the first types of tumors treated with the immunotherapy approach: more than 100 years ago American surgeon William B. Coley injected his patients with a mixture of heat-inactivated bacteria, achieving survival rates apparently higher than with surgery alone. In contrast to OS which exhibits recurrent somatic copy-number alterations, EwS possesses one of the lowest mutation rates among cancers, being driven by a single oncogenic fusion protein, most frequently EWS-FLI1. In spite these differences, both EwS and OS are allied with immune tolerance and low immunogenicity. We discuss here the potential mechanisms of immune escape in these tumors, including low representation of tumor-specific antigens, low expression levels of MHC-I antigen-presenting molecules, accumulation of immunosuppressive M2 macrophages and myeloid proinflammatory cells, and release of extracellular vesicles (EVs) which are capable of reprogramming host cells in the tumor microenvironment and systemic circulation. We also discuss the vulnerabilities of EwS and OS and potential novel strategies for their targeting.

Keywords: Ewing sarcoma; osteosarcoma; immunotherapy; William Coley; tumor microenvironment; extracellular vesicles; exosome; immunosuppression; retrotransposon; human endogenous retrovirus

Citation: Evdokimova, V.; Gassmann, H.; Radvanyi, L.; Burdach, S.E.G. Current State of Immunotherapy and Mechanisms of Immune Evasion in Ewing Sarcoma and Osteosarcoma. *Cancers* **2023**, *15*, 272. https://doi.org/10.3390/cancers15010272

Academic Editor: Maria Paola Paronetto

Received: 30 November 2022
Revised: 23 December 2022
Accepted: 27 December 2022
Published: 30 December 2022

1. Historical Perspectives and Modern Immunotherapy in Pediatric Sarcomas

1.1. William Bradley Coley (1862–1936): The First Focused Effort at Cancer Immunotherapy

Long before James Stephen Ewing, one of the leading US pathologists and the Medical Director of Memorial Hospital in New York, presented in 1920 the first case of a small round blue cell "endothelioma of bone" which since then carries his name [1], William B. Coley was treating hundreds of bone and soft-tissue cancer patients with a significant rate of success [2–4]. Being a renowned surgeon and the head of the Bone Tumor Service at the same hospital (nowadays Memorial Sloan Kettering Cancer Center), he pioneered the approach of stimulating the body's "resisting powers"/immunity with the attenuated bacterial mixtures, paving the road to modern immunotherapy.

By 1896, Dr. Coley reported 160 cases of patients with inoperable malignant tumors treated with the mixture of heat-inactivated bacteria, the Gram-positive *S. pyogenes* (erysipelas) and the Gram-negative *Bacillus prodigiosus* (*Serratia marcescens*) [3], which became known as "Coley's toxins". Starting from 1899, these toxins were commercially produced by Parke Davis & Co and were widely available to the physicians in North America and Europe until 1951 [2]. Importantly, among a large variety of cancers treated with Coley's toxins, the best response was achieved in patients with inoperable bone and soft tissue sarcomas, including EwS and OS, as well as in osteolytic types of rapidly growing malignant bone tumors (endothelioma, reticulum cell sarcoma and plasma cell myeloma), while carcinomas responded poorly to the treatment [5]. Coley himself concluded that the use of toxins should be limited to sarcomas [4].

Over the course of his 45-year-long career, Coley treated thousands of patients and published more than 150 monographs on the subject, and yet, his work came under scrutiny due to a number of reasons. These included variations in bacterial strain activity and lack of standard bioassays to test their efficacy prior to injecting into patients as well as a complexity and duration of their administration, making it difficult to standardize [5]. Furthermore, at the beginning of 20th century, radiation and then chemotherapy began to emerge as promising new cancer treatments (Figure 1). James Ewing, a big proponent of radiation therapy and Coley's boss, forbid Coley from using his toxin inside the Memorial Hospital. The final blow was delivered in 1963, when the Food and Drug Administration (FDA) refused to recognize Coley's toxin as a proven drug due to the lack of safety and efficacy data, despite over 70 years of its clinical use, hundreds of publications and, most importantly, remarkable improvements achieved by this treatment, including an independent controlled study in 1962 which showed a dramatic response in 20 of 93 cancer patients [6]. His legacy was saved by his children. Coley's son Bradley succeeded him as the head of the Bone Tumor Service at the Memorial Hospital and advocated Coley's toxin as adjunctive therapy for bone sarcomas after surgery to prevent metastasis [7]. His daughter, Helen Coley Nauts became one of the two founders of the Cancer Research Institute (CRI), which was established in 1953 and since then, is a leading institution in the field of cancer immunology and immunotherapy.

Ironically, the CRI-sponsored phase I clinical trial of Coley's toxins conducted in 2012 in Germany revealed both promise and obstacles associated with its use in modern times. While the main objective of the trial was to establish safety and determine optimal dosing, one of 12 patients with metastatic bladder cancer had a clear clinical response, experiencing a 50% reduction in his cancer [8]. The team of researchers and clinicians led by Dr. Jäger faithfully recapitulated Coley's treatment protocols, of which induction of fever is the key component of therapy. Patients received subcutaneous injections of Coley's toxins twice a week, and the dose was escalated in each patient until a body temperature reached 38 °C to 39.5 °C, followed by four additional doses. Remarkably, coincident with the highest body temperature, ten of 12 patients showed a consistent increase in serum IL-6 levels, and some (including a patient with metastatic bladder cancer) also manifested an increase in TNF-α, IFN-γ and IL-1β. The team concluded that fever-induced massive release of immunoregulatory cytokines may play an important role in tumor regression and suggested further exploration of Coley's vaccine as a potent immune modulator [8]. However, further testing of Coley's vaccine is under scrutiny again. The major obstacles include the cost and complexities associated with manufacturing a bacteria-derived product in accordance with Good Clinical Practice guidelines. Most importantly, no modern hospital institutional review board will permit maintaining fever as one of the study objectives. Nevertheless, growing number of pre-clinical and clinical studies into the use of bacteria to combat cancer provides additional rationale for implementing basic principles established by Dr. William Coley 100 years ago into the clinical practice [9].

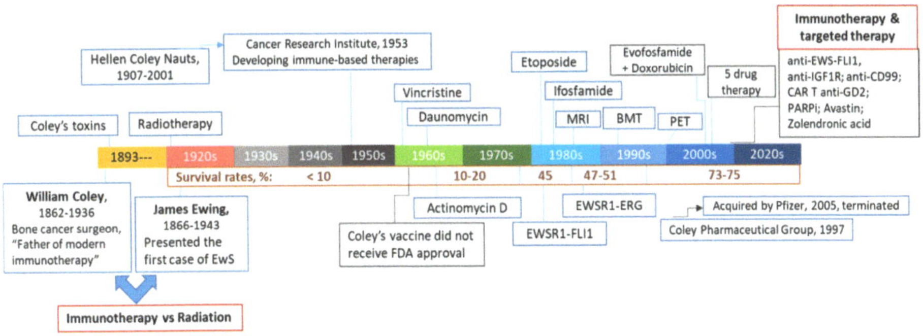

Figure 1. The 100-year-old battlefield between immunotherapy and radiotherapy in treating patients with bone and soft-tissues sarcomas. Milestones and major events are documented in a chronological order.

1.2. Conventional and Targeted Therapeutics in Treating EwS and OS

Nowadays, high-risk childhood bone cancers, including EwS and OS, range among the leading causes of cancer-related deaths in children. Despite a significant increase in the 5-year survival rates among EwS patients with localized disease from less than 10% in the pre-chemotherapy era to about 75% currently, outcomes have not significantly improved since late 1990s (Figure 1). Likewise, combination chemotherapy introduced in the 1970s for treatment of OS increased overall survival rates for patients with localized disease from 20% to approximately 70% today, however patients with metastatic or recurrent disease fare poorly with only 20% 5-year survival rates [10,11]. The standard protocols for both EwS and OS patients rely upon extensive surgical resections, high-dose multimodal chemotherapy and radiation [12–15], putting young survivors at risk of developing long-term disabilities and therapy-induced secondary cancers [16]. Furthermore, the relapse rates remain high and less than 30% of patients with metastasis survive for 5 years [12–14], urging the need for developing tailored therapy approaches, especially for these high-risk groups of patients.

In the past 20 years, multiple new-generation inhibitors of oncogenic driver proteins and pathways have been tested in clinical trials to enhance efficacy and overcome resistance to conventional therapies. Due to different reasons, targeting driver mutations in EwS and OS proved to be challenging. Unlike OS which has no clear driver and is uniformly characterized by high genomic instability and chromothripsis due to inactivating mutations in various DNA repair genes, EwS possess relatively stable genome with no recurrent mutations and gene amplifications [17]. Instead, it is driven by a single well-defined driver, the EWS-FLI1 fusion protein, which is found in ~85% of EwS cases [12,13]. Yet, EWS-FLI1 lacks enzymatic activity, and its direct chemical targeting was unsuccessful so far. Indirect EWS-FLI1 inhibition includes targeting downstream interacting proteins (e.g., RNA helicase A using YK-4-279, TK216 inhibitors), EWS-FLI1-driven epigenetic and transcriptional reprograming (e.g., using trabectedin and mithramycin), activated pathways (most notably, IGF1R and MEK), epigenetic modifiers (e.g., deacetylase, LSD, BRD4 and EZH2 inhibitors), anti-angiogenic tyrosine kinase inhibitors, and various combinatorial treatments; an overview of the past and current clinical trials is provided in [15,18,19]. Remarkably, despite their distinct biological and clinical properties, both EwS and OS exhibit gene expression profiles characteristic of BRCA1/2 mutant tumors (so-called "BRCAness") [20,21], and thus expected to be susceptible to the poly(ADP)-ribose polymerase (PARP) inhibitors olaparib and talazoparib. Unfortunately, adding these newer agents to the existing treatment protocols has not improved outcomes, with the acquired and intrinsic resistance to therapy impeding further progress [14,15]. It becomes increasingly evident that without re-activating the immune system and re-thinking the designs of clinical trials, conventional and targeted therapeutics may not substantially improve outcomes in EwS and OS, where standard-of-care has not been significantly changed in decades [10,11,22], as well as in the majority of other solid malignancies [23].

1.3. Immunotherapy of Pediatric Bone Cancers

Complete or partial tumor regression and prolonged remission in some sarcoma patients treated with Coley's vaccine provided one of the first clues that activating natural immunity may have anti-tumorigenic effects. Modern immunotherapies are aimed at building strong antitumor immunity by re-activating and amplifying pre-existing antitumor responses and inducing new responses against otherwise cryptic antigens (e.g., with checkpoint inhibitors, oncolytic viruses, and cancer vaccines) or by adoptive cell therapies, including engineered CAR T cells, TCR transgenic T cells or tumor-infiltrating lymphocytes (TIL) (Table 1). Their implementation in the past decade offered exceptional clinical benefits for patients with certain types of "immunogenic" cancers but even so there are unprecedented challenges, ranging from developing pre-clinical models, defining dominant drivers of anticancer immunity and immune escape to the off-target toxicities and the need to identify optimal combinations for any given patient [24,25]. The majority of solid tumors remain immune "cold" and at the outset unable to generate an appreciable de novo immune response that can be capitalized on by immunotherapy, owing to low mutational burden, lack of MHC expression and other factors.

For example, targeting overexpressed surface molecules, including HER2 [26], B7-H3 [27] and the gangliosides and disialogangliosides (GD2 and GD3) [28–30] using antibody-based therapy and CAR T cells is a promising therapeutic strategy in EwS and OS. Preclinical and clinical studies have indicated that targeting GD2 generates robust anti-tumor response in EwS and in some other tumor entities including neuroblastoma [30,31], however off-target neurotoxicity can occur [32]. The ongoing clinical trials will test its efficacy in patients with EwS and OS (NCT02107963, NCT04539366).

Similar to the majority of solid tumors in adults [23,33], the above immunotherapy strategies showed limited efficacy in pediatric cancers, including EwS and OS [19,22,34,35]. As discussed below, this is mainly because of the lack of tumor-specific antigens (TSAs) and neoantigens [36–38], low expression of the classical human major histocompatibility complex class I (MHC-I) and thus impaired MHC-peptide presentation [39], upregulation of checkpoints and the immunosuppressive tumor microenvironment (TME), which is mostly populated by pro-tumorigenic M2 macrophages. One of the emerging immune escape mechanisms in cancer also involves an increased secretion of TSAs, TSA-MHC complexes, oncogenic proteins and RNAs in extracellular vesicles (EVs) [40–42].

Table 1. Immunotherapy of pediatric bone cancers.

Approach	Goal	Target	Therapeutic Agent	Major Obstacles	Refs
Immune checkpoint inhibitors	Reactivating and amplifying preexisting antitumor immunity	PD-1	Nivolumab/OPDIVO® Pembrolizumab/ Keytruda®	Low expression, low mutational burden, Immunological cold TME	[19,43–55]
		PD-L1	Atezolizumab/Tecentriq®		
		CTLA-4	IpilimumabYERVOY®		
Tumor specific antigens (TSAs)	Direct tumor targeting	GD2	Dinutuximab/Unituxin® anti-GD2 CAR-T cells anti-GD2 CAR-engineered NK cells	Variable expression in EwS and OS Upregulation of HLA-G checkpoint	[28–30,56]
		IGF1R	Ganitumab, Dalotuzumab	Activation of compensatory mechanisms, Toxicity	[57]
		HER2	Trastuzumab/Herceptin®	Not expressed in EwS, no clinical benefit for OS	[18,45]
		B7-H3	Anti-B7-H3 CAR T cells		[27]

Table 1. *Cont.*

Approach	Goal	Target	Therapeutic Agent	Major Obstacles	Refs
Antitumor vaccines	Direct tumor targeting	Tumor TSAs or proteins	Dendritic cell vaccine	Need for autologous DCs, Labor-intensive and costly cell isolation	[58]
	Activation of DC responses	Multiple tumor antigens	Attenuated tumor cells, could be pulsed with GM-CSF, IL-2 or IL-7 or siRNAs	Immunosuppressive TME, low tumor immunogenicity	[22]
Oncolytic viruses	Increase tumor immunogenicity Induce immunogenic cell death	Tumor	Vaccinia virus/Pexa-Vec Reovirus/Reolysin HSV-1/HSV1716 Adenovirus X-Vir	Antiviral immunity, Low delivery efficacy, Immunosuppression, T cell exhaustion	[34,59–62]
Targeting immunosuppressive TME	Macrophage activation	TME	L-MTP-PE/Mifamurtide, BCG, Coley's toxins, oncolytic viruses		[63,64]
	Macrophage polarization	TME	All-trans retinoic acid (ATRA)	Low delivery efficacy	[65]
	Macrophage/MDSC depletion	TME	All-trans retinoic acid (ATRA) Trabectedin	Toxicity	[66–69]

2. Mechanisms of Immune Escape

A better understanding of immune evasion mechanisms employed by tumor to avoid recognition and killing by the immune system is a key to successful therapy aimed at combating cancer with minimally invasive and non-toxic modalities. To search for cure and not just for 5-year overall survival for pediatric patients, it is imperative to establish critical aspects of the interplay between the tumor and the immune system. Several key points are discussed below.

2.1. Lack of Tumor-Specific Antigens (TSAs)

Targeted therapeutics and immunotherapy strategies are critically dependent on identification of druggable TSAs and neoantigens. To achieve maximal clinical efficacy and minimal toxicity, the ideal target should be immunogenic, highly expressed and presented on the surface of the majority of tumor but not normal host cells, and play a role in tumorigenesis. A very few antigens in general (and none of them in pediatric sarcomas) meet such criteria. In contrast to adult cancers, pediatric tumors exhibit very low mutation rates and, consequently, much fewer TSAs and tumor neoantigens [22,34,35]. Even OS, which has relatively high level of copy-number variations and gene deletions [70], exhibits on average about 7 neoepitopes per tumor and only 2 of them are predicted to be expressed, while EwS has none [71]. With regard to cell surface proteins, the best studied targets include B7-H3, GD2, IGF1R (which are shared between OS and EwS) as well as HER2 (for OS) and CD99, endosialin/CD248, TRAIL-R and STEAP-1 (for EwS) [18,34,45]. These are being targeted using CAR T cell approaches, bispecific T-cell engagers and antibody-drug conjugates. While multiple pre-clinical development programs and clinical attempts are still ongoing [15,18,19], none of them has shown significant clinical benefit so far, in part due to toxicities associated with their expression in other tissues. Targeting intracellular TSAs is also problematic, mainly because of low expression of MHC-I molecules that are required for the intracellular peptide presentation to the effector $CD8^+$ T cells [39], see also Section 2.2. However, class I expression may pertain to advanced EwS and experimental approaches have been established to induce it [72–74].

One of the potential solutions to the lack of conventional TSAs in pediatric sarcomas may be hidden in noncoding regions of the genome, including introns, alternative splicing variants, gene fusions, endogenous retroelements and other unannotated open reading frames [75]. According to recent studies, noncoding regions could be the main source of targetable TSAs in human malignancies [76–79]. Activation of these regions mainly occurs due to demethylation of the genome and other epigenetic mechanisms that are highly dysregulated in tumor cells, rising a possibility that the resulting TSAs may be widely shared between different tumor types and absent from normal tissues [80,81]. In addition, pediatric sarcomas may also express unique neoantigens from noncoding regions, given that at least a third of them carry recurrent chromosomal translocations and express characteristic fusion proteins which act as transcriptional and epigenetic regulators [13,15,82]. For instance, in EwS, epigenetic changes are driven by EWS-FLI1, through its ability to bind to GGAA microsatellite sequences [13,14]. A recent report provided the first evidence that EWS-FLI1 and potentially other EWS fusions can drive transcription, processing and translation of neopeptides from the silent genomic regions including GGAA microsatellite repeats [83], although their presentation on MHC-I, immunogenicity, TCR addressability and druggability remains to be studied.

2.2. Low Expression of MHC-I and Upregulation of Immune Checkpoints

One of the mechanisms whereby pediatric sarcomas escape T cell-mediated immunosurveillance is impaired expression of MHC class I/Human Leukocyte Antigen (HLA) class I antigens [39]. About 48–79% of primary EwS and the majority of metastatic lesions, especially pulmonary metastasis, exhibit low-to-absent MHC class I and II expression [84,85]. In EwS, low levels of MHC-I correlate with reduced $CD8^+$ T cell infiltration and survival [85]. Likewise, ~25–52% of primary and 44–88% of metastatic OS manifest complete loss or downregulation of MHC-I, being strongly associated with decreased survival [46,86,87].

Upregulating MHC-I expression in pediatric sarcomas may thus be a promising strategy to activate $CD8^+$ T cell-mediated antitumor responses [88]. This can be achieved by stimulating proinflammatory pathways, such as TNF-TNF receptor-NFκB, type I IFNs-IFNAR1/2-STAT1/2/3 or type II IFN-IFNGR-STAT1 [89]. For example, MHC-I expression in EwS cell lines is induced by IFNγ or mediators of dendritic cell maturation, including TNF [84,90,91]. In particular, treatment with GM-CSF, IL-4, TNF, IL-6, IL-1β and PGE_2 upregulated MHC-I, ICAM-1 and CD83, and improved recognition of EwS cell lines by TSA-specific TCR transgenic T cells in vitro [74]. In clinical settings, certain treatments including irradiation and hyperthermia may induce TNF secretion and MHC-I expression and may thus render EwS cells more susceptible to MHC-I-dependent immunotherapies [74,92,93]. The MHC-I presentation on tumor cells can also be induced by adenovirus-based oncolytic virotherapy. Oncolytic adenoviruses selectively replicate in tumor cells, where they induce the cGAS-STING pathway [94], and promote MHC-I expression. One example is the YB-1-targeting adenovirus XVir-N-31, which demonstrated substantial antitumor activity in a murine EwS xenograft model in combination with CDK4/6 inhibitors [62]. It is thus likely that combinatorial treatments would be necessary to increase MHC-I expression and $CD8^+$ T cell-mediated antitumor immunity.

Concurrent with reduced MHC-I expression, upregulation of immunosuppressive receptors and checkpoints may contribute to the immune escape in pediatric sarcomas. For instance, HLA-G and HLA-E, the non-classical MHC-I molecules implicated in the protective maternal-fetal barrier in the placenta [95], are highly upregulated on tumor and myeloid cells in the EwS TME. HLA-G is expressed in ~34% of EwS biopsies, and can be further induced by proinflammatory signaling, including IFNγ and GD2-specific CAR-engineered NK cells [56,91,96]. When expressed on tumor cells, HLA-G and HLA-E were shown to interact with inhibitory receptors expressed on T cells and NK cells, negatively affecting cytotoxic functions of both $CD8^+$ T cells and NK cells [97]. Yet, in experimental in vitro model, ectopic expression of HLA-G by myeloid THP-1 cells but not by EwS tumor cells impeded functionality of CAR T cells, suggesting that immunosuppressive effects of HLA-G could be mediated by myeloid cells in the TME [96]. However, expression of various HLA-G isoforms and lack of specific antibodies are currently hampering the development of HLA-G targeting approaches in EwS and other tumors [97].

On a similar note, therapeutic targeting of PD-L1 and PD-1 immune checkpoints, which are expressed in ~20% of pediatric sarcoma patients in EwS and OS [46–52], has not shown clinical efficacy [43,53,54]. In line with low expression of immune checkpoints on EwS and OS tumor cells, only 35% of EwS- and OS-infiltrating immune cells express PD-L1 [47], whereby expression of PD-L1 or PD-1 on T cells is rare for the most part in EwS and OS [50,55,98,99], and is predominantly observed on macrophages [54]. Yet, ~10% of post-treatment OS tumors score in the top quartile of immune infiltration, which is comparable to other strongly immune-infiltrated malignancies, including lung cancer and renal clear cell carcinoma [100]. The respective groups of OS patients may potentially benefit from the immune checkpoint blockade.

2.3. *Immunosuppressive TME in EwS and OS*

2.3.1. Improving $CD8^+$ T Cell Infiltration and Antitumor Activity

Only 12–38% of EwS and ~52–68% OS tumors are infiltrated by cytotoxic $CD8^+$ T cells [48,55,91]. Poor $CD8^+$ T cell infiltration is a negative prognostic marker associated with metastatic progression and worse outcomes [85,99,101–104]. Key chemotactic mechanisms for the recruitment of TILs and activation of antitumor immune responses are the TME-derived C-X-C Motif Chemokine Ligand 9/10 (CXCL9/10) or stromal-derived Chemokine (C-C motif) ligand 5 (CCL5) and their respective receptors C-X-C Motif Chemokine Receptor 3 (CXCR3) and CCR5 [105,106]. In EwS, increased expression of CXCL9/10 and CCL5 correlates with infiltration of $CD8^+$ $CXCR3^+/CXCR5^+$ T cells [103]. However, in recurrent pediatric sarcomas, EwS-infiltrating macrophages express lower levels of CXCL9/10 compared to OS [107]. Inducing CXCL10 expression in the EwS TME in order to enhance T cell infiltration may thus be a promising therapeutic strategy. This can be achieved by intratumoral injection of attenuated pathogens or oncolytic viruses [108], especially those harboring CXCL10 transgene [109], as well as by using dipeptidylpeptidase 4 (DPP4) inhibitors to stabilize CXCL10 [110,111].

In spite of higher proportion of TILs in the OS TME compared to EwS, they exhibit terminally exhausted phenotypes, including expression of co-inhibitory receptors TIGIT, LAG3, PD-1 and TIM3 [107]. Apart from the PD-1/PD-L1 axis, their blockade in OS may thus enhance TILs cytolytic activity. This may be especially relevant in OS with pulmonary metastases, which show increased T cell infiltration at the interface between the adjacent healthy tissue and tumor stroma [100,112]. These interfaces are enriched with activated exhausted $CD8^+$ T cells positive for PD-1, LAG3 and IFNγ and with myeloid cells expressing M-MDSC and DC signatures. The core of pulmonary metastases is devoid of immune infiltrates, suggesting that myeloid cells may exclude TILs [112].

In contrast to OS, circulating lymphocytes in EwS exhibit mixed gene expression profiles associated with effector responses (granzyme A and B, and perforin) and intermediate exhausted phenotypes (TIM-3) [107]. This is partly in line with previous results observing increased frequencies of circulating $PD\text{-}1^+CD4^+$ and $CD8^+$ T cells [113], however immune checkpoint blockade failed to show clinical response, as discussed above.

2.3.2. Targeting Tumor-Associated Macrophages

The most abundant immune cells in the TME of EwS and OS are tumor-associated macrophages (TAMs) which exhibit immunosuppressive M2 signatures [104,114–116]. Based on recently published transcriptomic analysis, these M2 macrophages may be phenotypically and functionally distinct in EwS and OS [107]. In line with this, TAM infiltration in EwS was indicative of poorer survival [104,115,117], while opposite observations were made in OS, where infiltration with $CD14^+/CD163^+$ myeloid cells and M1/M2 macrophages correlated with improved outcomes [67,99,118]. However, infiltration with $CD68^+$ macrophages was associated with worse survival in OS [49], suggesting the existence of different TAM populations with opposite activities. Higher density of CD68$^+$ and CD163$^+$ macrophages in OS (the CD68$^+$ to TIL ratio is 5.9, compared to 2.5 in EwS) may contribute to OS aggressiveness [119].

Chemotactic signals from the TME recruit monocytes from the bone marrow into the tumor stroma [120], where they polarize into TAMs and pro-tumorigenic M2 macrophages [121] (Figure 2). Signaling in the TME promotes sarcoma progression by inducing angiogenesis [115,122,123], migration [124], extravasation [125] and chemotherapy resistance [126]. TAMs in pediatric bone sarcomas release pro-inflammatory cytokines [115], prevent T cells from entering the tumor core [112] and impede the activation and degranulation of T cells [127]. Targeting TAMs may therefore be a promising therapeutic strategy, which is currently being investigated in multiple pre-clinical studies and clinical trials. This involves the following directions:

- Macrophage activation using liposome-encapsulated muramyl tripeptide phosphatidyl ethanolamine (L-MTP-PE or mifamurtide), a constituent of the Mycobacterium cell wall originally purified from the attenuated *Mycoblasma bovis*, also known as Bacille Calmette-Guerin (BCG). BCG vaccine is currently used for treatment of certain types of cancer and may have a mechanism of action similar to Coley's toxins. The synthetic L-MTP-PE was shown to stimulate macrophages and to improve survival in OS [63,64,128,129]. Another approach to activate TAMs are oncolytic viruses, which are capable of inducing immunogenic tumor cell death (ICD). This is accompanied by release of TSAs and danger- and pathogen-associated molecular patterns (DAMPs and PAMPs), switching TAMs to antitumorigenic M1 macrophages [130].
- Blocking the immunosuppressive M2 polarization and depleting myeloid-derived suppressor cells (MDSCs) using all-trans retinoic acid (ATRA). Treatment with ATRA reduced metastasis in an OS mouse model [65] and improved the efficacy of CAR T cells against pediatric sarcomas in vivo [66].
- Depleting TAMs and MDSCs using chemotherapeutic agents. For example, trabectedin, which is a natural product from sea squirt shown to inhibit transcription factor bindings such as FUS-CHOP in myxoid liposarcoma or EWS-FLI1 in EwS [131], enhanced CD3$^+$ T cell infiltration in OS and other cancers as well as oncolytic virotherapy against EwS xenograft in a mouse model [67–69]. Trabectedin is currently tested in a clinical trial for EwS (NCT04067115).
- Other therapeutic options include block of recruitment and reprogramming metabolic switches.

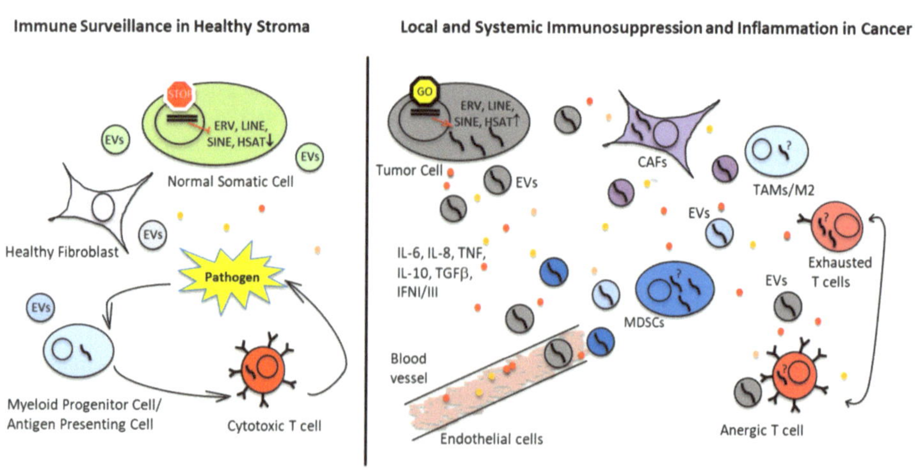

Figure 2. Immunosuppressive tumor environment and immune evasion mechanisms in EwS and OS. Oncogenic drivers and epigenetic changes in EwS, OS and other cancers activate proinflammatory

pathways and increase secretion of cytokines, chemokines, growth factors and EVs. Proinflammatory microenvironment attracts resident tissue macrophages and blood-circulating monocytes, while skewing their differentiation and promoting accumulation of M2 macrophages and immature myeloid cells with tolerogenic properties. Together with cancer-associated fibroblasts (CAFs), these cells (TAMs/M2 and MDSCs) constitute a majority in the TME. They fail to efficiently activate T cells and, instead, may induce T cell anergy and exhaustion. Additional contributing mechanisms may include re-expression of retroelements (LINEs and SINEs) and endogenous retroviruses including HERV-K in tumor cells. Dissemination of these virus-like RNAs in tumor EVs and their potential uptake by immune cells and CAFs may induce innate immune responses in these cells, leading to chronic inflammation. Tumor EVs bearing TSAs and pre-formed TSA-MHC complexes may also function as a decoy to divert antitumor immunity from cancer cells, especially when taken up by bystander host cells.

Therapeutic approaches directed at targeting TAMs and summary of clinical trials in sarcomas are discussed in detail elsewhere [132–134].

2.4. Immunogenicity and Response to Immunotherapy of EwS and OS in the Context of Bone and Soft Tissue Sarcomas

Bone and soft tissue sarcomas (STS) of children and adults comprise a heterogenous group of tumors with distinct biological properties, albeit they all share non-immunogenic properties and non-responsiveness to immunotherapy [22,43,135–137]. The most common bone tumors include EwS, OS and chondrosarcoma, while fibrosarcoma, gastrointestinal stromal tumors (GIST), leiomyosarcoma, liposarcoma, rhabdomyosarcoma (RMS), undifferentiated pleomorphic sarcoma and synovial sarcoma are the most frequent STS [138].

Immune infiltrates are heterogenous between sarcoma entities and age-dependent [139]. Sarcomas driven by mutations and copy number alteration tend to be T cell-inflamed, while translocation-driven sarcomas are immunologically cold [101,140,141]. Remarkably, mutation rates among sarcomas are low compared to other tumor entities [142], with pediatric sarcomas exhibiting the lowest numbers of mutations per Mb (EwS 0.24; OS 0.38 and RMS 0.33, the most frequent pediatric STS, compared to adult STS 1.06) [141,143]. Expression of immune checkpoints and clinical response to immune checkpoint blockade in sarcomas is dynamic, variable, and dependent on the histologic subtype. PD-L1 expression is sparse on the pediatric sarcomas EwS, OS and RMS, similar to the majority of adult STS (~20% PD-L1 expression) [144–146]. Consistent with this, the majority of pediatric and adult sarcomas do not respond to immune checkpoint blockade and predictive markers for responsiveness to ICI are yet to be determined [54,147]. The exceptions are the undifferentiated pleomorphic sarcoma and alveolar soft-part sarcoma, which are inflamed tumors with higher mutational burden [43,142], positive for PD-L1 in up to 40% of cases and partially responsive to immune checkpoint blockade [145,148,149]. Specific immunotherapeutic approaches in sarcomas have been reviewed elsewhere [22,26,137,150,151].

2.5. Extracellular Vesicles (EVs) as Means of Immune Escape

Communications between tumor and host cells in local and distant tumor sites are mediated by diffusible molecules such as cytokines, chemokines and lipids as well as to a large extent by extracellular vesicles (EVs) that create permissive environment for tumorigenic progression (Figure 2). EVs do so by transferring nucleic acids, proteins, lipids and various metabolites from tumor to various host cells, and vice versa [152–154]. As such, the EV cargo reflects the cell of origin and its physiological conditions, representing an important source of cancer-associated biomarkers. Most importantly, the EV cargo is encapsulated into lipid bilayer membrane and thus protected from degradation. When taken up by bystander normal and malignant cells, the EV cargo and is capable of functionally reprogramming the acceptor cells.

The EVs are comprised of highly heterogeneous populations of vesicles, whose secretion and composition are influenced by environmental conditions and tissue homeostasis. Most studies are focused on the nanosized vesicles (30–200 nm) originating from endosomal compartments (exosomes) or plasma membrane (ectosomes), which are believed to be important players in extracellular communications in healthy and diseased states [154–156]. The original hypothesis proposed in early 1980s implicated exosomes (EVs) as garbage bags for removal of unwanted proteins or harmful metabolites from the cells [157,158]. Indeed, secretion of cellular waste in EVs (or in specialized EV subsets) may be important for maintaining cellular homeostasis in normal and cancer cells. Recent findings have indicated that secretion in exosomes is essential for removal of damaged DNA and that blocking exosomal pathways provokes innate immune responses and induces senescence-like phenotype or apoptosis in normal cells due to accumulation of nuclear DNA in the cytosol [159]. Packaging in EVs is also required for expulsion of chemotherapy drugs and cellular toxins [160,161]. The EV-mediated waste management may be especially important for cancer cells, given their high proliferation and metabolic rates, and deficiencies in DNA repair pathways.

As discussed in numerous comprehensive reviews, tumor-derived EVs influence all major hallmarks of cancer, including immune evasion, tumor-promoted inflammation, angiogenesis, metabolic and epigenetic reprograming of the recipient cells, extracellular matrix remodeling, cancer metastasis and drug resistance [40,153,162]. In bone sarcomas and in cancers that preferentially metastasize to bones (such as prostate and breast carcinomas), EVs secreted by tumor cells are also capable of interfering with osteogenesis to promote tumor-supporting microenvironment inside the bone [163–166]. Particularly, OS-derived EVs enhanced angiogenic activities of endothelial cells, triggered macrophage dedifferentiation and increased a number of osteoclast-like cells and cancer-associated fibroblasts (CAFs) in local TME and metastatic sites [167,168]. Using murine NIH3T3 fibroblasts which are more susceptible to oncogenic transformation, it was also shown that OS EVs may induce tumor-like phenotype in non-transformed cells [169]. Moreover, OS EVs promoted epigenetic changes in mesenchymal stem cells (MSCs) but not in pre-osteoblasts, indicating that MSCs are highly susceptible to the EV-mediated epigenetic transformation [170]. Given that MSCs of osteogenic lineage are believed to be potential cells of origin for OS [171], their reprogramming by OS EVs can be an early event during OS development. Other mechanisms may involve a membrane-associated form of TGFβ which is transported in OS EVs to MSCs, leading to enhanced production of the proinflammatory cytokine IL-6 [172]. Likewise, IL-6 secreted by stromal cells in EwS TME was shown to contribute to EwS progression by protecting from apoptosis and promoting migration [173]. Using IL6- and TGFβ-blocking agents may thus be a viable therapeutic option for OS and EwS patients [172].

Lack of TSAs and low expression of MHC molecules have been described in previous sections as one of the major impediments for therapeutic targeting of EwS and OS cells. The available evidence suggests that their release in EVs could be one of the mechanisms employed by tumor cells to eliminate their specific antigens and MHCs and to reduce their recognition by cytotoxic T cells. Indeed, presence of tumor-derived MHCs and antigens (including pre-formed functional TSA-MHC complexes) in EVs is a well-documented phenomenon [40–42], albeit its role in EwS and OS remains to be elucidated. Dissemination of tumor EVs harboring TSAs and their subsequent acquisition and cross-presentation by bystander immune and non-immune cells may also act as a decoy to divert antitumor immunity from cancer cells. Remarkably, EVs carrying MHC-peptide complexes can directly activate cognate receptors on T cells, but the efficacy of antigen presentation is increased when EVs are attached to the surface of mature DCs [41,42]. However, this mechanism known as cross-dressing [174] may be compromised in tumor settings, given accumulation of MDSCs and immature DCs.

EwS EVs may be directly involved in generation of immature proinflammatory myeloid cells in local TME and systemic circulation. We showed that EwS EVs induced secretion of IL-6, IL-8 and tumor necrosis factor (TNF) by primary $CD33^+$ myeloid cells and $CD14^+$ monocytes, and inhibited their maturation into antigen-presenting DCs [175]. In particular, $CD14^+$ cells differentiated in the presence of EwS EVs exhibited a semi-mature phenotype and immunosuppressive activity, including reduced expression of co-stimulatory molecules CD80, CD86 and HLA-DR, activation of the innate immune response gene expression programs, and the ability to interfere with activation of CD4$^+$ and CD8$^+$ T cells. Therefore, EwS EVs may contribute to systemic inflammation and immunosuppression by skewing differentiation and maturation of blood-circulating and tumor-infiltrating myeloid cells.

Mechanistically, induction of immunosuppressive myeloid cells is primarily mediated by various protein and RNA constituents present in tumor EVs [176]. EwS EVs carry multiple mRNAs encoding oncogenic drivers, including EWS-FLI1, EZH2 and stem cell-associated proteins [177,178], some of which can be transferred to the neighboring mesenchymal stem cells [179]. Whether or not EV-derived RNAs are actually capable of driving a sustainable protein expression in the recipient cells is an open question, given that the majority of these RNAs, including mRNAs and microRNAs, are severely fragmented and present in less than one copy per EV [180,181]. Additionally, given that the RNA lifetime is relatively short, massive production of RNA-containing EVs and their highly selective uptake might be required to reprogram the recipient cells by a particular RNA in EVs [180].

Of further consideration, our recent whole transcriptome analysis of EVs isolated from plasma of EwS patients and cell lines showed that the vast majority of RNAs in these EVs (up to 70–90%) are derived from satellite repeats, endogenous retroelements and retroviruses (ERVs), including *HSAT2*, *LINEs*, *LTR/ERVs*, *SINE/Alu*, and *7SL RNA* [182]. Similar results were reported for EVs from low-passage brain cancer cell lines [183] and patient-derived glioma stem cell-like cultures [180], as well as from co-cultures of breast cancer cells with stromal fibroblasts [184]. Compared to the respective parental cells, these EV preparations exhibited a significant enrichment with *ERV*, *LINE*, *SINE* and other repeat RNAs. Transcriptional activation of the respective heterochromatic genomic regions could be attributed to demethylation of the genome and other epigenetic changes characteristic of many if not all human malignancies [80,81]. In turn, their release in EVs could be a protective mechanism directed at preventing activation of innate immune responses in tumor cells, given highly immunogenic virus-like features of these transcripts. Moreover, we [182] and others [183] provided preliminary experimental evidence that at least some of these RNAs may be transferred in tumor EVs to normal cells, essentially mimicking viral infection (Figure 2). In contrast to other RNAs in EVs, retroelement-derived transcripts including *LINE-1* and *HERV-K* may retain some coding and replicative potential and may potentially be able to propagate in the "infected" cells via reverse transcription mechanisms [185,186]. Further investigation into these mechanisms may provide unique insights into cancer-associated inflammation, evasion of antitumor immunity and immunosuppression, and open novel directions for therapeutic targeting.

3. Conclusions and Perspectives

The development of efficacious immunotherapies in immune-inert pediatric sarcomas requires addressing both the tumor cell itself and the TME. Immunogenicity of the tumor cell can be increased by upregulating MHCs and/or immune stimulatory molecules such as CD83 and ICAM-1 on the tumor cell surface, or by increasing its sensitivity to immune checkpoint inhibitors. A combination of these approaches is warranted. In turn, immunogenicity of the tumor environment can be enhanced by altering macrophage differentiation and polarization or by administering activating cytokines. Additional tumor microenvironment-directed approaches could be designed to interfere with the immunosuppressive mechanisms active in the immunological deserts, e.g., blocking immunosuppressive EVs or immunosuppressive metabolic mechanisms, or probably both. This can be achieved by engineering bifunctional TCR or CAR transgenic T cells that could simultaneously manipulate the TME and target tumor-specific cell surface antigens. Moreover, epigenetic activation of gene expression from non-coding sequences may provide targetable neo-epitopes even in immune inert malignancies. Finally, expression of immunogenic neo-epitopes from non-coding sequences can be combined with repression of non-coding sequences associated with immune suppression or tolerance. Multifunctional T cell engineering can be envisioned, where transgenic TCR or CAR T, or NK cells recognizing tumor cells are designed to co-express immune stimulatory and TME manipulating gene products, such as those affecting macrophage polarization and/or molecules repressing EVs. Lastly, a better understanding of the role of sarcoma EVs in mediating immune dysfunction and ways of alleviating this are needed. We also need to establish whether EV-induced immune disruption (e.g., via releasing retroelement RNAs and inducing systemic inflammation) is a more widespread phenomenon across other cancers.

Author Contributions: Conceptualization, V.E., writing—original draft preparation, V.E., H.G., L.R. and S.E.G.B.; writing—review and editing, V.E., H.G., L.R. and S.E.G.B.; visualization, V.E. and S.E.G.B.; supervision, L.R. and S.E.G.B.; funding acquisition, V.E., H.G., L.R. and S.E.G.B. All authors have read and agreed to the published version of the manuscript.

Funding: This work was supported by funding provided by the Government of Ontario (to V.E. and L.R.), the Cura Placida Children's Cancer Research Foundation grants (to V.E. and S.E.G.B.), the St. Baldrick's Foundation (Martha's BEST Grant for all #663113) and the 'Du musst kämpfen' Wohltätigkeitsinitiative (Access to Systems Biology-Based Individualized Cell Therapies for Adolescents with Refractory Pediatric Cancer, AdoRe, #200310stb), to S.E.G.B.), and the Dr. Robert-Pfleger Foundation and the KKF clinician-scientist program of the School of Medicine, Technical University of Munich (to H.G.).

Acknowledgments: This article is dedicated to children with cancer and their families in hopes to induce progress and find cure.

Conflicts of Interest: S.E.G.B. has an ownership interest in PDL BioPharma and has had US and EU intellectual properties in gene expression analysis. He served as a consultant to EOS Biotechnology Inc. and serves as an advisor to Bayer AG and Swedish Orphan Biovitrum AB. The other authors declare no conflict of interest.

References

1. Ewing, J. Classics in oncology. Diffuse endothelioma of bone. James Ewing. Proceedings of the New York Pathological Society, 1921. *CA Cancer J. Clin.* **1972**, *22*, 95–98. [CrossRef] [PubMed]
2. McCarthy, E.F. The toxins of William B. Coley and the treatment of bone and soft-tissue sarcomas. *Iowa Orthop. J.* **2006**, *26*, 154–158. [PubMed]
3. Coley, W.B. The therapeutic value of the mixed toxins of the streptococcus of erysipelas and bacillus prodigiosus in the treatment of inoperable malignant tumors: With a report of one hundred and sixty cases. *Am. J. Med. Sci.* **1896**, *112*, 251. [CrossRef]
4. Coley, W.B. The Treatment of Inoperable Sarcoma by Bacterial Toxins (the Mixed Toxins of the Streptococcus erysipelas and the Bacillus prodigiosus). *Proc. R. Soc. Med.* **1910**, *3*, 1–48. [CrossRef] [PubMed]
5. Nauts, H.C.; Fowler, G.A.; Bogatko, F.H. A review of the influence of bacterial infection and of bacterial products (Coley's toxins) on malignant tumors in man; a critical analysis of 30 inoperable cases treated by Coley's mixed toxins, in which diagnosis was confirmed by microscopic examination selected for special study. *Acta Med. Scandinavica Suppl.* **1953**, *276*, 1–103.
6. Johnston, B.J.; Novales, E.T. Clinical effect of Coley's toxin. II. A seven-year study. *Cancer Chemother. Rep.* **1962**, *21*, 43–68.
7. Coley, B.L. *Neoplasms of Bone*; Medical Book Departmnent of Harper & Brothers: New York, NY, USA, 1949; pp. 565–570.
8. Karbach, J.; Neumann, A.; Brand, K.; Wahle, C.; Siegel, E.; Maeurer, M.; Ritter, E.; Tsuji, T.; Gnjatic, S.; Old, L.J.; et al. Phase I clinical trial of mixed bacterial vaccine (Coley's toxins) in patients with NY-ESO-1 expressing cancers: Immunological effects and clinical activity. *Clin. Cancer Res.* **2012**, *18*, 5449–5459. [CrossRef]
9. Sedighi, M.; Zahedi Bialvaei, A.; Hamblin, M.R.; Ohadi, E.; Asadi, A.; Halajzadeh, M.; Lohrasbi, V.; Mohammadzadeh, N.; Amiriani, T.; Krutova, M.; et al. Therapeutic bacteria to combat cancer; current advances, challenges, and opportunities. *Cancer Med.* **2019**, *8*, 3167–3181. [CrossRef]
10. Lilienthal, I.; Herold, N. Targeting Molecular Mechanisms Underlying Treatment Efficacy and Resistance in Osteosarcoma: A Review of Current and Future Strategies. *Int. J. Mol. Sci.* **2020**, *21*, 6885. [CrossRef]
11. Harrison, D.J.; Geller, D.S.; Gill, J.D.; Lewis, V.O.; Gorlick, R. Current and future therapeutic approaches for osteosarcoma. *Expert Rev. Anticancer Ther.* **2018**, *18*, 39–50. [CrossRef]
12. Grunewald, T.G.P.; Cidre-Aranaz, F.; Surdez, D.; Tomazou, E.M.; de Alava, E.; Kovar, H.; Sorensen, P.H.; Delattre, O.; Dirksen, U. Ewing sarcoma. *Nat Rev Dis Prim.* **2018**, *4*, 5. [CrossRef]
13. Riggi, N.; Suvà, M.L.; Stamenkovic, I. Ewing's Sarcoma. *N. Engl. J. Med.* **2021**, *384*, 154–164. [CrossRef]
14. Grünewald, T.G.; Alonso, M.; Avnet, S.; Banito, A.; Burdach, S.; Cidre-Aranaz, F.; Di Pompo, G.; Distel, M.; Dorado-Garcia, H.; Garcia-Castro, J.; et al. Sarcoma treatment in the era of molecular medicine. *EMBO Mol. Med.* **2020**, *12*, e11131. [CrossRef]

15. Zöllner, S.K.; Amatruda, J.F.; Bauer, S.; Collaud, S.; de Álava, E.; DuBois, S.G.; Hardes, J.; Hartmann, W.; Kovar, H.; Metzler, M.; et al. Ewing Sarcoma-Diagnosis, Treatment, Clinical Challenges and Future Perspectives. *J. Clin. Med.* **2021**, *10*, 1685. [CrossRef]
16. Ginsberg, J.P.; Goodman, P.; Leisenring, W.; Ness, K.K.; Meyers, P.A.; Wolden, S.L.; Smith, S.M.; Stovall, M.; Hammond, S.; Robison, L.L.; et al. Long-term survivors of childhood Ewing sarcoma: Report from the childhood cancer survivor study. *J. Natl. Cancer Inst.* **2010**, *102*, 1272–1283. [CrossRef]
17. Crompton, B.D.; Stewart, C.; Taylor-Weiner, A.; Alexe, G.; Kurek, K.C.; Calicchio, M.L.; Kiezun, A.; Carter, S.L.; Shukla, S.A.; Mehta, S.S.; et al. The genomic landscape of pediatric Ewing sarcoma. *Cancer Discov.* **2014**, *4*, 1326–1341. [CrossRef]
18. Bailey, K.; Cost, C.; Davis, I.; Glade-Bender, J.; Grohar, P.; Houghton, P.; Isakoff, M.; Stewart, E.; Laack, N.; Yustein, J.; et al. Emerging novel agents for patients with advanced Ewing sarcoma: A report from the Children's Oncology Group (COG) New Agents for Ewing Sarcoma Task Force. *F1000Research* **2019**, *8*. [CrossRef]
19. Gupta, A.; Cripe, T.P. Immunotherapies for Pediatric Solid Tumors: A Targeted Update. *Paediatr. Drugs* **2022**, *24*, 1–12. [CrossRef]
20. Kovac, M.; Blattmann, C.; Ribi, S.; Smida, J.; Mueller, N.S.; Engert, F.; Castro-Giner, F.; Weischenfeldt, J.; Kovacova, M.; Krieg, A.; et al. Exome sequencing of osteosarcoma reveals mutation signatures reminiscent of BRCA deficiency. *Nat. Commun.* **2015**, *6*, 8940. [CrossRef]
21. Gorthi, A.; Romero, J.C.; Loranc, E.; Cao, L.; Lawrence, L.A.; Goodale, E.; Iniguez, A.B.; Bernard, X.; Masamsetti, V.P.; Roston, S.; et al. EWS-FLI1 increases transcription to cause R-loops and block BRCA1 repair in Ewing sarcoma. *Nature* **2018**, *555*, 387–391. [CrossRef]
22. Dyson, K.A.; Stover, B.D.; Grippin, A.; Mendez-Gomez, H.R.; Lagmay, J.; Mitchell, D.A.; Sayour, E.J. Emerging trends in immunotherapy for pediatric sarcomas. *J. Hematol. Oncol.* **2019**, *12*, 78. [CrossRef] [PubMed]
23. Maeda, H.; Khatami, M. Analyses of repeated failures in cancer therapy for solid tumors: Poor tumor-selective drug delivery, low therapeutic efficacy and unsustainable costs. *Clin. Transl. Med.* **2018**, *7*, 11. [CrossRef] [PubMed]
24. Hegde, P.S.; Chen, D.S. Top 10 Challenges in Cancer Immunotherapy. *Immunity* **2020**, *52*, 17–35. [CrossRef] [PubMed]
25. Johnson, D.B.; Nebhan, C.A.; Moslehi, J.J.; Balko, J.M. Immune-checkpoint inhibitors: Long-term implications of toxicity. *Nat. Rev. Clin. Oncol.* **2022**, *19*, 254–267. [CrossRef] [PubMed]
26. Ahmed, N.; Brawley, V.S.; Hegde, M.; Robertson, C.; Ghazi, A.; Gerken, C.; Liu, E.; Dakhova, O.; Ashoori, A.; Corder, A.; et al. Human Epidermal Growth Factor Receptor 2 (HER2) -Specific Chimeric Antigen Receptor-Modified T Cells for the Immunotherapy of HER2-Positive Sarcoma. *J. Clin. Oncol.* **2015**, *33*, 1688–1696. [CrossRef]
27. Majzner, R.G.; Theruvath, J.L.; Nellan, A.; Heitzeneder, S.; Cui, Y.; Mount, C.W.; Rietberg, S.P.; Linde, M.H.; Xu, P.; Rota, C.; et al. CAR T Cells Targeting B7-H3, a Pan-Cancer Antigen, Demonstrate Potent Preclinical Activity Against Pediatric Solid Tumors and Brain Tumors. *Clin. Cancer Res.* **2019**, *25*, 2560–2574. [CrossRef]
28. Dobrenkov, K.; Ostrovnaya, I.; Gu, J.; Cheung, I.Y.; Cheung, N.K. Oncotargets GD2 and GD3 are highly expressed in sarcomas of children, adolescents, and young adults. *Pediatr. Blood Cancer* **2016**, *63*, 1780–1785. [CrossRef]
29. Poon, V.I.; Roth, M.; Piperdi, S.; Geller, D.; Gill, J.; Rudzinski, E.R.; Hawkins, D.S.; Gorlick, R. Ganglioside GD2 expression is maintained upon recurrence in patients with osteosarcoma. *Clin. Sarcoma Res.* **2015**, *5*, 4. [CrossRef]
30. Kailayangiri, S.; Altvater, B.; Meltzer, J.; Pscherer, S.; Luecke, A.; Dierkes, C.; Titze, U.; Leuchte, K.; Landmeier, S.; Hotfilder, M.; et al. The ganglioside antigen GD2 is surface-expressed in Ewing sarcoma and allows for MHC-independent immune targeting. *Br. J. Cancer* **2012**, *106*, 1123–1133. [CrossRef]
31. Nazha, B.; Inal, C.; Owonikoko, T.K. Disialoganglioside GD2 Expression in Solid Tumors and Role as a Target for Cancer Therapy. *Front. Oncol.* **2020**, *10*, 1000. [CrossRef]
32. Richman, S.A.; Nunez-Cruz, S.; Moghimi, B.; Li, L.Z.; Gershenson, Z.T.; Mourelatos, Z.; Barrett, D.M.; Grupp, S.A.; Milone, M.C. High-Affinity GD2-Specific CAR T Cells Induce Fatal Encephalitis in a Preclinical Neuroblastoma Model. *Cancer Immunol. Res.* **2018**, *6*, 36–46. [CrossRef]
33. Bagchi, S.; Yuan, R.; Engleman, E.G. Immune Checkpoint Inhibitors for the Treatment of Cancer: Clinical Impact and Mechanisms of Response and Resistance. *Annu. Rev. Pathol.* **2021**, *16*, 223–249. [CrossRef]
34. Morales, E.; Olson, M.; Iglesias, F.; Dahiya, S.; Luetkens, T.; Atanackovic, D. Role of immunotherapy in Ewing sarcoma. *J. Immunother. Cancer* **2020**, *8*, e000653. [CrossRef]
35. Roberts, S.S.; Chou, A.J.; Cheung, N.K. Immunotherapy of Childhood Sarcomas. *Front. Oncol.* **2015**, *5*, 181. [CrossRef]
36. Corre, I.; Verrecchia, F.; Crenn, V.; Redini, F.; Trichet, V. The Osteosarcoma Microenvironment: A Complex But Targetable Ecosystem. *Cells* **2020**, *9*, 976. [CrossRef]
37. Morales, E.; Olson, M.; Iglesias, F.; Luetkens, T.; Atanackovic, D. Targeting the tumor microenvironment of Ewing sarcoma. *Immunotherapy* **2021**, *13*, 1439–1451. [CrossRef]
38. Cersosimo, F.; Lonardi, S.; Bernardini, G.; Telfer, B.; Mandelli, G.E.; Santucci, A.; Vermi, W.; Giurisato, E. Tumor-Associated Macrophages in Osteosarcoma: From Mechanisms to Therapy. *Int. J. Mol. Sci.* **2020**, *21*, 5207. [CrossRef]
39. Haworth, K.B.; Leddon, J.L.; Chen, C.Y.; Horwitz, E.M.; Mackall, C.L.; Cripe, T.P. Going back to class I: MHC and immunotherapies for childhood cancer. *Pediatr. Blood Cancer* **2015**, *62*, 571–576. [CrossRef]
40. Buzas, E.I. The roles of extracellular vesicles in the immune system. *Nat. Rev. Immunol.* **2022**, 1–15. [CrossRef]
41. Robbins, P.D.; Morelli, A.E. Regulation of immune responses by extracellular vesicles. *Nat. Rev. Immunol.* **2014**, *14*, 195–208. [CrossRef]

42. Zeng, F.; Morelli, A.E. Extracellular vesicle-mediated MHC cross-dressing in immune homeostasis, transplantation, infectious diseases, and cancer. *Semin. Immunopathol.* **2018**, *40*, 477–490. [CrossRef] [PubMed]
43. Tawbi, H.A.; Burgess, M.; Bolejack, V.; Van Tine, B.A.; Schuetze, S.M.; Hu, J.; D'Angelo, S.; Attia, S.; Riedel, R.F.; Priebat, D.A.; et al. Pembrolizumab in advanced soft-tissue sarcoma and bone sarcoma (SARC028): A multicentre, two-cohort, single-arm, open-label, phase 2 trial. *Lancet Oncol.* **2017**, *18*, 1493–1501. [CrossRef] [PubMed]
44. Keung, E.Z.; Burgess, M.; Salazar, R.; Parra, E.R.; Rodrigues-Canales, J.; Bolejack, V.; Van Tine, B.A.; Schuetze, S.M.; Attia, S.; Riedel, R.F.; et al. Correlative Analyses of the SARC028 Trial Reveal an Association Between Sarcoma-Associated Immune Infiltrate and Response to Pembrolizumab. *Clin. Cancer Res.* **2020**, *26*, 1258–1266. [CrossRef] [PubMed]
45. Gill, J.; Gorlick, R. Advancing therapy for osteosarcoma. *Nat. Rev. Clin. Oncol.* **2021**, *18*, 609–624. [CrossRef] [PubMed]
46. Sundara, Y.T.; Kostine, M.; Cleven, A.H.; Bovée, J.V.; Schilham, M.W.; Cleton-Jansen, A.M. Increased PD-L1 and T-cell infiltration in the presence of HLA class I expression in metastatic high-grade osteosarcoma: A rationale for T-cell-based immunotherapy. *Cancer Immunol. Immunother. CII* **2017**, *66*, 119–128. [CrossRef]
47. Majzner, R.G.; Simon, J.S.; Grosso, J.F.; Martinez, D.; Pawel, B.R.; Santi, M.; Merchant, M.S.; Geoerger, B.; Hezam, I.; Marty, V.; et al. Assessment of programmed death-ligand 1 expression and tumor-associated immune cells in pediatric cancer tissues. *Cancer* **2017**, *123*, 3807–3815. [CrossRef]
48. Machado, I.; López-Guerrero, J.A.; Scotlandi, K.; Picci, P.; Llombart-Bosch, A. Immunohistochemical analysis and prognostic significance of PD-L1, PD-1, and CD8+ tumor-infiltrating lymphocytes in Ewing's sarcoma family of tumors (ESFT). *Virchows Arch. Int. J. Pathol.* **2018**, *472*, 815–824. [CrossRef]
49. Koirala, P.; Roth, M.E.; Gill, J.; Piperdi, S.; Chinai, J.M.; Geller, D.S.; Hoang, B.H.; Park, A.; Fremed, M.A.; Zang, X.; et al. Immune infiltration and PD-L1 expression in the tumor microenvironment are prognostic in osteosarcoma. *Sci. Rep.* **2016**, *6*, 30093. [CrossRef]
50. Spurny, C.; Kailayangiri, S.; Jamitzky, S.; Altvater, B.; Wardelmann, E.; Dirksen, U.; Hardes, J.; Hartmann, W.; Rossig, C. Programmed cell death ligand 1 (PD-L1) expression is not a predominant feature in Ewing sarcomas. *Pediatr. Blood Cancer* **2018**, *65*. [CrossRef]
51. Pinto, N.; Park, J.R.; Murphy, E.; Yearley, J.; McClanahan, T.; Annamalai, L.; Hawkins, D.S.; Rudzinski, E.R. Patterns of PD-1, PD-L1, and PD-L2 expression in pediatric solid tumors. *Pediatr. Blood Cancer* **2017**, *64*, e26613. [CrossRef]
52. Raj, S.; Bui, M.; Gonzales, R.; Letson, D.; Antonia, S.J. Impact of Pdl1 Expression on Clinical Outcomes in Subtypes of Sarcoma. *Ann. Oncol.* **2014**, *25*, iv498. [CrossRef]
53. D'Angelo, S.P.; Mahoney, M.R.; Van Tine, B.A.; Atkins, J.; Milhem, M.M.; Jahagirdar, B.N.; Antonescu, C.R.; Horvath, E.; Tap, W.D.; Schwartz, G.K.; et al. Nivolumab with or without ipilimumab treatment for metastatic sarcoma (Alliance A091401): Two open-label, non-comparative, randomised, phase 2 trials. *Lancet Oncol.* **2018**, *19*, 416–426. [CrossRef]
54. Davis, K.L.; Fox, E.; Merchant, M.S.; Reid, J.M.; Kudgus, R.A.; Liu, X.; Minard, C.G.; Voss, S.; Berg, S.L.; Weigel, B.J.; et al. Nivolumab in children and young adults with relapsed or refractory solid tumours or lymphoma (ADVL1412): A multicentre, open-label, single-arm, phase 1-2 trial. *Lancet Oncol.* **2020**, *21*, 541–550. [CrossRef]
55. Van Erp, A.E.M.; Versleijen-Jonkers, Y.M.H.; Hillebrandt-Roeffen, M.H.S.; van Houdt, L.; Gorris, M.A.J.; van Dam, L.S.; Mentzel, T.; Weidema, M.E.; Savci-Heijink, C.D.; Desar, I.M.E.; et al. Expression and clinical association of programmed cell death-1, programmed death-ligand-1 and CD8(+) lymphocytes in primary sarcomas is subtype dependent. *Oncotarget* **2017**, *8*, 71371–71384. [CrossRef]
56. Kailayangiri, S.; Altvater, B.; Spurny, C.; Jamitzky, S.; Schelhaas, S.; Jacobs, A.H.; Wiek, C.; Roellecke, K.; Hanenberg, H.; Hartmann, W.; et al. Targeting Ewing sarcoma with activated and GD2-specific chimeric antigen receptor-engineered human NK cells induces upregulation of immune-inhibitory HLA-G. *Oncoimmunology* **2017**, *6*, e1250050. [CrossRef]
57. Amin, H.M.; Morani, A.C.; Daw, N.C.; Lamhamedi-Cherradi, S.E.; Subbiah, V.; Menegaz, B.A.; Vishwamitra, D.; Eskandari, G.; George, B.; Benjamin, R.S.; et al. IGF-1R/mTOR Targeted Therapy for Ewing Sarcoma: A Meta-Analysis of Five IGF-1R-Related Trials Matched to Proteomic and Radiologic Predictive Biomarkers. *Cancers* **2020**, *12*, 1768. [CrossRef]
58. Merchant, M.S.; Bernstein, D.; Amoako, M.; Baird, K.; Fleisher, T.A.; Morre, M.; Steinberg, S.M.; Sabatino, M.; Stroncek, D.F.; Venkatasan, A.M.; et al. Adjuvant Immunotherapy to Improve Outcome in High-Risk Pediatric Sarcomas. *Clin. Cancer Res.* **2016**, *22*, 3182–3191. [CrossRef]
59. Cripe, T.P.; Ngo, M.C.; Geller, J.I.; Louis, C.U.; Currier, M.A.; Racadio, J.M.; Towbin, A.J.; Rooney, C.M.; Pelusio, A.; Moon, A.; et al. Phase 1 study of intratumoral Pexa-Vec (JX-594), an oncolytic and immunotherapeutic vaccinia virus, in pediatric cancer patients. *Mol. Ther. J. Am. Soc. Gene Ther.* **2015**, *23*, 602–608. [CrossRef]
60. Kolb, E.A.; Sampson, V.; Stabley, D.; Walter, A.; Sol-Church, K.; Cripe, T.; Hingorani, P.; Ahern, C.H.; Weigel, B.J.; Zwiebel, J.; et al. A phase I trial and viral clearance study of reovirus (Reolysin) in children with relapsed or refractory extra-cranial solid tumors: A Children's Oncology Group Phase I Consortium report. *Pediatr. Blood Cancer* **2015**, *62*, 751–758. [CrossRef]
61. Streby, K.A.; Currier, M.A.; Triplet, M.; Ott, K.; Dishman, D.J.; Vaughan, M.R.; Ranalli, M.A.; Setty, B.; Skeens, M.A.; Whiteside, S.; et al. First-in-Human Intravenous Seprehvir in Young Cancer Patients: A Phase 1 Clinical Trial. *Mol. Ther. J. Am. Soc. Gene Ther.* **2019**, *27*, 1930–1938. [CrossRef]
62. Koch, J.; Schober, S.J.; Hindupur, S.V.; Schöning, C.; Klein, F.G.; Mantwill, K.; Ehrenfeld, M.; Schillinger, U.; Hohnecker, T.; Qi, P.; et al. Targeting the Retinoblastoma/E2F repressive complex by CDK4/6 inhibitors amplifies oncolytic potency of an oncolytic adenovirus. *Nat. Commun.* **2022**, *13*, 4689. [CrossRef] [PubMed]

63. Meyers, P.A.; Schwartz, C.L.; Krailo, M.D.; Healey, J.H.; Bernstein, M.L.; Betcher, D.; Ferguson, W.S.; Gebhardt, M.C.; Goorin, A.M.; Harris, M.; et al. Osteosarcoma: The addition of muramyl tripeptide to chemotherapy improves overall survival—A report from the Children's Oncology Group. *J. Clin. Oncol. Off. J. Am. Soc. Clin. Oncol.* **2008**, *26*, 633–638. [CrossRef] [PubMed]
64. Pahl, J.H.W.; Kwappenberg, K.M.C.; Varypataki, E.M.; Santos, S.J.; Kuijjer, M.L.; Mohamed, S.; Wijnen, J.T.; van Tol, M.J.D.; Cleton-Jansen, A.-M.; Egeler, R.M.; et al. Macrophages inhibit human osteosarcoma cell growth after activation with the bacterial cell wall derivative liposomal muramyl tripeptide in combination with interferon-γ. *J. Exp. Clin. Cancer Res.* **2014**, *33*, 27. [CrossRef] [PubMed]
65. Zhou, Q.; Xian, M.; Xiang, S.; Xiang, D.; Shao, X.; Wang, J.; Cao, J.; Yang, X.; Yang, B.; Ying, M.; et al. All-Trans Retinoic Acid Prevents Osteosarcoma Metastasis by Inhibiting M2 Polarization of Tumor-Associated Macrophages. *Cancer Immunol. Res.* **2017**, *5*, 547–559. [CrossRef] [PubMed]
66. Long, A.H.; Highfill, S.L.; Cui, Y.; Smith, J.P.; Walker, A.J.; Ramakrishna, S.; El-Etriby, R.; Galli, S.; Tsokos, M.G.; Orentas, R.J.; et al. Reduction of MDSCs with All-trans Retinoic Acid Improves CAR Therapy Efficacy for Sarcomas. *Cancer Immunol. Res.* **2016**, *4*, 869–880. [CrossRef]
67. Deng, C.; Xu, Y.; Fu, J.; Zhu, X.; Chen, H.; Xu, H.; Wang, G.; Song, Y.; Song, G.; Lu, J.; et al. Reprograming the tumor immunologic microenvironment using neoadjuvant chemotherapy in osteosarcoma. *Cancer Sci.* **2020**, *111*, 1899–1909. [CrossRef]
68. Germano, G.; Frapolli, R.; Belgiovine, C.; Anselmo, A.; Pesce, S.; Liguori, M.; Erba, E.; Uboldi, S.; Zucchetti, M.; Pasqualini, F.; et al. Role of macrophage targeting in the antitumor activity of trabectedin. *Cancer Cell* **2013**, *23*, 249–262. [CrossRef]
69. Denton, N.L.; Chen, C.-Y.; Hutzen, B.; Currier, M.A.; Scott, T.; Nartker, B.; Leddon, J.L.; Wang, P.-Y.; Srinivas, R.; Cassady, K.A.; et al. Myelolytic Treatments Enhance Oncolytic Herpes Virotherapy in Models of Ewing Sarcoma by Modulating the Immune Microenvironment. *Mol. Ther.—Oncolytics* **2018**, *11*, 62–74. [CrossRef]
70. Alexandrov, L.B.; Nik-Zainal, S.; Wedge, D.C.; Aparicio, S.A.; Behjati, S.; Biankin, A.V.; Bignell, G.R.; Bolli, N.; Borg, A.; Børresen-Dale, A.L.; et al. Signatures of mutational processes in human cancer. *Nature* **2013**, *500*, 415–421. [CrossRef]
71. Chang, T.C.; Carter, R.A.; Li, Y.; Li, Y.; Wang, H.; Edmonson, M.N.; Chen, X.; Arnold, P.; Geiger, T.L.; Wu, G.; et al. The neoepitope landscape in pediatric cancers. *Genome Med.* **2017**, *9*, 78. [CrossRef]
72. Thiel, U.; Schober, S.J.; Einspieler, I.; Kirschner, A.; Thiede, M.; Schirmer, D.; Gall, K.; Blaeschke, F.; Schmidt, O.; Jabar, S.; et al. Ewing sarcoma partial regression without GvHD by chondromodulin-I/HLA-A*02:01-specific allorestricted T cell receptor transgenic T cells. *Oncoimmunology* **2017**, *6*, e1312239. [CrossRef]
73. Staege, M.S.; Gorelov, V.; Bulankin, A.; Fischer, U.; Dumon, K.; Hohndorf, L.; Hattenhorst, U.; Kramm, C.; Burdach, S. Stable transgenic expression of IL-2 and HSV1-tk by single and fusion tumor cell lines bearing EWS/FLI-1 chimeric genes. *Pediatr. Hematol. Oncol.* **2003**, *20*, 119–140. [CrossRef]
74. Biele, E.; Schober, S.J.; Prexler, C.; Thiede, M.; Heyking, K.V.; Gassmann, H.; Eck, J.; Xue, B.; Burdach, S.; Thiel, U. Monocyte Maturation Mediators Upregulate CD83, ICAM-1 and MHC Class 1 Expression on Ewing's Sarcoma, Enhancing T Cell Cytotoxicity. *Cells* **2021**, *10*, 3070. [CrossRef]
75. Smith, C.C.; Selitsky, S.R.; Chai, S.; Armistead, P.M.; Vincent, B.G.; Serody, J.S. Alternative tumour-specific antigens. *Nat. Rev. Cancer* **2019**, *19*, 465–478. [CrossRef]
76. Laumont, C.M.; Vincent, K.; Hesnard, L.; Audemard, E.; Bonneil, E.; Laverdure, J.P.; Gendron, P.; Courcelles, M.; Hardy, M.P.; Cote, C.; et al. Noncoding regions are the main source of targetable tumor-specific antigens. *Sci. Transl. Med.* **2018**, *10*, eaau5516. [CrossRef]
77. Bonaventura, P.; Alcazer, V.; Mutez, V.; Tonon, L.; Martin, J.; Chuvin, N.; Michel, E.; Boulos, R.E.; Estornes, Y.; Valladeau-Guilemond, J.; et al. Identification of shared tumor epitopes from endogenous retroviruses inducing high-avidity cytotoxic T cells for cancer immunotherapy. *Sci. Adv.* **2022**, *8*, eabj3671. [CrossRef]
78. Ouspenskaia, T.; Law, T.; Clauser, K.R.; Klaeger, S.; Sarkizova, S.; Aguet, F.; Li, B.; Christian, E.; Knisbacher, B.A.; Le, P.M.; et al. Unannotated proteins expand the MHC-I-restricted immunopeptidome in cancer. *Nat. Biotechnol.* **2022**, *40*, 209–217. [CrossRef]
79. Camp, F.A.; Slansky, J.E. Implications of Antigen Selection on T Cell-Based Immunotherapy. *Pharmaceuticals* **2021**, *14*, 993. [CrossRef]
80. Ishak, C.A.; Carvalho, D.D.D. Reactivation of Endogenous Retroelements in Cancer Development and Therapy. *Annu. Rev. Cancer Biol.* **2020**, *4*, 159–176. [CrossRef]
81. Nishiyama, A.; Nakanishi, M. Navigating the DNA methylation landscape of cancer. *Trends Genet. TIG* **2021**, *37*, 1012–1027. [CrossRef]
82. Perry, J.A.; Seong, B.K.A.; Stegmaier, K. Biology and Therapy of Dominant Fusion Oncoproteins Involving Transcription Factor and Chromatin Regulators in Sarcomas. *Annu. Rev. Cancer Biol.* **2019**, *3*, 299–321. [CrossRef]
83. Vibert, J.; Saulnier, O.; Collin, C.; Petit, F.; Borgman, K.J.E.; Vigneau, J.; Gautier, M.; Zaidi, S.; Pierron, G.; Watson, S.; et al. Oncogenic chimeric transcription factors drive tumor-specific transcription, processing, and translation of silent genomic regions. *Mol. Cell* **2022**, *82*, 2458–2471.e2459. [CrossRef]
84. Berghuis, D.; de Hooge, A.S.; Santos, S.J.; Horst, D.; Wiertz, E.J.; van Eggermond, M.C.; van den Elsen, P.J.; Taminiau, A.H.; Ottaviano, L.; Schaefer, K.L.; et al. Reduced human leukocyte antigen expression in advanced-stage Ewing sarcoma: Implications for immune recognition. *J. Pathol.* **2009**, *218*, 222–231. [CrossRef] [PubMed]

85. Yabe, H.; Tsukahara, T.; Kawaguchi, S.; Wada, T.; Torigoe, T.; Sato, N.; Terai, C.; Aoki, M.; Hirose, S.; Morioka, H.; et al. Prognostic significance of HLA class I expression in Ewing's sarcoma family of tumors. *J. Surg. Oncol.* **2011**, *103*, 380–385. [CrossRef] [PubMed]
86. Tsukahara, T.; Kawaguchi, S.; Torigoe, T.; Asanuma, H.; Nakazawa, E.; Shimozawa, K.; Nabeta, Y.; Kimura, S.; Kaya, M.; Nagoya, S.; et al. Prognostic significance of HLA class I expression in osteosarcoma defined by anti-pan HLA class I monoclonal antibody, EMR8-5. *Cancer Sci.* **2006**, *97*, 1374–1380. [CrossRef]
87. Nada, O.H.; Ahmed, N.S.; Abou Gabal, H.H. Prognostic significance of HLA EMR8-5 immunohistochemically analyzed expression in osteosarcoma. *Diagn. Pathol.* **2014**, *9*, 72. [CrossRef]
88. Morrison, B.J.; Steel, J.C.; Morris, J.C. Reduction of MHC-I expression limits T-lymphocyte-mediated killing of Cancer-initiating cells. *BMC Cancer* **2018**, *18*, 469. [CrossRef]
89. Cornel, A.M.; Mimpen, I.L.; Nierkens, S. MHC Class I Downregulation in Cancer: Underlying Mechanisms and Potential Targets for Cancer Immunotherapy. *Cancers* **2020**, *12*, 1760. [CrossRef]
90. Borowski, A.; van Valen, F.; Ulbrecht, M.; Weiss, E.H.; Blasczyk, R.; Jorgens, H.; Gobel, U.; Schneider, E.M. Monomorphic HLA class I-(non-A, non-B) expression on Ewing's tumor cell lines, modulation by TNF-α and IFN-γ. *Immunobiology* **1999**, *200*, 1–20. [CrossRef]
91. Spurny, C.; Kailayangiri, S.; Altvater, B.; Jamitzky, S.; Hartmann, W.; Wardelmann, E.; Ranft, A.; Dirksen, U.; Amler, S.; Hardes, J.; et al. T cell infiltration into Ewing sarcomas is associated with local expression of immune-inhibitory HLA-G. *Oncotarget* **2018**, *9*, 6536–6549. [CrossRef]
92. Reits, E.A.; Hodge, J.W.; Herberts, C.A.; Groothuis, T.A.; Chakraborty, M.; Wansley, E.K.; Camphausen, K.; Luiten, R.M.; de Ru, A.H.; Neijssen, J.; et al. Radiation modulates the peptide repertoire, enhances MHC class I expression, and induces successful antitumor immunotherapy. *J. Exp. Med.* **2006**, *203*, 1259–1271. [CrossRef]
93. Rübe, C.E.; van Valen, F.; Wilfert, F.; Palm, J.; Schuck, A.; Willich, N.; Winkelmann, W.; Jürgens, H.; Rübe, C. Ewing's sarcoma and peripheral primitive neuroectodermal tumor cells produce large quantities of bioactive tumor necrosis factor-alpha (TNF-alpha) after radiation exposure. *Int. J. Radiat. Oncol. Biol. Phys.* **2003**, *56*, 1414–1425. [CrossRef]
94. Lam, E.; Stein, S.; Falck-Pedersen, E. Adenovirus detection by the cGAS/STING/TBK1 DNA sensing cascade. *J. Virol.* **2014**, *88*, 974–981. [CrossRef]
95. Rouas-Freiss, N.; Gonçalves, R.M.; Menier, C.; Dausset, J.; Carosella, E.D. Direct evidence to support the role of HLA-G in protecting the fetus from maternal uterine natural killer cytolysis. *Proc. Natl. Acad. Sci. USA* **1997**, *94*, 11520–11525. [CrossRef]
96. Altvater, B.; Kailayangiri, S.; Pérez Lanuza, L.F.; Urban, K.; Greune, L.; Flügge, M.; Meltzer, J.; Farwick, N.; König, S.; Görlich, D.; et al. HLA-G and HLA-E Immune Checkpoints Are Widely Expressed in Ewing Sarcoma but Have Limited Functional Impact on the Effector Functions of Antigen-Specific CAR T Cells. *Cancers* **2021**, *13*, 2857. [CrossRef]
97. Loustau, M.; Anna, F.; Dréan, R.; Lecomte, M.; Langlade-Demoyen, P.; Caumartin, J. HLA-G Neo-Expression on Tumors. *Front. Immunol.* **2020**, *11*, 1685. [CrossRef]
98. Chowdhury, F.; Dunn, S.; Mitchell, S.; Mellows, T.; Ashton-Key, M.; Gray, J.C. PD-L1 and CD8+PD1+ lymphocytes exist as targets in the pediatric tumor microenvironment for immunomodulatory therapy. *Oncoimmunology* **2015**, *4*, e1029701. [CrossRef]
99. Gomez-Brouchet, A.; Illac, C.; Gilhodes, J.; Bouvier, C.; Aubert, S.; Guinebretiere, J.M.; Marie, B.; Larousserie, F.; Entz-Werlé, N.; de Pinieux, G.; et al. CD163-positive tumor-associated macrophages and CD8-positive cytotoxic lymphocytes are powerful diagnostic markers for the therapeutic stratification of osteosarcoma patients: An immunohistochemical analysis of the biopsies fromthe French OS2006 phase 3 trial. *Oncoimmunology* **2017**, *6*, e1331193. [CrossRef]
100. Wu, C.-C.; Beird, H.C.; Andrew Livingston, J.; Advani, S.; Mitra, A.; Cao, S.; Reuben, A.; Ingram, D.; Wang, W.-L.; Ju, Z.; et al. Immuno-genomic landscape of osteosarcoma. *Nat. Commun.* **2020**, *11*, 1008. [CrossRef]
101. Brohl, A.S.; Sindiri, S.; Wei, J.S.; Milewski, D.; Chou, H.-C.; Song, Y.K.; Wen, X.; Kumar, J.; Reardon, H.V.; Mudunuri, U.S.; et al. Immuno-transcriptomic profiling of extracranial pediatric solid malignancies. *Cell Rep.* **2021**, *37*, 110047. [CrossRef]
102. Trieb, K.; Lechleitner, T.; Lang, S.; Windhager, R.; Kotz, R.; Dirnhofer, S. Evaluation of HLA-DR expression and T-lymphocyte infiltration in osteosarcoma. *Pathol. Res. Pract.* **1998**, *194*, 679–684. [CrossRef] [PubMed]
103. Berghuis, D.; Santos, S.J.; Baelde, H.J.; Taminiau, A.H.M.; Maarten Egeler, R.; Schilham, M.W.; Hogendoorn, P.C.W.; Lankester, A.C. Pro-inflammatory chemokine–chemokine receptor interactions within the Ewing sarcoma microenvironment determine CD8+ T-lymphocyte infiltration and affect tumour progression. *J. Pathol.* **2011**, *223*, 347–357. [CrossRef] [PubMed]
104. Stahl, D.; Gentles, A.J.; Thiele, R.; Gütgemann, I. Prognostic profiling of the immune cell microenvironment in Ewing's Sarcoma Family of Tumors. *Oncoimmunology* **2019**, *8*, e1674113. [CrossRef] [PubMed]
105. House, I.G.; Savas, P.; Lai, J.; Chen, A.X.Y.; Oliver, A.J.; Teo, Z.L.; Todd, K.L.; Henderson, M.A.; Giuffrida, L.; Petley, E.V.; et al. Macrophage-Derived CXCL9 and CXCL10 Are Required for Antitumor Immune Responses Following Immune Checkpoint Blockade. *Clin. Cancer Res.* **2020**, *26*, 487–504. [CrossRef] [PubMed]
106. Zumwalt, T.J.; Arnold, M.; Goel, A.; Boland, C.R. Active secretion of CXCL10 and CCL5 from colorectal cancer microenvironments associates with GranzymeB+ CD8+ T-cell infiltration. *Oncotarget* **2015**, *6*, 2981–2991. [CrossRef]
107. Cillo, A.R.; Mukherjee, E.; Bailey, N.G.; Onkar, S.; Daley, J.; Salgado, C.; Li, X.; Li, D.; Ranganathan, S.; Burgess, M.; et al. Ewing sarcoma and osteosarcoma have distinct immune signatures and intercellular communication networks. *Clin. Cancer Res.* **2022**, *28*, 4968–4982. [CrossRef]

108. Liu, M.; Guo, S.; Hibbert, J.M.; Jain, V.; Singh, N.; Wilson, N.O.; Stiles, J.K. CXCL10/IP-10 in infectious diseases pathogenesis and potential therapeutic implications. *Cytokine Growth Factor Rev.* **2011**, *22*, 121–130. [CrossRef]
109. Li, X.; Lu, M.; Yuan, M.; Ye, J.; Zhang, W.; Xu, L.; Wu, X.; Hui, B.; Yang, Y.; Wei, B.; et al. CXCL10-armed oncolytic adenovirus promotes tumor-infiltrating T-cell chemotaxis to enhance anti-PD-1 therapy. *Oncoimmunology* **2022**, *11*, 2118210. [CrossRef]
110. Barreira da Silva, R.; Laird, M.E.; Yatim, N.; Fiette, L.; Ingersoll, M.A.; Albert, M.L. Dipeptidylpeptidase 4 inhibition enhances lymphocyte trafficking, improving both naturally occurring tumor immunity and immunotherapy. *Nat. Immunol.* **2015**, *16*, 850–858. [CrossRef]
111. Decalf, J.; Tarbell, K.V.; Casrouge, A.; Price, J.D.; Linder, G.; Mottez, E.; Sultanik, P.; Mallet, V.; Pol, S.; Duffy, D.; et al. Inhibition of DPP4 activity in humans establishes its in vivo role in CXCL10 post-translational modification: Prospective placebo-controlled clinical studies. *EMBO Mol. Med.* **2016**, *8*, 679–683. [CrossRef]
112. Ligon, J.A.; Choi, W.; Cojocaru, G.; Fu, W.; Hsiue, E.H.-C.; Oke, T.F.; Siegel, N.; Fong, M.H.; Ladle, B.; Pratilas, C.A.; et al. Pathways of immune exclusion in metastatic osteosarcoma are associated with inferior patient outcomes. *J. ImmunoTherapy Cancer* **2021**, *9*, e001772. [CrossRef]
113. Altvater, B.; Kailayangiri, S.; Theimann, N.; Ahlmann, M.; Farwick, N.; Chen, C.; Pscherer, S.; Neumann, I.; Mrachatz, G.; Hansmeier, A.; et al. Common Ewing sarcoma-associated antigens fail to induce natural T cell responses in both patients and healthy individuals. *Cancer Immunol. Immunother. CII* **2014**, *63*, 1047–1060. [CrossRef]
114. Gentles, A.J.; Newman, A.M.; Liu, C.L.; Bratman, S.V.; Feng, W.; Kim, D.; Nair, V.S.; Xu, Y.; Khuong, A.; Hoang, C.D.; et al. The prognostic landscape of genes and infiltrating immune cells across human cancers. *Nat. Med.* **2015**, *21*, 938–945. [CrossRef]
115. Fujiwara, T.; Fukushi, J.; Yamamoto, S.; Matsumoto, Y.; Setsu, N.; Oda, Y.; Yamada, H.; Okada, S.; Watari, K.; Ono, M.; et al. Macrophage infiltration predicts a poor prognosis for human ewing sarcoma. *Am. J. Pathol.* **2011**, *179*, 1157–1170. [CrossRef]
116. Vakkila, J.; Jaffe, R.; Michelow, M.; Lotze, M.T. Pediatric cancers are infiltrated predominantly by macrophages and contain a paucity of dendritic cells: A major nosologic difference with adult tumors. *Clin. Cancer Res.* **2006**, *12*, 2049–2054. [CrossRef]
117. Handl, M.; Hermanova, M.; Hotarkova, S.; Jarkovsky, J.; Mudry, P.; Shatokhina, T.; Vesela, M.; Sterba, J.; Zambo, I. Clinicopathological correlation of tumor-associated macrophages in Ewing sarcoma. *Biomed. Pap.* **2018**, *162*, 54–60. [CrossRef]
118. Buddingh, E.P.; Kuijjer, M.L.; Duim, R.A.J.; Bürger, H.; Agelopoulos, K.; Myklebost, O.; Serra, M.; Mertens, F.; Hogendoorn, P.C.W.; Lankester, A.C.; et al. Tumor-Infiltrating Macrophages Are Associated with Metastasis Suppression in High-Grade Osteosarcoma: A Rationale for Treatment with Macrophage Activating Agents. *Clin. Cancer Res.* **2011**, *17*, 2110–2119. [CrossRef]
119. Dancsok, A.R.; Gao, D.; Lee, A.F.; Steigen, S.E.; Blay, J.-Y.; Thomas, D.M.; Maki, R.G.; Nielsen, T.O.; Demicco, E.G. Tumor-associated macrophages and macrophage-related immune checkpoint expression in sarcomas. *Oncoimmunology* **2020**, *9*, 1747340. [CrossRef]
120. Qian, B.-Z.; Li, J.; Zhang, H.; Kitamura, T.; Zhang, J.; Campion, L.R.; Kaiser, E.A.; Snyder, L.A.; Pollard, J.W. CCL2 recruits inflammatory monocytes to facilitate breast-tumour metastasis. *Nature* **2011**, *475*, 222–225. [CrossRef]
121. Wynn, T.A.; Chawla, A.; Pollard, J.W. Macrophage biology in development, homeostasis and disease. *Nature* **2013**, *496*, 445–455. [CrossRef]
122. Dumars, C.; Ngyuen, J.-M.; Gaultier, A.; Lanel, R.; Corradini, N.; Gouin, F.; Heymann, D.; Heymann, M.-F. Dysregulation of macrophage polarization is associated with the metastatic process in osteosarcoma. *Oncotarget* **2016**, *7*, 78343–78354. [CrossRef] [PubMed]
123. Ségaliny, A.I.; Mohamadi, A.; Dizier, B.; Lokajczyk, A.; Brion, R.; Lanel, R.; Amiaud, J.; Charrier, C.; Boisson-Vidal, C.; Heymann, D. Interleukin-34 promotes tumor progression and metastatic process in osteosarcoma through induction of angiogenesis and macrophage recruitment. *Int. J. Cancer* **2015**, *137*, 73–85. [CrossRef] [PubMed]
124. Han, Y.; Guo, W.; Ren, T.; Huang, Y.; Wang, S.; Liu, K.; Zheng, B.; Yang, K.; Zhang, H.; Liang, X. Tumor-associated macrophages promote lung metastasis and induce epithelial-mesenchymal transition in osteosarcoma by activating the COX-2/STAT3 axis. *Cancer Lett.* **2019**, *440–441*, 116–125. [CrossRef] [PubMed]
125. Hesketh, A.J.; Maloney, C.; Behr, C.A.; Edelman, M.C.; Glick, R.D.; Al-Abed, Y.; Symons, M.; Soffer, S.Z.; Steinberg, B.M. The Macrophage Inhibitor CNI-1493 Blocks Metastasis in a Mouse Model of Ewing Sarcoma through Inhibition of Extravasation. *PLoS ONE* **2016**, *10*, e0145197. [CrossRef] [PubMed]
126. Liang, X.; Guo, W.; Ren, T.; Huang, Y.; Sun, K.; Zhang, H.; Yu, Y.; Wang, W.; Niu, J. Macrophages reduce the sensitivity of osteosarcoma to neoadjuvant chemotherapy drugs by secreting Interleukin-1 beta. *Cancer Lett.* **2020**, *480*, 4–14. [CrossRef]
127. Han, Q.; Shi, H.; Liu, F. CD163+ M2-type tumor-associated macrophage support the suppression of tumor-infiltrating T cells in osteosarcoma. *Int. Immunopharmacol.* **2016**, *34*, 101–106. [CrossRef]
128. MacEwen, E.G.; Kurzman, I.D.; Rosenthal, R.C.; Smith, B.W.; Manley, P.A.; Roush, J.K.; Howard, P.E. Therapy for osteosarcoma in dogs with intravenous injection of liposome-encapsulated muramyl tripeptide. *J. Natl. Cancer Inst.* **1989**, *81*, 935–938. [CrossRef]
129. Kurzman, I.D.; Shi, F.; Vail, D.M.; MacEwen, E.G. In vitro and in vivo enhancement of canine pulmonary alveolar macrophage cytotoxic activity against canine osteosarcoma cells. *Cancer Biother. Radiopharm.* **1999**, *14*, 121–128. [CrossRef]
130. Workenhe, S.T.; Mossman, K.L. Oncolytic virotherapy and immunogenic cancer cell death: Sharpening the sword for improved cancer treatment strategies. *Mol. Ther. J. Am. Soc. Gene Ther.* **2014**, *22*, 251–256. [CrossRef]
131. Grohar, P.J.; Griffin, L.B.; Yeung, C.; Chen, Q.R.; Pommier, Y.; Khanna, C.; Khan, J.; Helman, L.J. Ecteinascidin 743 interferes with the activity of EWS-FLI1 in Ewing sarcoma cells. *Neoplasia* **2011**, *13*, 145–153. [CrossRef]

132. Fujiwara, T.; Healey, J.; Ogura, K.; Yoshida, A.; Kondo, H.; Hata, T.; Kure, M.; Tazawa, H.; Nakata, E.; Kunisada, T.; et al. Role of Tumor-Associated Macrophages in Sarcomas. *Cancers* **2021**, *13*, 1086. [CrossRef]
133. Koo, J.; Hayashi, M.; Verneris, M.R.; Lee-Sherick, A.B. Targeting Tumor-Associated Macrophages in the Pediatric Sarcoma Tumor Microenvironment. *Front. Oncol.* **2020**, *10*, 581107. [CrossRef]
134. Cassetta, L.; Pollard, J.W. Targeting macrophages: Therapeutic approaches in cancer. *Nat. Rev. Drug Discov.* **2018**, *17*, 887–904. [CrossRef]
135. Toulmonde, M.; Penel, N.; Adam, J.; Chevreau, C.; Blay, J.Y.; Le Cesne, A.; Bompas, E.; Piperno-Neumann, S.; Cousin, S.; Grellety, T.; et al. Use of PD-1 Targeting, Macrophage Infiltration, and IDO Pathway Activation in Sarcomas: A Phase 2 Clinical Trial. *JAMA Oncol.* **2018**, *4*, 93–97. [CrossRef]
136. Merchant, M.S.; Wright, M.; Baird, K.; Wexler, L.H.; Rodriguez-Galindo, C.; Bernstein, D.; Delbrook, C.; Lodish, M.; Bishop, R.; Wolchok, J.D.; et al. Phase I Clinical Trial of Ipilimumab in Pediatric Patients with Advanced Solid Tumors. *Clin. Cancer Res.* **2016**, *22*, 1364–1370. [CrossRef]
137. Heymann, M.-F.; Schiavone, K.; Heymann, D. Bone sarcomas in the immunotherapy era. *Br. J. Pharmacol.* **2021**, *178*, 1955–1972. [CrossRef]
138. WHO. *WHO Editorial Board WHO Classification of Tumours: Soft Tissue and Bone Tumours*, 5th ed.; WHO: Geneva, Switzerland, 2020; ISBN 978-92-8324502-5.
139. Smolle, M.A.; Herbsthofer, L.; Goda, M.; Granegger, B.; Brcic, I.; Bergovec, M.; Scheipl, S.; Prietl, B.; El-Heliebi, A.; Pichler, M.; et al. Influence of tumor-infiltrating immune cells on local control rate, distant metastasis, and survival in patients with soft tissue sarcoma. *Oncoimmunology* **2021**, *10*, 1896658. [CrossRef]
140. Dancsok, A.R.; Setsu, N.; Gao, D.; Blay, J.-Y.; Thomas, D.; Maki, R.G.; Nielsen, T.O.; Demicco, E.G. Expression of lymphocyte immunoregulatory biomarkers in bone and soft-tissue sarcomas. *Mod. Pathol.* **2019**, *32*, 1772–1785. [CrossRef]
141. Abeshouse, A.; Adebamowo, C.; Adebamowo, S.N.; Akbani, R.; Akeredolu, T.; Ally, A.; Anderson, M.L.; Anur, P.; Appelbaum, E.L.; Armenia, J.; et al. Comprehensive and Integrated Genomic Characterization of Adult Soft Tissue Sarcomas. *Cell* **2017**, *171*, 950–965.e928. [CrossRef]
142. Chalmers, Z.R.; Connelly, C.F.; Fabrizio, D.; Gay, L.; Ali, S.M.; Ennis, R.; Schrock, A.; Campbell, B.; Shlien, A.; Chmielecki, J.; et al. Analysis of 100,000 human cancer genomes reveals the landscape of tumor mutational burden. *Genome Med.* **2017**, *9*, 34. [CrossRef]
143. Gröbner, S.N.; Worst, B.C.; Weischenfeldt, J.; Buchhalter, I.; Kleinheinz, K.; Rudneva, V.A.; Johann, P.D.; Balasubramanian, G.P.; Segura-Wang, M.; Brabetz, S.; et al. The landscape of genomic alterations across childhood cancers. *Nature* **2018**, *555*, 321–327. [CrossRef] [PubMed]
144. Orth, M.F.; Buecklein, V.L.; Kampmann, E.; Subklewe, M.; Noessner, E.; Cidre-Aranaz, F.; Romero-Pérez, L.; Wehweck, F.S.; Lindner, L.; Issels, R.; et al. A comparative view on the expression patterns of PD-L1 and PD-1 in soft tissue sarcomas. *Cancer Immunol. Immunother. CII* **2020**, *69*, 1353–1362. [CrossRef] [PubMed]
145. Boxberg, M.; Steiger, K.; Lenze, U.; Rechl, H.; von Eisenhart-Rothe, R.; Wörtler, K.; Weichert, W.; Langer, R.; Specht, K. PD-L1 and PD-1 and characterization of tumor-infiltrating lymphocytes in high grade sarcomas of soft tissue—Prognostic implications and rationale for immunotherapy. *Oncoimmunology* **2018**, *7*, e1389366. [CrossRef] [PubMed]
146. Bertolini, G.; Bergamaschi, L.; Ferrari, A.; Renne, S.L.; Collini, P.; Gardelli, C.; Barisella, M.; Centonze, G.; Chiaravalli, S.; Paolino, C.; et al. PD-L1 assessment in pediatric rhabdomyosarcoma: A pilot study. *BMC Cancer* **2018**, *18*, 652. [CrossRef] [PubMed]
147. Siozopoulou, V.; Domen, A.; Zwaenepoel, K.; Van Beeck, A.; Smits, E.; Pauwels, P.; Marcq, E. Immune Checkpoint Inhibitory Therapy in Sarcomas: Is There Light at the End of the Tunnel? *Cancers* **2021**, *13*, 360. [CrossRef]
148. Saerens, M.; Brusselaers, N.; Rottey, S.; Decruyenaere, A.; Creytens, D.; Lapeire, L. Immune checkpoint inhibitors in treatment of soft-tissue sarcoma: A systematic review and meta-analysis. *Eur. J. Cancer* **2021**, *152*, 165–182. [CrossRef]
149. Somaiah, N.; Conley, A.P.; Parra, E.R.; Lin, H.; Amini, B.; Solis Soto, L.; Salazar, R.; Barreto, C.; Chen, H.; Gite, S.; et al. Durvalumab plus tremelimumab in advanced or metastatic soft tissue and bone sarcomas: A single-centre phase 2 trial. *Lancet Oncol.* **2022**, *23*, 1156–1166. [CrossRef]
150. Nakata, E.; Fujiwara, T.; Kunisada, T.; Ito, T.; Takihira, S.; Ozaki, T. Immunotherapy for sarcomas. *Jpn. J. Clin. Oncol.* **2021**, *51*, 523–537. [CrossRef]
151. Nathenson, M.J.; Conley, A.P.; Sausville, E. Immunotherapy: A New (and Old) Approach to Treatment of Soft Tissue and Bone Sarcomas. *Oncol.* **2017**, *23*, 71–83. [CrossRef]
152. Mathieu, M.; Martin-Jaular, L.; Lavieu, G.; Théry, C. Specificities of secretion and uptake of exosomes and other extracellular vesicles for cell-to-cell communication. *Nat. Cell Biol.* **2019**, *21*, 9–17. [CrossRef]
153. Wortzel, I.; Dror, S.; Kenific, C.M.; Lyden, D. Exosome-Mediated Metastasis: Communication from a Distance. *Dev. Cell* **2019**, *49*, 347–360. [CrossRef]
154. Pegtel, D.M.; Gould, S.J. Exosomes. *Annu. Rev. Biochem.* **2019**, *88*, 487–514. [CrossRef]
155. Kalluri, R.; LeBleu, V.S. The biology, function, and biomedical applications of exosomes. *Science* **2020**, *367*, eaau6977. [CrossRef]
156. Van Niel, G.; D'Angelo, G.; Raposo, G. Shedding light on the cell biology of extracellular vesicles. *Nat. Rev. Mol. Cell Biol.* **2018**, *19*, 213–228. [CrossRef]
157. Harding, C.; Heuser, J.; Stahl, P. Receptor-mediated endocytosis of transferrin and recycling of the transferrin receptor in rat reticulocytes. *J. Cell Biol.* **1983**, *97*, 329–339. [CrossRef]

158. Pan, B.T.; Johnstone, R.M. Fate of the transferrin receptor during maturation of sheep reticulocytes in vitro: Selective externalization of the receptor. *Cell* **1983**, *33*, 967–978. [CrossRef]
159. Takahashi, A.; Okada, R.; Nagao, K.; Kawamata, Y.; Hanyu, A.; Yoshimoto, S.; Takasugi, M.; Watanabe, S.; Kanemaki, M.T.; Obuse, C.; et al. Exosomes maintain cellular homeostasis by excreting harmful DNA from cells. *Nat. Commun.* **2017**, *8*, 15287. [CrossRef]
160. Fontana, F.; Carollo, E.; Melling, G.E.; Carter, D.R.F. Extracellular Vesicles: Emerging Modulators of Cancer Drug Resistance. *Cancers* **2021**, *13*, 749. [CrossRef]
161. Ab Razak, N.S.; Ab Mutalib, N.S.; Mohtar, M.A.; Abu, N. Impact of Chemotherapy on Extracellular Vesicles: Understanding the Chemo-EVs. *Front. Oncol.* **2019**, *9*, 1113. [CrossRef]
162. Paskeh, M.D.A.; Entezari, M.; Mirzaei, S.; Zabolian, A.; Saleki, H.; Naghdi, M.J.; Sabet, S.; Khoshbakht, M.A.; Hashemi, M.; Hushmandi, K.; et al. Emerging role of exosomes in cancer progression and tumor microenvironment remodeling. *J. Hematol. Oncol.* **2022**, *15*, 83. [CrossRef]
163. Cappariello, A.; Rucci, N. Tumour-Derived Extracellular Vesicles (EVs): A Dangerous "Message in A Bottle" for Bone. *Int. J. Mol. Sci.* **2019**, *20*, 4805. [CrossRef] [PubMed]
164. Chicon-Bosch, M.; Tirado, O.M. Exosomes in Bone Sarcomas: Key Players in Metastasis. *Cells* **2020**, *9*, 241. [CrossRef] [PubMed]
165. Pachva, M.C.; Lai, H.; Jia, A.; Rouleau, M.; Sorensen, P.H. Extracellular Vesicles in Reprogramming of the Ewing Sarcoma Tumor Microenvironment. *Front. Cell Dev. Biol.* **2021**, *9*, 726205. [CrossRef] [PubMed]
166. Ye, H.; Hu, X.; Wen, Y.; Tu, C.; Hornicek, F.; Duan, Z.; Min, L. Exosomes in the tumor microenvironment of sarcoma: From biological functions to clinical applications. *J. Nanobiotechnol.* **2022**, *20*, 403. [CrossRef] [PubMed]
167. Raimondi, L.; De Luca, A.; Gallo, A.; Costa, V.; Russelli, G.; Cuscino, N.; Manno, M.; Raccosta, S.; Carina, V.; Bellavia, D.; et al. Osteosarcoma cell-derived exosomes affect tumor microenvironment by specific packaging of microRNAs. *Carcinogenesis* **2020**, *41*, 666–677. [CrossRef] [PubMed]
168. Mazumdar, A.; Urdinez, J.; Boro, A.; Migliavacca, J.; Arlt, M.J.E.; Muff, R.; Fuchs, B.; Snedeker, J.G.; Gvozdenovic, A. Osteosarcoma-Derived Extracellular Vesicles Induce Lung Fibroblast Reprogramming. *Int. J. Mol. Sci.* **2020**, *21*, 5451. [CrossRef]
169. Urciuoli, E.; Giorda, E.; Scarsella, M.; Petrini, S.; Peruzzi, B. Osteosarcoma-derived extracellular vesicles induce a tumor-like phenotype in normal recipient cells. *J. Cell. Physiol.* **2018**, *233*, 6158–6172. [CrossRef]
170. Mannerström, B.; Kornilov, R.; Abu-Shahba, A.G.; Chowdhury, I.M.; Sinha, S.; Seppänen-Kaijansinkko, R.; Kaur, S. Epigenetic alterations in mesenchymal stem cells by osteosarcoma-derived extracellular vesicles. *Epigenetics* **2019**, *14*, 352–364. [CrossRef]
171. Sarhadi, V.K.; Daddali, R.; Seppänen-Kaijansinkko, R. Mesenchymal Stem Cells and Extracellular Vesicles in Osteosarcoma Pathogenesis and Therapy. *Int. J. Mol. Sci.* **2021**, *22*, 1035. [CrossRef]
172. Baglio, S.R.; Lagerweij, T.; Pérez-Lanzón, M.; Ho, X.D.; Léveillé, N.; Melo, S.A.; Cleton-Jansen, A.M.; Jordanova, E.S.; Roncuzzi, L.; Greco, M.; et al. Blocking Tumor-Educated MSC Paracrine Activity Halts Osteosarcoma Progression. *Clin. Cancer Res.* **2017**, *23*, 3721–3733. [CrossRef]
173. Lissat, A.; Joerschke, M.; Shinde, D.A.; Braunschweig, T.; Meier, A.; Makowska, A.; Bortnick, R.; Henneke, P.; Herget, G.; Gorr, T.A.; et al. IL6 secreted by Ewing sarcoma tumor microenvironment confers anti apoptotic and cell-disseminating paracrine responses in Ewing sarcoma cells. *BMC Cancer* **2015**, *15*, 552. [CrossRef]
174. Dolan, B.P.; Gibbs, K.D., Jr.; Ostrand-Rosenberg, S. Dendritic cells cross-dressed with peptide MHC class I complexes prime CD8+ T cells. *J. Immunol.* **2006**, *177*, 6018–6024. [CrossRef]
175. Gassmann, H.; Schneider, K.; Evdokimova, V.; Ruzanov, P.; Schober, S.J.; Xue, B.; von Heyking, K.; Thiede, M.; Richter, G.H.S.; Pfaffl, M.W.; et al. Ewing Sarcoma-Derived Extracellular Vesicles Impair Dendritic Cell Maturation and Function. *Cells* **2021**, *10*, 2081. [CrossRef]
176. Arkhypov, I.; Lasser, S.; Petrova, V.; Weber, R.; Groth, C.; Utikal, J.; Altevogt, P.; Umansky, V. Myeloid Cell Modulation by Tumor-Derived Extracellular Vesicles. *Int. J. Mol. Sci.* **2020**, *21*, 6319. [CrossRef]
177. Miller, I.V.; Raposo, G.; Welsch, U.; Prazeres da Costa, O.; Thiel, U.; Lebar, M.; Maurer, M.; Bender, H.U.; von Luettichau, I.; Richter, G.H.; et al. First identification of Ewing's sarcoma-derived extracellular vesicles and exploration of their biological and potential diagnostic implications. *Biol. Cell* **2013**, *105*, 289–303. [CrossRef]
178. Tsugita, M.; Yamada, N.; Noguchi, S.; Yamada, K.; Moritake, H.; Shimizu, K.; Akao, Y.; Ohno, T. Ewing sarcoma cells secrete EWS/Fli-1 fusion mRNA via microvesicles. *PLoS ONE* **2013**, *8*, e77416. [CrossRef]
179. Villasante, A.; Marturano-Kruik, A.; Ambati, S.R.; Liu, Z.; Godier-Furnemont, A.; Parsa, H.; Lee, B.W.; Moore, M.A.; Vunjak-Novakovic, G. Recapitulating the Size and Cargo of Tumor Exosomes in a Tissue-Engineered Model. *Theranostics* **2016**, *6*, 1119–1130. [CrossRef]
180. Wei, Z.; Batagov, A.O.; Schinelli, S.; Wang, J.; Wang, Y.; El Fatimy, R.; Rabinovsky, R.; Balaj, L.; Chen, C.C.; Hochberg, F.; et al. Coding and noncoding landscape of extracellular RNA released by human glioma stem cells. *Nat. Commun.* **2017**, *8*, 1145. [CrossRef]
181. Chevillet, J.R.; Kang, Q.; Ruf, I.K.; Briggs, H.A.; Vojtech, L.N.; Hughes, S.M.; Cheng, H.H.; Arroyo, J.D.; Meredith, E.K.; Gallichotte, E.N.; et al. Quantitative and stoichiometric analysis of the microRNA content of exosomes. *Proc. Natl. Acad. Sci. USA* **2014**, *111*, 14888–14893. [CrossRef]
182. Evdokimova, V.; Ruzanov, P.; Gassmann, H.; Zaidi, S.H.; Peltekova, V.; Heisler, L.E.; McPherson, J.D.; Orlic-Milacic, M.; Specht, K.; Steiger, K.; et al. Exosomes transmit retroelement RNAs to drive inflammation and immunosuppression in Ewing Sarcoma. *bioRxiv* **2019**, 806851. [CrossRef]

183. Balaj, L.; Lessard, R.; Dai, L.; Cho, Y.J.; Pomeroy, S.L.; Breakefield, X.O.; Skog, J. Tumour microvesicles contain retrotransposon elements and amplified oncogene sequences. *Nat. Commun.* **2011**, *2*, 180. [CrossRef] [PubMed]
184. Boelens, M.C.; Wu, T.J.; Nabet, B.Y.; Xu, B.; Qiu, Y.; Yoon, T.; Azzam, D.J.; Twyman-Saint Victor, C.; Wiemann, B.Z.; Ishwaran, H.; et al. Exosome transfer from stromal to breast cancer cells regulates therapy resistance pathways. *Cell* **2014**, *159*, 499–513. [CrossRef] [PubMed]
185. Kassiotis, G.; Stoye, J.P. Immune responses to endogenous retroelements: Taking the bad with the good. *Nat. Rev. Immunol.* **2016**, *16*, 207–219. [CrossRef] [PubMed]
186. Burns, K.H. Transposable elements in cancer. *Nat. Rev. Cancer* **2017**, *17*, 415–424. [CrossRef]

cancers

MDPI

Article

Exploiting the Stemness and Chemoresistance Transcriptome of Ewing Sarcoma to Identify Candidate Therapeutic Targets and Drug-Repurposing Candidates

Elizabeth Ann Roundhill [1,*], Pan Pantziarka [2], Danielle E. Liddle [1], Lucy A. Shaw [1], Ghadeer Albadrani [1] and Susan Ann Burchill [1,*]

1 Children's Cancer Research Group, Leeds Institute of Medical Research, St James's University Hospital, Beckett Street, Leeds LS9 7TF, UK
2 Anticancer Fund, Brusselsesteenweg 11, 1860 Meise, Belgium
* Correspondence: e.a.roundhill@leeds.ac.uk (E.A.R.); s.a.burchill@leeds.ac.uk (S.A.B.)

Simple Summary: Ewing sarcoma is a cancer arising most frequently in teenagers and young adults. For many patients, outcomes are the same today as they were 30 years ago, emphasising the need for more effective treatments to eradicate the cells responsible for progression and relapse. These cells responsible for progression and relapse have been identified using assays that evaluate functional characteristics and expression of cell surface markers. For the first time, we reveal ABCG1 as an additional potential cell surface marker of progression. In rare cancers like Ewing sarcoma, commercial development of new drugs is seldom a priority, reflecting the small number of patients and lack of well-characterised molecular subtypes. Therefore, rather than creating new drugs, which can take 20 to 30 years, repurposing of existing drugs may be an efficient cost-effective strategy to accelerate novel molecularly targeted therapy into clinical trials for patients with Ewing sarcoma. To identify candidate molecular targets, we have used a combination of functional assays and transcriptomic analyses to characterize the cells responsible for progression and relapse. We have then applied a bespoke in silico pipeline to find drugs with known safety profiles that bind to these targets. In the future, after preclinical validation of efficacy and specificity in Ewing sarcoma, some of these drugs may be assessed as combination treatments in clinical trials, with the goal of improving outcomes.

Abstract: Outcomes for most patients with Ewing sarcoma (ES) have remained unchanged for the last 30 years, emphasising the need for more effective and tolerable treatments. We have hypothesised that using small-molecule inhibitors to kill the self-renewing chemotherapy-resistant cells (Ewing sarcoma cancer stem-like cells; ES-CSCs) responsible for progression and relapse could improve outcomes and minimise treatment-induced morbidities. For the first time, we demonstrate that ABCG1, a potential oncogene in some cancers, is highly expressed in ES-CSCs independently of CD133. Using functional models, transcriptomics and a bespoke in silico drug-repurposing pipeline, we have prioritised a group of tractable small-molecule inhibitors for further preclinical studies. Consistent with the cellular origin of ES, 21 candidate molecular targets of pluripotency, stemness and chemoresistance were identified. Small-molecule inhibitors to 13 of the 21 molecular targets (62%) were identified. POU5F1/OCT4 was the most promising new therapeutic target in Ewing sarcoma, interacting with 10 of the 21 prioritised molecular targets and meriting further study. The majority of small-molecule inhibitors (72%) target one of two drug efflux proteins, p-glycoprotein (n = 168) or MRP1 (n = 13). In summary, we have identified a novel cell surface marker of ES-CSCs and cancer/non-cancer drugs to targets expressed by these cells that are worthy of further preclinical evaluation. If effective in preclinical models, these drugs and drug combinations might be repurposed for clinical evaluation in patients with ES.

Keywords: Ewing sarcoma; stemness; pluripotency; CD133; multidrug resistance; ABCG1; POU5F1/CT-4; drug repurposing

Citation: Roundhill, E.A.; Pantziarka, P.; Liddle, D.E.; Shaw, L.A.; Albadrani, G.; Burchill, S.A. Exploiting the Stemness and Chemoresistance Transcriptome of Ewing Sarcoma to Identify Candidate Therapeutic Targets and Drug-Repurposing Candidates. *Cancers* **2023**, *15*, 769. https://doi.org/10.3390/cancers15030769

Academic Editors: Franck Verrecchia, Stefan Burdach, Uta Dirksen and Poul H. Sorensen

Received: 7 November 2022
Revised: 8 January 2023
Accepted: 17 January 2023
Published: 26 January 2023

1. Introduction

Ewing sarcoma (ES) arises in the bone or soft tissue [1], most frequently presenting in young people aged 10–25 years [2]. Standard of care treatment, including a combination of multi-agent chemotherapy, surgery and radiotherapy, has improved outcomes for some patients [1]. However, around 40% of these patients develop multidrug-resistant (MDR) metastatic disease [2–5], leading to relapse and survival rates normally associated with metastatic disease (5-year survival 10% [5–7]). Late relapses and chemotherapy-induced morbidities are additional enduring burdens for patients, their families and carers. A major unmet clinical need is therefore the introduction of molecularly targeted therapy to minimise treatment-related toxicity and improve outcomes.

Progression and relapse are driven by subpopulations of cells capable of self-renewal and migration within tumours that are resistant to current treatments. These cancer stem-like cells (CSCs) have been identified in a range of adult and paediatric solid tumours based on expression of cell surface markers, most frequently CD133 (also known as prominim-1) [8–12]. Ewing sarcoma cancer stem-like cells (ES-CSCs) have been isolated based on expression of CD133 [13–15]. However, distinct CD133-negative CSCs are present in some cancers [8–12], including Ewing sarcoma [15]. Therefore, additional approaches, including formation of clones from single cells and generation of three-dimensional (3D) spheroids, have been used to improve the identification of ES-CSC [16,17] and CSC in other cancer types [18–20].

In this study, we have investigated the completeness of CD133 as a cell surface marker of ES-CSCs and identified ABCG1 as an additional marker of these cells using 3D spheroids and self-renewal from single cells. The expression and prognostic potential of this ABC transporter protein has been evaluated in patient-derived cells and tumours in the online dataset GSE17618. Genes that regulate pluripotency, stemness and the MDR ABC transporter proteins have been identified by comparing the transcriptomes of substrate adherent two-dimensional (2D) non-CSC and 3D spheroid-derived ES-CSCs. We combined genes that were differentially expressed in ES cells grown in 3D spheroids and cells grown in 2D with genes previously reported in patient-derived ES-CSCs [17] to pinpoint candidate molecular targets expressed by ES-CSCs. We then developed a bespoke pipeline to identify small-molecule inhibitors of these targets for further preclinical studies. If these drugs are effective in preclinical models of ES, they may in the future be accelerated into clinical trials for evaluation in patients. This repurposing strategy seeks to reuse existing licensed drugs to treat new indications and is a complementary strategy to de novo drug development [21,22].

2. Materials and Methods

2.1. Cell Culture

ES cell lines (A673, RD-ES, SKES-1, SK-N-MC, TC-32 and TTC-466) were cultured as previously described [23] and purchased from the American Type Culture Collection, Manassas, VA, except for the following cells that were kind gifts: TC-32 cells from Dr. J. Toretsky (Division of Pediatrics, University of Maryland, Baltimore, MD, USA), the TTC-466 cells from Dr. P. Sorenson (British Columbia Children's Hospital, Vancouver, BC, Canada). Primary ES cell cultures and daughter ES-CSCs were cultured as previously described [17]. The embryonic stem cell (ESC) culture SHEF-4 (RRID:CVCL_9791) was a gift from Professor P. Andrews, Centre for Stem Cell Biology, University of Sheffield, Sheffield, UK [24] and used as a positive control for CD133 and ATP binding cassette subfamily G member 1 (ABCG1). The glioblastoma cell line (T98G) was a gift from Professor M. Knowles, University of Leeds, and was a positive control for multidrug resistance-associated protein 1 (MRP1, gene name = ABCC1) [23]. The human embryonic fibroblast line KMST-6 (cultured in MEM containing 10% FBS and 2 mM glutamine) and Jurkat cells (cultured in RPMI 1640 containing 10% FBS and 2 mM glutamine) were gifts from Dr E. Morrison, University of Leeds, and used as the positive control and to generate the calibration curve, respectively, for the measurement of telomere length.

2.2. Western Blotting

Western blotting (WB) was performed as previously described [23]. Equal loading of proteins was confirmed using α-tubulin [23] or β-actin (0.4 μg/mL, A5441, Sigma-Aldrich, Paisley, UK). For the detection of CD133, protein extracts were heated at 95 °C for 5 min before cooling on ice and WB (CD133, 1 μg/mL, W6B3C1, Miltenyi Biotech, Surrey, UK). After incubation with primary antibodies (MRP1 [23] or ABCG1 (1:1000, ab36969; Abcam Plc., Cambridge, UK)) and secondary antibodies [23], proteins were visualised by image capture (at different exposure times depending on the intensity of signal) and quantified using the Li-cor Odyssey infrared imaging system (Li-cor Biosciences, Lincoln, NE, USA).

2.3. Flow Cytometry

2.3.1. Cell Surface Expression of CD133

Cells (5×10^5) were incubated in FcR blocking buffer (Miltenyi Biotech) and either anti-CD133/2 (4.5 μg/mL, clone 293C3, Miltenyi Biotech) or the isotype control IgG2b-PE (4.5 μg/mL, Miltenyi Biotech) antibodies in the dark at 4 °C for 10 min. Cells (10,000 per sample) were then analysed by flow cytometry using the FACSCalibur (BD Biosciences, Berkshire, UK).

2.3.2. Co-Expression of ABCG1 and CD133

SK-N-MC cells were incubated in normal goat serum (1:10, Dako, Agilent Technologies, Santa Clara, CA, USA) in BD Perm/Wash™ Buffer for 30 min at 4 °C. Cells were then incubated with CD133 (anti-CD133/2, 4.5 μg/mL, clone 293C3) and/or ABCG1 rabbit polyclonal antibody (100 μg/mL, PA5-13462, Thermo-Fisher Scientific, Paisley, UK) in BD Perm/Wash™ Buffer for 1h at 4 °C in the dark. Control cells were incubated with IgG2b-PE (4.5 μg/mL, Miltenyi Biotech) or rabbit IgG isotype control antibody (100 μg/mL, Dako). Cells were then incubated with the secondary antibody (2 μg/mL goat anti rabbit IgG FITC, A31556, Thermo-Fisher Scientific) in BD Perm/Wash™ Buffer for 30 min at 4 °C in the dark. Cells (10,000 per sample) were analysed by flow cytometry using the CytoFLEX (Beckman Coulter, High Wycombe, UK).

2.4. Self-Renewing Ability

Growth of progeny from a single cell as an adherent culture was determined as previously described [17,18]. A single cell (Poisson distribution probability of $\lambda < 1 = 0.9$) was seeded into each well of 10 Primaria™ 96-well plates (Corning) and the number of wells containing ≥5 cells was recorded after 21 days by light microscopy (Olympus CKX41). Where possible, single cell self-renewing cell populations were propagated to establish daughter cell cultures; these are subsequently referred to as ES-CSCs. To examine colony forming efficiency in soft agar, a single cell suspension (1.8×10^5 cells) in cell-specific media containing 0.3% agar (Aldrich, Poole, UK), was overlaid on a solid agar bed (0.5% agar in media). After 21 days, colonies were stained with 8 mM iodonitrotetrazolium chloride (made up in ddH_2O (*w*/*v*); Sigma-Aldrich) for 16 h and colony number counted by light microscopy. Colony forming efficiency = [the number of colonies formed in the field of view/number of cells seeded] × 100.

For spheroid formation, a single cell was seeded into each well of an ultra-low attachment plate (Corning, UK) in cell line-specific media. The number of spheroids at 21 days was recorded to calculate the spheroid forming efficiency (SFE); SFE = [number of wells containing a spheroid/the number of wells seeded with a single cell] × 100. Spheroids were imaged by light microscopy (Olympus CKX41). For reverse transcriptase quantitative polymerase chain reaction (RTqPCR), Western blot, immunocytochemistry (ICC) and flow cytometry, spheroids formed from single cells were collected at 21 days.

2.5. Magnetically Activated Cell Sorting for the Enrichment of CD133-Positive Cells

ES cells (1×10^8) were incubated with 300 μL of CD133 microbeads and 100 μL of FcR blocking buffer (Miltenyi Biotech). CD133-positive cells were isolated using LS columns

(Miltenyi Biotech). CD133-negative cells were depleted of labelled (CD133-positive) cells by passing the cells through two LD columns (Miltenyi Biotech). Cell surface expression of CD133 was confirmed by flow cytometry immediately after separation: >90% of the CD133-positive selection expressed CD133, and CD133 was expressed by <5% of the CD133-negative selected cells. CD133-positive A673 and TC-32 cells remained positive over 15 passages.

2.6. RNA Expression of Markers of Pluripotency and Differentiation, the Wnt Signalling Pathway and ABC Transporter Proteins

RNA was extracted using the RNeasy Micro Kit (Qiagen, Düsseldorf, Germany) and RNA quantity and quality measured using the Nanodrop and Agilent 2100 bioanalyzer. RNA (with a RIN > 8) was converted to cDNA by reverse transcription [23]. The mRNA expression was evaluated using the TaqMan® Human Stem Cell Pluripotency Array, TaqMan® Human Wnt Pathway and TaqMan® Human ABC Transporter Array (Applied Biosystems, Waltham, MA, USA) [23]. To allow direct comparison across the 3 array platforms target Ct values were normalised using the global mean [25–27]. ABC transporter and pluripotency mRNAs (n = 140, excluding endogenous control mRNAs) more highly expressed than other mRNAs (Ct values < 25), expressed (Ct values 25–35) and not expressed (Ct values >35, [18,28,29]) in 3D spheroids from TTC 466 and SK-N-MC cells were identified. Unique and shared mRNAs in each group were analysed using the Search Tool for Retrieval of Interacting Genes/Proteins (STRING) database (http://string-db.org, [17,30]) to identify GO terms. Significant differences in mRNA expression were determined using Linear Models for Microarray Data (LIMMA) [18]. Target mRNAs were validated using individual RTqPCR assays if the Q value was <0.1, there was at least a mean Ct difference of >2 between the compared populations and Ct values were <35 [18]. For mRNA validation, total RNA (20 ng) was reverse-transcribed and cDNA added to the PCR mix containing sequence specific reverse and forward primers and probe for PPIA (the housekeeping gene) or ABCG1 (Thermo Fisher Scientific; ABCG1 Hs00245154_m1) and 1 × TaqMan® Universal PCR Master Mix (Thermo Fisher Scientific). RNA expression was calculated using the comparative Ct method [31].

ABCG1 transcripts were characterised in total RNA extracted from SK-N-MC cells grown as 2D cultures or 3D spheroids, sequenced and aligned as previously described [17]. Normalised read counts were identified in total RNA sequencing data using DESeq2 [32], adjusted p values of <0.01 were considered significant.

2.7. Immunohistochemistry (IHC)

Formalin-fixed paraffin-embedded (FFPE) sections (4 μm) of spheroids were deparaffinised in xylene (2 min) and rehydrated before antigen retrieval in citric acid buffer (10 mM in ddH_2O, pH6, [33]). Endogenous peroxidases were blocked using 3% hydrogen peroxide for 5 min and endogenous biotin, biotin receptors and avidin-binding sites blocked using an avidin/biotin blocking kit (Invitrogen, Waltham, MA, USA). Sections were incubated for 1 h with ABCG1 primary antibody (10 μg/mL, PA5-13462, Thermo-Fisher Scientific) or rabbit IgG control (10 μg/mL, Dako) at room temperature, followed by incubation with the secondary antibody. Sections were then incubated with streptavidin–peroxidase (Abcam Plc.), followed by DAB substrate (Dako) for 10 min and nuclei counterstained with haematoxylin.

2.8. siRNA Knockdown of ABCG1

Cells (5×10^4) were seeded and allowed to adhere overnight. Media were replaced with Accell siRNA Delivery Media (B-005000-500, Dharmacon, Lafayette, CO, USA) alone or containing ABCG1 siRNA (1.5 μM, E-008615-00-0005, SMARTpool:Accell ABCG1 siRNA, containing 4 siRNAs targeting exon 6, present in all canonical and novel transcripts (Dharmacon) or non-targeting control siRNAs (1.5 μM, D-001910-10-20, Dharmacon) for 72 h. ABCG1 knockdown was confirmed by RTqPCR.

2.9. Apoptosis

Cells (5×10^4) were harvested following treatment with ABCG1 and control siRNA for 24–72 h and apoptosis measured by annexin V and propidium iodide (PI) labelling of cells (annexin V–FITC apoptosis detection kit, BD Biosciences, [34]).

2.10. Statistical Analyses

Differences in proliferation and viable cell numbers were log transformed and analysed by linear regression. Differences in the gradients of each plot were compared using the extra sum of squares F test. For all other experiments, data were analysed by analysis of variance (ANOVA followed by a Bonferroni's or Tukey's post hoc test), or a non-parametric Mann–Whitney or unpaired two-tailed *t*-test. Correlations were determined using a Pearson's correlation coefficient (r). p values of <0.05 were considered significantly different. Statistical analyses were performed using GraphPad PRISM 7.03 (GraphPad Software, San Diego, CA, USA).

2.11. Identification of Candidate Drugs for Prioritised Molecular Targets

Gene lists were analysed using the STRING database. To identify drugs reported to target these molecular candidates, we interrogated DrugBank (version 5.1.9 [35]) and DGI database (DGIdb, version v4.2.0 [36]) to access information on FDA-approved drugs and their molecular targets. Additional drug-target information was derived from the literature-based repurposing drugs in oncology (ReDO) database [37]. A polypharmacology approach identified targets for each drug candidate. Those targets previously associated with ES using the Open Targets Platform [38] and DisGeNET [39] were employed as the identification source. Data from DGIdb and Open Targets characterise the strength of target–disease and drug–target associations. Using DGIdb, the interaction score is computed from: (publication count + data source count) × (average known gene partners for all drugs/known gene partners for candidate drug) × (average known drug partners for all genes/known drug partners for target gene). The Dir DGI score therefore incorporates data on the number of molecular targets of the drug, the number of drugs that target a gene and the number of publications and data sources supporting the association. The Dir Assoc score is generated from the Open Targets database and is a measure of the relationship between the molecular target and "cancer" as the disease term.

Licensed cancer drugs were identified from the list of drug candidates addressing >1 ES molecular target using the Cancer Drugs database [40]. ES trials were identified using clinicaltrials.gov, EU Clinical Trials Register and WHO International Clinical Trials Registry Platform. For non-cancer drugs previously identified as oncological repurposing candidates (ReDO database), clinical trial activity in any cancer was extracted using the ReDO_Trials database of active repurposing trials in oncology [41]. For these repurposing candidates, a support score was calculated from data in the ReDO database to characterise the range of data available illustrating the anticancer effects of the drugs (e.g., in vitro, in vivo, case reports, observational data and clinical trial data).

3. Results

3.1. ES Cell Lines Produce Spheroids and Clones from a Single Cell

All ES cell lines (6/6) formed clones or spheroids from a single cell when cultured in soft agar or on ultra-low attachment plates (Table 1, Figure 1), consistent with previous reports that ES cell cultures contain a self-renewing cell population [8–14,17].

Table 1. Self-renewing ability of ES cell lines. A single cell suspension of each cell line was seeded into soft agar or into a 96 well ultra-low adherence plate at a density of 1–10,000 cells. The number of colonies was counted after 21 days. For studies in soft agar, the mean (±SEM) percentage colony formation is expressed as the number of colonies counted at 21 days relative to the number of cells seeded. For cells seeded onto an ultra-low adherence plate, the mean (±SEM) percentage colony formation is expressed as the number of wells containing a colony of >5 cells relative to the total number of wells with cells seeded. Results are representative of three independent experiments Percentage colony formation was compared using ANOVA and Bonferroni's post hoc test.

Cell Line	Clone or Spheroid Formation (%)					
	Clone Formation in Soft Agar from a Single Cell	Ultra-Low Attachment Plate				
		1 Cell	10 Cells	100 Cells	1000 Cells	10,000 Cells
A673	17 ±1	18 ± 5	88 ± 10	100	100	100
RD-ES	17 ±1	5 ± 1	45 ± 13	100	100	100
SKES-1	29 ± 2 ($p < 0.05$)	75 ± 7 ($p < 0.05$)	98 ± 2	100	100	100
SK-N-MC	12 ± 1	64 ± 1	97 ± 3	100	100	100
TC-32	7 ± 0.5 ($p < 0.05$)	15 ± 3	93 ± 9	100	100	100
TTC 466	13 ± 1	45 ± 2	96 ± 6	100	100	100

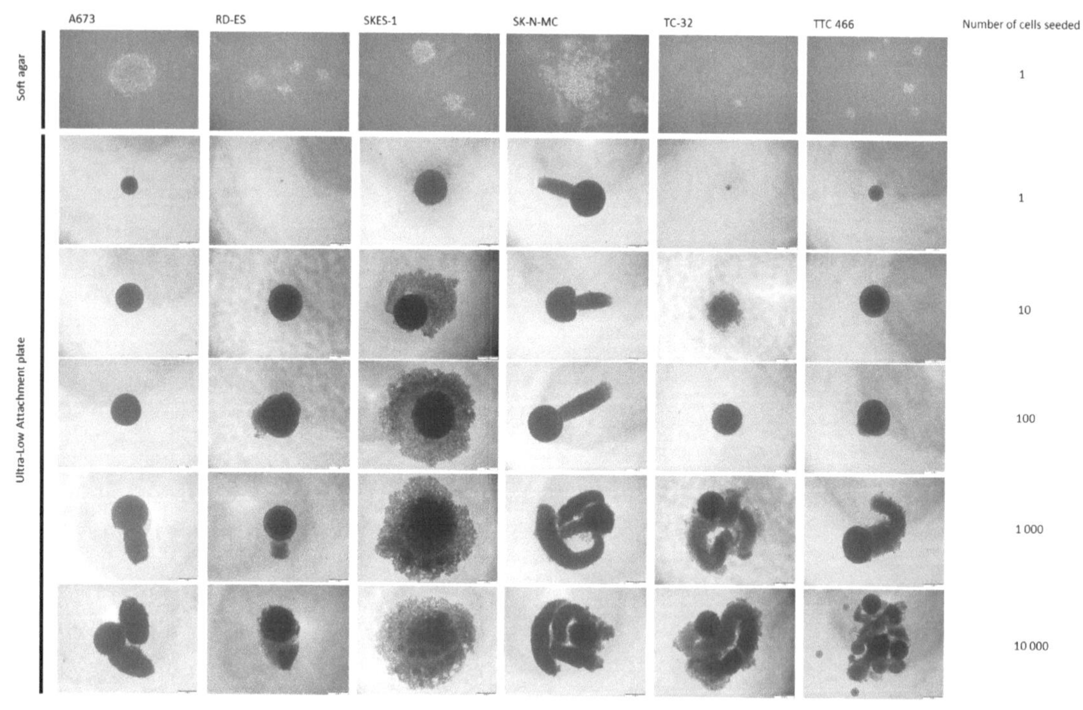

Figure 1. Self-renewing ability of ES cell lines. Single cell suspensions of ES cells were prepared in 0.3% soft agar or 1–10,000 cells seeded into each well of an ultra-low attachment plate (96 wells). Images show colonies formed at 21 days and are representative of three independent experiments.

The formation of single cell-derived clones and spheroids in soft agar (29 ± 2%, $p < 0.05$) and ultra-low attachment conditions (75 ± 7%, $p < 0.05$; Table 1, Figure 1) was most efficient in the SKES-1 cell line. Spheroid formation was 100% in all cell lines when combining 100 cells or more (Table 1). Larger spheroids (>400 µm), produced cell line-specific morphology (Figure 1). SKES-1 spheroids produced a relatively uniform sphere of disseminated or dissociated cells, consistent with spheroid formation in a range of solid tumour cell types [42–46]. For the first time, we identified 3D projections developing from the central core in five of six ES cell lines (Figure 1). The biological significance of these projections requires further investigation.

3.2. CD133 Identifies Some Self-Renewing Drug-Resistant ES-CSCs

CD133 protein was detected in all ES cell lines, with the exception of SK-N-MC cells (Figure 2A and Supplementary Data S1). A673 and TC-32 CD133-positive cells formed significantly more colonies in soft agar than the CD133-negative SK-N-MC cells (Supplementary Data S1–S3). However, there was no difference in proliferation, cell cycle status or telomere length across the two cell lines (Supplementary Data S1–S3), phenotypes frequently associated with CSCs and self-renewing ability [47]. Moreover, expression and activity of the multidrug-resistant protein MRP1 was increased in CD133-positive TC-32 cells, but not the A673 CD133-positive population (Supplementary Data S1–S3). The shared CSC phenotype of both CD133-positive and CD133-negative ES populations demonstrates that CD133 expression alone is not sufficient to enrich for all the ES-CSC population.

A

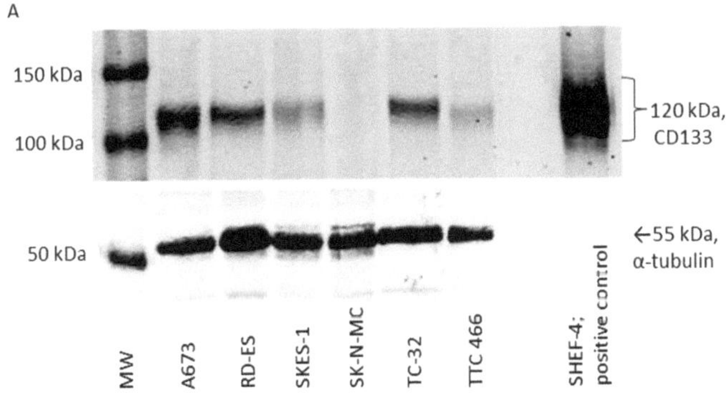

B

Gene	Mean Ct CD133 negative	Mean Ct CD133 positive	Difference in ΔCt	Fold change in expression comparing CD133 positive and CD133 negative cells	Increased (I) or decreased (D) expression in CD133 positive cells	Q value
CRABP2	28.37	25.68	0.70	1.40	I	0.1176
CXXC4	29.81	26.57	3.40	6.80	I	0.1176
NES	31.81	32.98	0.46	-0.92	D	0.1176
SFRP2	24.54	23.38	0.24	0.48	I	0.1176
ABCA4	35.86	34.25	0.40	0.81	I	0.1176
COL2A1	31.32	28.09	0.87	1.74	I	0.1176
FOXD3	32.05	29.05	0.80	[illegible]	I	[illegible]
GBX2	37.61	33.26	1.20	2.40	I	0.1176
GATA6	35.37	31.87	0.94	1.88	I	0.1176
BCRP	32.69	38.09	-1.70	-3.40	D	0.1176
NR5A2	34.58	33.02	0.36	0.72	I	0.1232
EOMES	33.70	32.28	0.32	0.64	I	0.1232
Xist	25.11	22.31	0.73	1.46	I	0.1467
PODXL	26.46	24.87	0.37	0.74	I	0.1467
NOG	33.80	33.41	0.01	[illegible]	I	0.1680
SLC9A3R1	29.71	28.22	0.92	1.84	I	0.1721

C

Gene	Mean Ct CD133 negative	Mean Ct CD133 positive	Difference in ΔCt	Fold change in expression comparing CD133 positive and CD133 negative cells	Increased (I) or decreased (D) expression in CD133 positive cells	Q value
TFCP2L1	34.36	31.70	2.57	5.14	I	0.4438
DKK1	30.10	29.30	0.10	0.20	I	0.4438
EEF1A1	25.43	24.25	1.16	2.32	I	0.4438
ABCA13	29.02	33.07	-4.1	-8.20	D	0.4438
GCG	40.00	40.00	0.01	0.02	I	0.4438
TLE2	24.50	24.70	-0.013	-0.03	D	0.4438
BCRP	29.70	29.16	0.73	1.46	I	0.5145
CDH5	33.76	31.27	2.49	4.98	I	[illegible]
POU5F1	29.81	27.96	1.83	3.66	I	0.5145
LIFR	34.96	31.90	3.05	6.10	I	0.5145
LAMB1	22.47	21.39	1.07	2.14	I	0.5145
ABCA6	30.29	31.33	-2	-4.00	D	0.5145
SEMA3A	26.20	23.94	2.25	4.50	I	0.5145
GFAP	40.00	40.00	0.01	0.02	I	0.5145
[illegible]	[illegible]	[illegible]	-1.4	-2.80	D	0.5145
ABCC9	40.00	38.70	1.12	2.24	I	0.5145
TERT	27.81	27.13	0.66	1.32	I	0.5145
TLE6	29.70	29.90	-0.00001	0.00	D	0.5145
BRIX	22.70	22.20	0.57	1.14	I	0.5145
NOG	30.81	29.65	1.15	2.30	I	0.5145
WNT10A	34.20	34.70	-0.0002	0.00	D	0.5145
LAMC1	23.06	22.06	0.99	1.98	I	0.5145
FRZB	24.90	24.60	0.15	0.30	I	0.5145

Figure 2. Characterisation of CD133 cells in ES cell lines. (**A**) Protein expression of CD133 in ES cell lines (n = 6) was evaluated by Western blot. Equal protein loading was confirmed by probing the blots for α-tubulin. The top differentially expressed genes comparing (**B**) TC-32 and (**C**) A673 CD133-positive and -negative cells using LIMMA were not significant. The difference in ΔCt between the two groups and corresponding Q values are shown. Genes were considered significantly differentially expressed if the Q value was <0.1 and the difference in ΔCt > 2. Genes with Ct values of >35 in both cell populations were excluded. Dark grey = ABC transporter proteins, grey = pluripotency associated genes, white = Wnt signalling pathway genes. Results of two independent experiments. SHEF-4 embryonic stem cells were included as a positive control for CD133 expression throughout.

Consistently with the cellular origin of ES and the high level of stemness markers expressed by these tumour cells [48–50], no significant differentially expressed genes associated with pluripotency, ABC transporter and Wnt signalling pathways were identified in CD133-positive and CD133-negative cells (Figure 2B (TC-32) and Figure 2C (A673), Supplementary Data S4). These observations are also consistent with the premise that CD133 can be used to identify some cells with characteristics of ES-CSCs, although pathways classically associated with the CSC phenotype in other cancer types are also expressed in CD133-negative ES cells. To investigate the CD133-independent ES-CSC phenotype, we went on to investigate ES-CSCs enriched through spheroid formation in ES cells with no or low CD133 expression.

3.3. Gene Expression Profile of CD133 Low or Negative ES 3D Spheroids and 2D Cultures

As TTC 466 and SK-N-MC cells had low or no detectable CD133 protein (Figure 2A), we compared the transcriptome of TTC 466 and SK-N-MC 3D spheroids and 2D cultures (Figure 3A). Consistently with the premise that spheroid formation is a feature of CSCs, expression of the stemness markers LEFTB and LIN28 were significantly increased in SK-N-MC spheroids compared to cells in 2D culture, whereas expression of LAMA1, COL2A1, ACTC, GCG, SEMA3A and PODXL were significantly decreased (Q value < 0.1, Figure 3B,D, Supplementary Data S5A). In TTC 466 spheroids, no markers were significantly increased; however, expression of FLT1 and GATA6 was significantly decreased compared to cells in 2D culture (Figure 3C,E, Supplementary Data S5A).

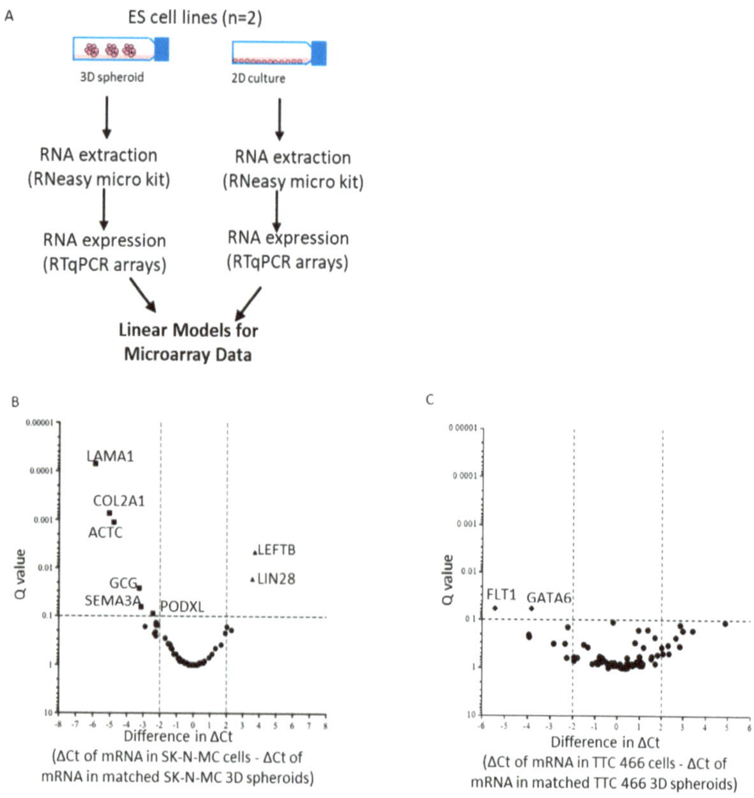

Figure 3. *Cont.*

D

Gene	Mean Ct 3D spheroids	Mean Ct cells	Difference in Δ Ct	Fold change in expression comparing SK-N-MC cells and 3D spheroids	Increased (I) or decreased (D) expression in 3D spheroids	Q value
LAMA1	40.00	33.87	-5.88	-11.77	D	0.0001
COL2A1	39.17	33.93	-5.07	-10.14	D	0.0007
ACTC	37.90	32.96	-4.79	-9.58	D	0.0011
LEFTB	30.47	34.10	3.72	7.44	I	0.0046
LIN28	30.78	34.34	3.57	7.15	I	0.0163
GCG	31.06	27.81	-3.25	-6.51	D	0.0261
SEMA3A	36.27	33.21	-3.14	-6.28	D	0.0650
PODXL	27.25	24.85	-2.42	-4.83	D	0.0892
SOX2	29.09	27.05	-2.23	-4.45	D	0.1361
DNMT3B	28.41	26.27	-2.13	-4.27	D	0.1532
KIT	26.19	23.93	-2.23	-4.46	D	0.1532
DES	35.23	32.39	-2.89	-5.77	D	0.1650

E

Gene	Mean Ct 3D spheroids	Mean Ct cells	Difference in ΔCt	Fold change in expression comparing TTC466 cells and 3D spheroids	Increased (I) or decreased (D) expression in 3D spheroids	Q value
FLT1	36.34	30.50	-5.84	-11.69	D	0.0587
GATA6	34.59	30.36	-4.23	-8.46	D	0.0587
SOX17	32.34	36.82	4.48	8.96	I	0.1120
LAMC1	27.47	24.84	-2.62	-5.25	D	0.1420
TDGF1	31.98	34.57	2.58	5.17	I	0.1662
PODXL	31.05	26.71	-4.34	-8.69	D	0.2058
DNMT3B	28.91	26.81	-2.09	-4.19	D	0.2303
FGF5	38.14	33.82	-4.32	-8.65	D	0.2303
IMP2	26.50	23.76	-2.74	-5.48	D	0.3122
OLIG2	37.53	34.29	-3.24	-6.49	D	0.3122
SEMA3A	29.05	27.14	-1.91	-3.82	D	0.3284
HBZ	31.90	33.81	1.91	3.82	I	0.3328
TERT	29.48	27.76	-1.72	-3.45	D	0.3692

Figure 3. Expression of pluripotency genes in SK-N-MC and TTC 466 cells in 2D and 3D spheroid cultures. (**A**) Summary of strategy to identify shared markers of stemness and MDR in ES cell lines. RNA (1 µg) from (**B**) SK-N-MC and (**C**) TTC 466 cells in 2D and 3D cultures were analysed by RTqPCR using the TaqMan® Human Stem Cell Pluripotency array. The level of mRNA expression is reported using the comparative Ct method, after normalisation of target Ct values to the global mean of all mRNAs. A volcano plot summarising the differentially expressed mRNAs in 2D and 3D spheroids, displaying the Q value (level of significance) and the difference in ΔCt between the cells in 2D and 3D spheroids is shown. The vertical dashed lines indicate the ΔCt thresholds of ±ΔCt of >2 (*x*-axis) and the horizontal line a significant Q value < 0.1 (*y*-axis). Black squares = mRNAs significantly decreased in SK-N-MC 3D spheroids and with a change in ΔCt > 2, Black diamonds = mRNAs significantly decreased in TTC 466 3D spheroids and with a change in ΔCt > 2, black triangles = mRNAs significantly increased in 3D spheroids and with a change in ΔCt >2, black circles = mRNAs with a ΔCt of ±<2 and Q value > 0.1. Genes with Ct values of >35 in 2D cells were excluded. The top differentially expressed genes with a Q value of <0.2 comparing (**D**) SK-N-MC and (**E**) TTC 466 cells in 2D and 3D cultures using LIMMA are shown. The difference in ΔCt between the two groups and corresponding Q values are shown. Genes were considered significantly differentially expressed if the Q value was <0.1 and the difference in ΔCt > 2.

Expression of 20 ABC transporter genes was significantly different in SK-N-MC spheroids compared to the cells in 2D cultures, although only ABCG1 and CFTR were differentially expressed greater than twofold (Figure 4A–C). However, ABC transporter genes, including ABCG1, were not significantly differentially expressed in TTC 466 cells grown in 2D or as 3D spheroids (Figure 4D–F). Comparison of the mRNAs in SK-N-MC and TTC 466 spheroids revealed that 35% of genes were undetected in both populations and an additional seven (5%) and six (4%) unique mRNAs were undetected in TTC 466 and SK-N-MC spheroids respectively (Figure 4G). Fifty-six percent of mRNAs were detected in both spheroid populations and 8% of these were highly expressed (Figure 4G). The six mRNAs uniquely expressed in TTC 466 spheroids are involved in cell differentiation (GO:0030154), which may explain why increased expression of pluripotency mRNAs was not observed in TTC 466 spheroids. Previous studies have shown knockdown of *EWSR1-FLI1* decreases the expression of stemness markers and stem-like properties of ES cells [51]. Whether the lack of shared pluripotency mRNAs increased in the TTC 466 and SK-N-MC spheroids reflects the different transcriptional activators of TTC 466 (*EWSR1-ERG)* and SK-N-MC cells (*EWSR1-FLI1*) remains to be seen.

A

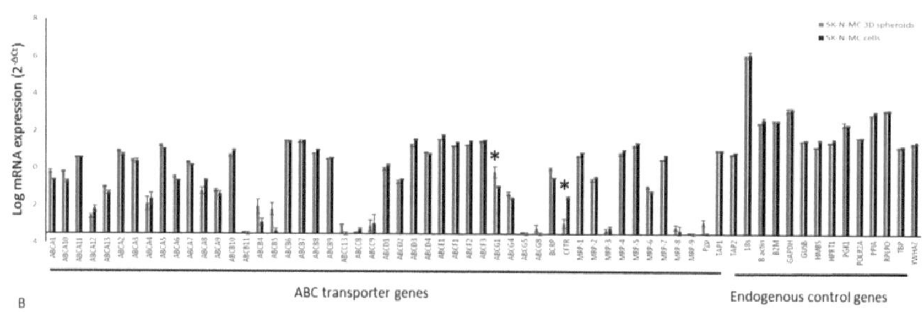

B

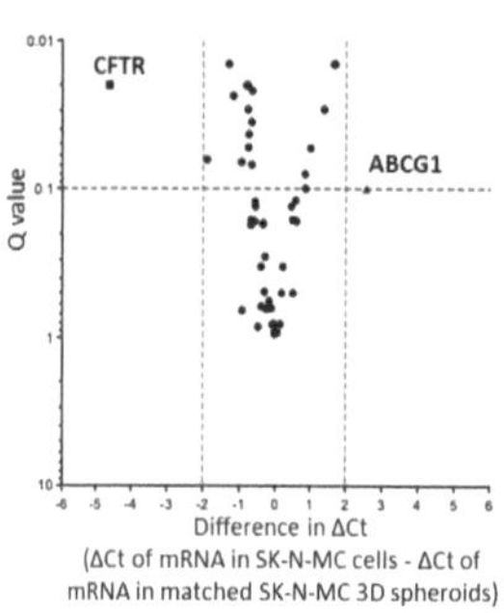

C

Gene	Mean Ct 3D spheroids	Mean Ct cells	Difference in ΔCt	Fold change in expression comparing SK-N-MC cells and 3D spheroids	Increased (I) or decreased (D) expression in 3D spheroids	Q value
HMBS	24.69	23.17	-1.3	-2.60	D	0.014
ABCA10	29.32	30.76	1.66	3.32	I	0.014
BCRP	28.66	30.13	1.7	3.40	I	0.014
CFTR	38.32	33.48	-4.61	-9.22	D	0.020
MRP-7	27.00	25.96	-1.04	-2.08	D	0.020
ABCF2	24.46	23.46	-0.78	-1.56	D	0.020
ABCB8	25.98	25.09	-0.65	-1.30	D	0.022
ABCD3	24.55	23.14	-1.18	-2.36	D	0.023
ABCF1	24.69	23.69	-0.77	-1.54	D	0.029
ABCA1	29.32	30.49	1.39	2.78	I	0.029
MRP-1	26.46	25.56	-0.67	-1.34	D	0.036
ABCE1	23.36	22.38	-0.75	-1.50	D	0.043
MRP-4	26.03	25.03	-0.76	-1.52	D	0.053
ABCA13	31.99	32.78	1.01	2.02	I	0.053
ABCA8	32.70	30.58	-1.89	-3.78	D	0.064
ABCB10	26.43	25.25	-0.95	-1.90	D	0.066
HPRT1	23.96	23.07	-0.66	-1.32	D	0.069
MRP-6	31.84	32.47	0.85	1.70	I	0.080
ABCG4	33.06	33.70	0.86	1.72	I	0.099
ABCG1	29.24	31.58	2.56	5.12	I	0.099
MRP-5	24.61	23.80	-0.8	-1.60	D	0.118
ABCA6	30.18	30.54	0.58	1.16	I	0.120

D

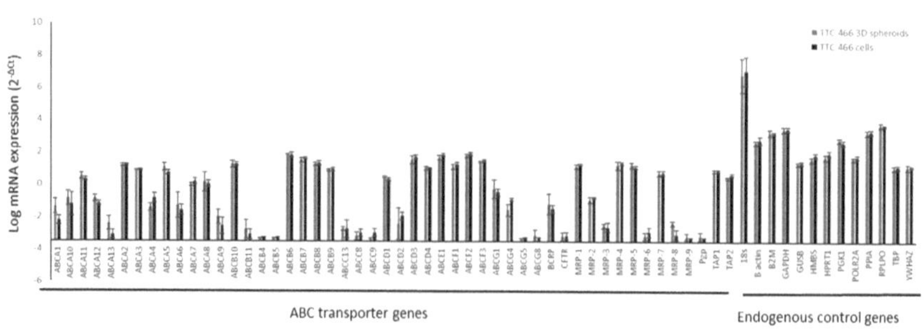

Figure 4. *Cont.*

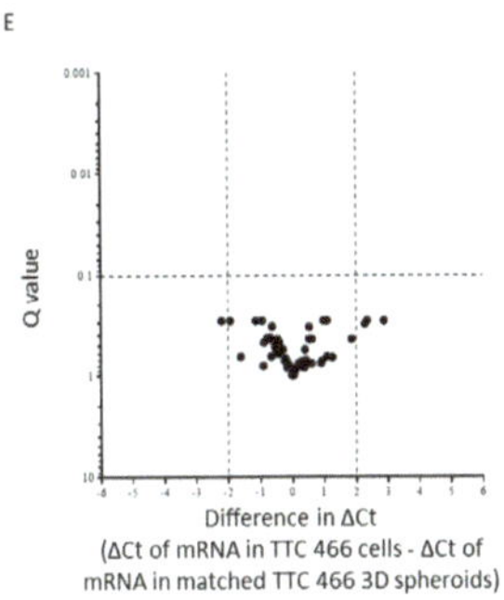

F

Gene	Mean Ct spheroids	Mean Ct cells	Difference in ΔCt	Fold change in expression comparing TTC 466 cells and 3D spheroids	Increased (I) or decreased (D) expression in 3D spheroids	Q value
ABCA1	33.5	36.6	3.1	6.2	I	0.2856915
ABCA12	31.9	33.1	1.2	2.4	I	0.2856915
ABCA4	33.7	32.0	-1.7	-3.4	D	0.2856915
ABCA5	25.5	26.8	1.3	2.6	I	0.2856915
ABCC9	40.0	39.1	-0.9	-1.9	D	0.2856915
ABCG4	34.2	32.2	-2.0	-4.0	D	0.2856915
HMBS	23.7	23.0	-0.7	-1.5	D	0.2856915
MRP-8	36.8	39.4	2.6	5.1	I	0.2856915
ABCA13	37.0	39.5	2.5	5.0	I	0.3076924
ABCE1	23.5	23.1	-0.4	-0.8	D	0.3268183
ABCD1	27.4	28.1	0.7	1.5	I	0.3304234
HPRT1	23.3	22.7	-0.6	-1.2	D	0.4274480
TAP2	27.5	27.1	-0.4	-0.9	D	0.4274480
ABCA11	27.3	28.2	0.8	1.7	I	0.4348587
ABCA7	29.1	28.6	-0.5	-0.9	D	0.4348587
ABCA9	35.6	37.7	2.1	4.1	I	0.4348587
MRP-2	32.1	31.8	-0.3	-0.5	D	0.4348587
PGK1	19.6	20.3	0.7	1.5	I	0.4348587
ABCB7	24.0	23.8	-0.2	-0.4	D	0.4684602
ABCF2	23.1	22.9	-0.2	-0.4	D	0.4684602
MRP-1	25.2	25.0	-0.2	-0.5	D	0.4684602
MRP-6	39.6	38.9	-0.7	-1.4	D	0.4684602
ABCF1	25.3	25.1	-0.3	-0.6	D	0.4986656
ABCD3	23.9	23.6	-0.3	-0.7	D	0.5142617
ABCF3	24.4	24.3	-0.1	-0.2	D	0.5511673

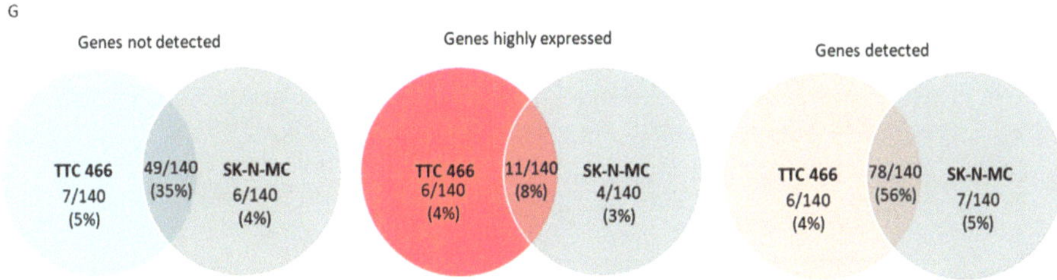

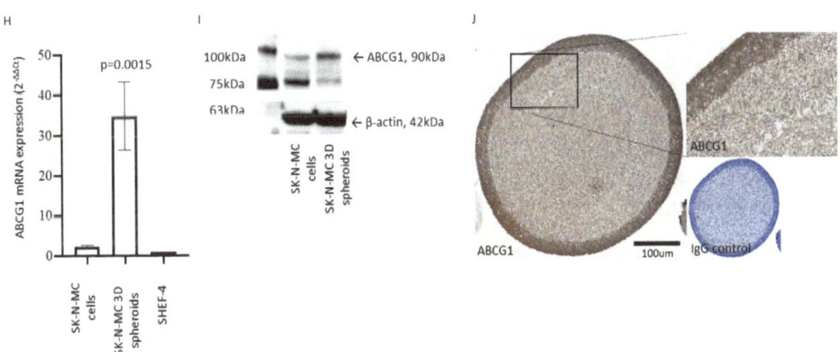

Figure 4. Expression of ABC transporter proteins by SK-N-MC and TTC 466 cells in 2D and 3D spheroid cultures. RNA (1 µg) from (**A**) SK-N-MC and (**D**) TTC 466 cells in 2D and 3D cultures were analysed by RTqPCR using the TaqMan® Human ABC Transporter Array. The level of mRNA expression was reported using the comparative Ct method as mean ± SEM, after normalisation of target Ct values to the global mean of all mRNAs. Results are the mean of two independent experiments. * = Genes with change in ΔCt > 2 and Q value < 0.1 comparing SK-N-MC cells in 2D and 3D cultures. Genes with Ct values of >35 in SK-N-MC cells grown in 2D were excluded. Volcano plot summarising the differentially expressed ABC transporter mRNAs in (**B**) SK-N-MC and (**E**) TTC 466 2D and 3D spheroids, displaying the Q value (level of significance) and the difference in ΔCt between the cells in 2D and 3D spheroids. The vertical dashed lines indicate the ΔCt thresholds of +/−ΔCt of >2 (*x*-axis) and the horizontal line a significant Q value < 0.1 (*y*-axis). Black square = CFTR which was significantly decreased and black triangle = ABCG1 which was significantly increased in 3D spheroids compared to 2D cultures. Black circles = ABC transporter proteins with a ΔCt of ±<2. The differentially expressed genes all have a Q value < 0.1. Mean Ct values of ABC transporter mRNAs in (**C**) SK-N-MC and (**F**) TTC 466 cells grown in 2D and 3D, the difference in ΔCt between the two groups and corresponding Q values are shown. Grey = genes with change in ΔCt > 2 and Q value < 0.1. (**G**) ABC transporter and pluripotency mRNAs (*n* = 140) detected (Ct

values < 35), highly expressed (Ct values < 25) and not detected (Ct values > 35) in 3D spheroids from TTC 466 and SK-N-MC cells. Venn diagrams show the number of mRNAs not detected, unique or shared for each cell line. (**H**) Validation of ABCG1 mRNA expression by RTqPCR and reported as $2^{-\Delta\Delta Ct}$ in SK-N-MC cells grown in 2D and 3D culture; expression of ABCG1 is normalised to the endogenous control gene PPIA and the control cell line, SHEF-4. RNA expression was compared between 2D and 3D SK-N-MC cells, using a two-tailed *t*-test. The results are representative of 2 independent experiments. (**I**) Increased expression of ABCG1 protein in SK-N-MC cells grown as 3D spheroids compared to 2D cultures was validated by Western blot. Equal loading of each protein was confirmed by probing the Western blot for ß-actin. The results are representative of 2 independent experiments. (**J**) High expression of ABCG1 protein expression in the outer 50 μm region of SK-N-MC spheroids detected by IHC; nuclei are labelled with haematoxylin. Black scale bar = 100 μm. IgG control = SK-N-MC spheroid section incubated with the isotype control antibody (4 μg/mL, Negative Control Mouse IgG1, X0931 (Dako) and 20 μg/μL, Normal Rabbit Serum Control Ig mix, 086199 (Life Technologies), stained with haematoxylin.

The decrease in expression of the CFTR gene was not validated using RTqPCR or Western blot (results not shown). However, the increase in ABCG1 expression was confirmed in SK-N-MC spheroids at the mRNA ($p = 0.0015$, Figure 4H) and protein (Figure 4I) level. ABCG1 is heterogeneously expressed in all ES cell lines (6/6; Supplementary Data S5B). By IHC, ABCG1 expression was confirmed in the outer proliferating region of 3D SK-N-MC spheroids but not the hypoxic region or inner necrotic core (Figure 4J; [34]), suggesting expression of ABCG1 may be regulated by hypoxia and have a functional role in ES cells under these conditions.

3.4. Functional Role and Characterisation of ABCG1 in ES-CSCs

Knockdown of ABCG1 using siRNA had no effect on the viability or apoptosis of SK-N-MC cells in culture (10 ± 0.8% in non-targeting and 10 ± 1.1% in ABCG1 siRNA-treated cells, $p > 0.05$; Figure 5A). There was also no effect on proliferation (0.93 ± 0.1 and 0.94 ± 0.05 in non-targeting and ABCG1 siRNA treated cells respectively, $p > 0.05$) or self-renewing ability from a single cell in soft agar (60% in non-targeting and 58% in ABCG1 siRNA-treated cells). However, after knockdown of ABCG1 spheroid production was significantly reduced; 65 ± 3% in non-targeting and 33 ± 5% in ABCG1 siRNA-treated cells ($p < 0.0001$, $n = 10$). These data suggest that although ABCG1 may not have a role in homeostasis of ES cells in 2D, it may affect cell–cell interactions and components of the tumour microenvironment important in the development of 3D spheroids and possibly tumours. This hypothesis requires further investigation.

Analysis of ABCG1 RNA in SK-N-MC spheroids and 2D cultures using total RNA sequencing, revealed expression of 11 previously reported transcripts (www.ensembl.org, accessed on 27 April 2016; Figure 5B,C) and two novel ABCG1 RNA species (Figure 5C). Canonical ABCG1 generates the multiple transcripts by alternative splicing (Figure 5C). Novel transcript 1 is most similar to transcript 4 (ENST00000450121.5), both sequences missing exon 5 (Figure 5C). However, novel transcript 1 included exons 8–15 and had an extended 3′UTR region predicted to produce a larger molecular weight protein than transcript 4. The RNA sequence of novel transcript 2 is most similar to transcript 5 (ENST00000361802.6), the additional 3′UTR sequence unlikely to produce a unique protein product. Expression of both novel transcripts at the RNA level was increased above canonical protein producing sequences (transcripts 1, 3–8) in SK-N-MC cells grown as spheroids compared to 2D cultures, this was most significant for novel transcript 1 (Figure 5D). The increase in ABCG1 gene expression in SK-N-MC cells from spheroids (Figure 4G,H), might then reflect expression of novel transcript 1. Further studies are required to investigate this novel ABCG1 transcript and its role in ES-CSCs.

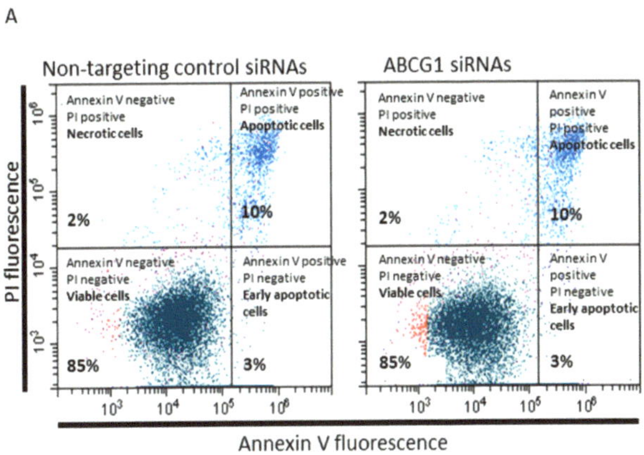

B

Ensembl listed ABCG1 transcript	Paired RNA sequencing identified transcript
ENST00000398457.6	Transcript 1
ENST00000462050.5*	Transcript 2
ENST00000347800.6	Transcript 3
ENST00000450121.5	Transcript 4
ENST00000361802.6	Transcript 5
ENST00000398449.7	Transcript 6
ENST00000343687.7	Transcript 7
ENST00000398437.1	Transcript 8
ENST00000472587.5#	Transcript 9
ENST00000467818.1#	Transcript 10
ENST00000496783.1#	Transcript 11

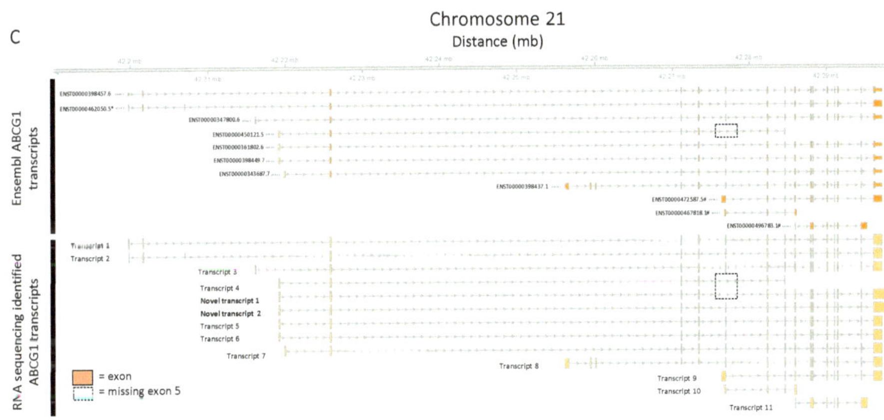

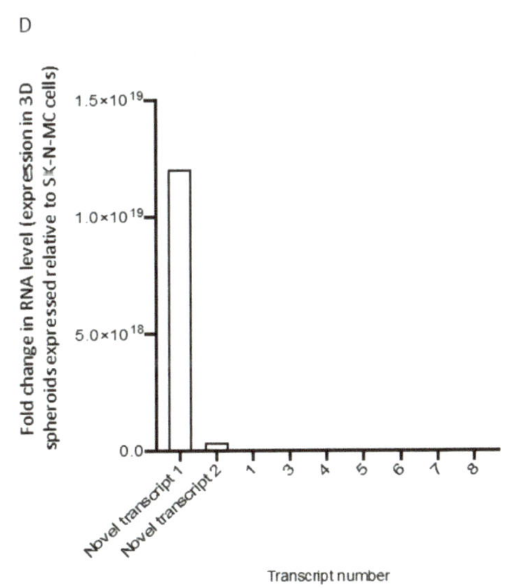

Figure 5. Functional role and characterisation of ABCG1 in ES. (**A**) Representative dot plot of annexin V and PI labelling of SK-N-MC cells analysed by flow cytometry following knockdown of ABCG1 using

SMARTpool:Accell ABCG1 siRNA and non-targeting control siRNAs. The upper left quadrant shows the mean percentage of necrotic cells, the upper right = apoptotic cells, lower left = viable cells and the lower right early apoptotic cells. The percentage of apoptotic cells was compared using a non-parametric Mann–Whitney two-tailed *t*-test. (**B**) Canonical ABCG1 RNA transcripts were downloaded from Ensembl.gov and labelled with the prefix ENST and unique transcript number. * = No protein produced from this transcript (www.ensembl.org, accessed on 27 April 2016), # = no protein produced from this transcript, retained intron (www.ensembl.org, accessed on 27 April 2016) (**C**) Canonical transcripts 1 to 11 were detected in SK-N-MC cells in 2D and 3D spheroids. In addition, two novel transcripts (transcript 1 and 2) were identified in 3D spheroids. Orange box = exon, grey arrows = direction of transcription, dotted line box = missing exon 5. (**D**) Novel transcript 1 was the most highly differentially expressed ABCG1 transcript in 3D spheroids compared to cells grown in 2D.

Since ABCG1 mRNA was increased in ES-CSCs from SK-N-MC spheroids, we examined its co-expression with CD133. There was no significant difference in the percentage of cells expressing ABCG1 or CD133 in SK-N-MC cells from 3D spheroids or 2D cultures (Figure 6A,B). Consistently with the increase in ABCG1 in protein extracts from 3D spheroids (Figure 4H), the level of ABCG1 protein per cell was greater in cells from 3D spheroids compared to 2D cultures (2.5 $\pm$ 0.8-fold increase, $p < 0.05$, Figure 6C). Levels of CD133 per cell were also increased (6.3 $\pm$ 1 fold increase, $p < 0.05$, Figure 6C), consistent with the higher expression of ABCG1 protein in TC-32 CD133-positive cells compared to CD133-negative cells (Figure 6D). Expression of ABCG1 RNA was confirmed at the protein level in primary patient-derived ES cells (Figure 6E). However, there was no correlation between the percentage of CD133-positive cell lines or patient-derived cells and progeny-producing ability measured using the colony formation assay in soft agar ($R^2 < 0.1$) or self-renewing ability from a single cell seeded on low adherent or adherent plates ($R^2 = 0.002$; Figure 6F), suggesting in patient-derived cells CD133 may not identify cell populations with a self-renewing phenotype.

Interrogation of the publicly available GSE17618 RNA dataset revealed high expression of CD133, but not ABCG1, was associated with a more than threefold risk of an event and poor outcome (Supplementary Data S6), consistent with the hypothesis that high expression of CD133 identifies ES-CSCs that may be responsible for progression and relapse in some patients (Figure 6G).

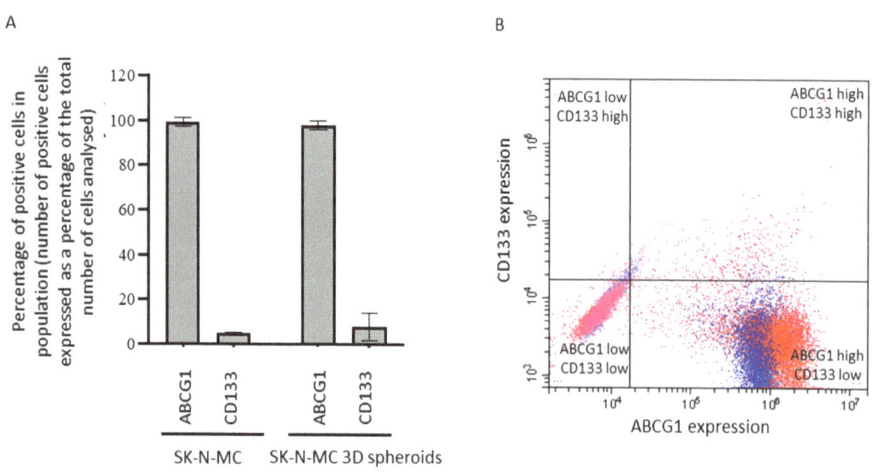

Figure 6. *Cont.*

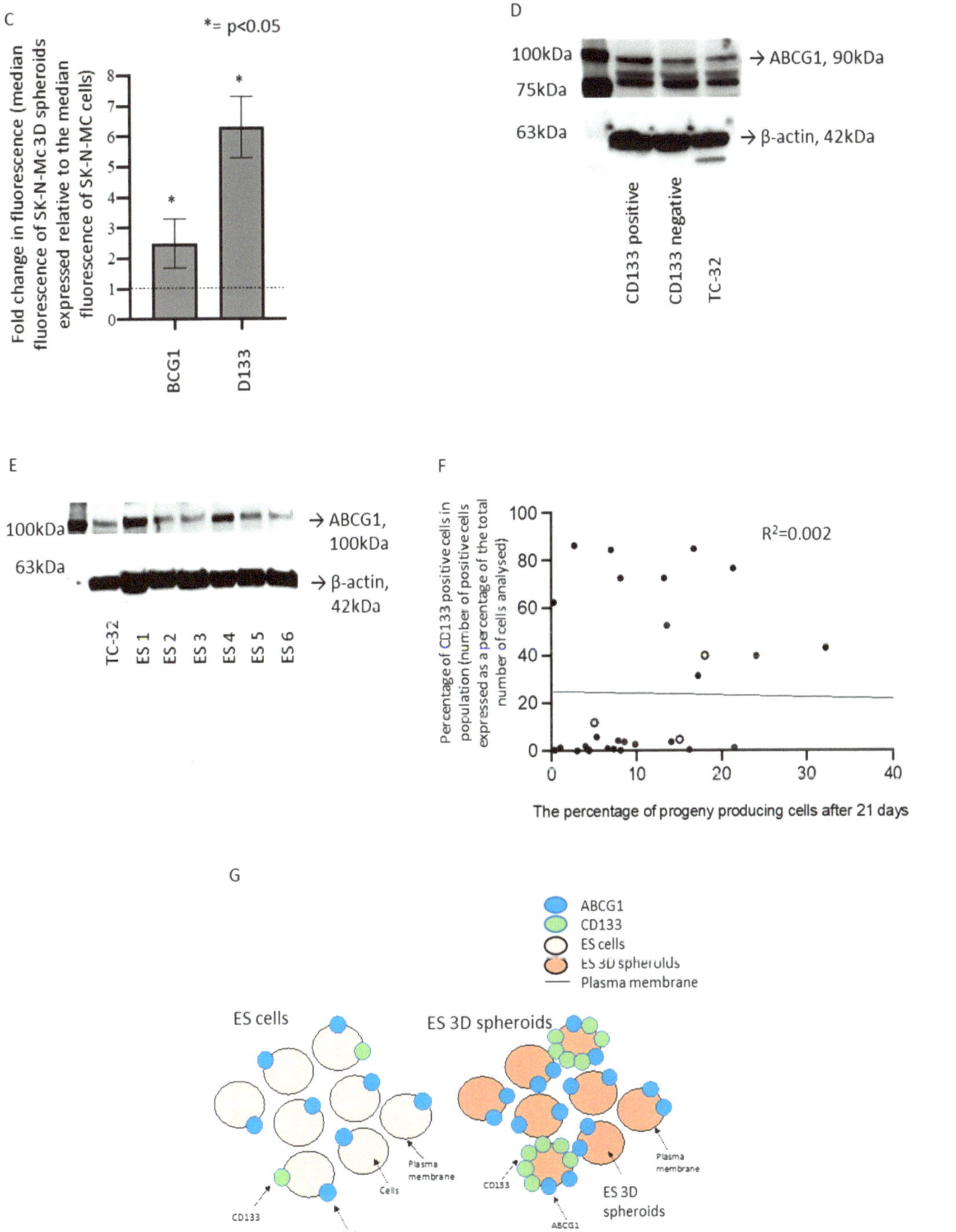

Figure 6. Expression profile of ABCG1, CD133 and MRP1 in ES cell lines and patient-derived cells. (**A**) The percentage of SK-N-MC cells grown in 2D and 3D cultures expressing ABCG1 and CD133 was

was quantified by flow cytometry; 10,000 events were examined for each condition. The percentage of positive cells is presented as the mean ± SEM (n = 6). ABCG1 and CD133 expression in SK-N-MC cells grown in 2D and 3D cultures was compared using a non-parametric Mann–Whitney two-tailed t-test. (**B**) Representative dot plot of ABCG1 and CD133 expression of SK-N-MC cells grown in 2D and disaggregated 3D cultures, analysed by flow cytometry. Quadrants represent cells with ABCG1 low CD133 high (**upper left**), ABCG1 high CD133 high (**upper right**), ABCG1 low CD133 low (**lower left**) and ABCG1 high CD133 low (**lower right**) levels of expression. Red spots = cells in 3D culture, blue spots = cells in 2D culture. (**C**) Protein expression of ABCG1 and CD133 in SK-N-MC cells grown in 2D and 3D culture was quantified by flow cytometry. Data are presented as the fold change in fluorescence, expressing the median target protein fluorescence in SK-N-MC cells grown in 3D relative to cells grown in 2D. Results show the mean ± SEM for two independent experiments, 3 replicates per experiment (n = 6). ABCG1 and CD133 expression in 2D and 3D cultures was compared using an unpaired two-tailed t-test. (**D**) ABCG1 protein expression in CD133-positive and CD133-negative TC-32 cells detected by Western blot. Equal protein loading was confirmed by probing the blots for ß-actin. Results are representative of 2 independent experiments. (**E**) ABCG1 protein expression in patient-derived ES cells, determined by Western blot. Equal protein loading was confirmed by probing the blots for ß-actin. Results are representative of 2 independent experiments. (**F**) There was no correlation between the level of CD133 expression and ability to produce progeny from a single cell. The level of CD133 expression was quantified by flow cytometry and the ability to produce progeny from a single cell evaluated in 2D adherent culture at 21 days. Correlation was examined using linear regression. Open circles = cell lines, filled circles = patient-derived cells. (**G**) Summary of ABCG1 and MRP1 protein expression in ES cells grown in 2D and 3D culture. Salmon-pink circle = ES cells grown in 2D, orange circle = ES cells grown in 3D (spheroids), blue circle = ABCG1 protein, green circle = CD133 protein, black line = plasma membrane.

3.5. *Identification of Candidate Drugs for ES-CSC Molecular Targets*

Candidate small-molecule inhibitors to 62% (13/21) of the unique molecular targets increased in 3D spheroids from TTC 466 and SK-N-MC cells and ES-CSCs (Figure 7A,B) were identified using our bespoke pipeline (Figure 7C). Of the 279 drugs directed against the targets, 66 (24%) are licensed cancer drugs and 213 (76%) are non-cancer drugs (Figure 7C and Supplementary Data S7). The majority of the drugs (202/279; 72%) target one of the two ATP-binding cassette (ABC) drug efflux proteins p-glycoprotein (n = 168, includes n = 10 targeting p-glycoprotein and additional proteins) or MRP1 (n = 13), while 21 drugs target both p-glycoprotein and MRP1. Forty eight of the 279 (17%) drugs hitting the targets have been evaluated in oncology trials, whereas the majority of drugs have been evaluated in non-cancer drug trials (n = 986; Figure 7C, Table 2, Supplementary Data S8).

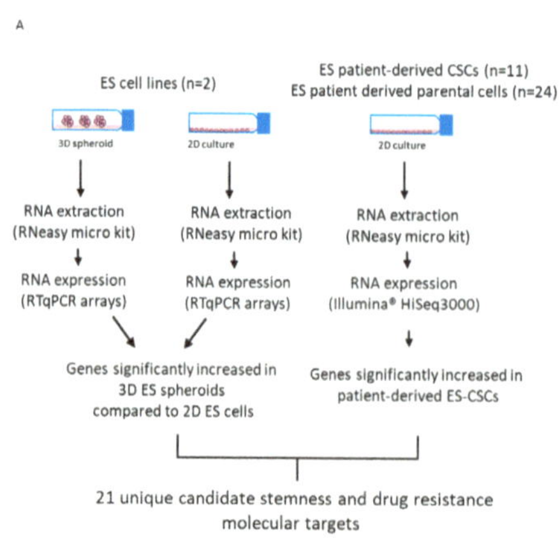

Figure 7. *Cont.*

B

Full or Alternative gene name	Gene	Source of gene
Activated leukocyte cell adhesion molecule, CD166	ALCAM	1
Caveolin 1	CAV1	1
Leukocyte antigen MIC3	CD9	1
Stro-1	CD34	1
Epican	CD44	1
DNA (cytosine-5)-methyltransferase 3B	DNMT3B	1
Endoglin	ENG	1
Gamma-amiobutyric acid receptor subunit beta-3	GABRB3	1
Intracelluair adhesion molecule 1, CD54	ICAM1	1
Integrin beta-1	ITGB1	1
Mast/stem cell growth factor receptor Kit	KIT	1
Podocalyxin	PODXL	1
Octamer-binding protein 4, OCT-4	POU5F1	1
Thy-1 membrane glycoprotein	THY1	1
Vimentin	VIM	1
Aldehyde dehydrogenase E3 isozyme	ALDH	1
Prominin-1	CD133	1
P-glycoprotein, MDR1 protein, ABCB1	Pgp	1
Multi-drug resistance associated protein 1, MRP1, ABCC1	ABCC1	1
Lin-28, Zinc Finger CCHC Domain-Containing prtoein 1	LIN28	2
LEFTY1, Left-Right Determination Factor 1	LEFTB	2

C

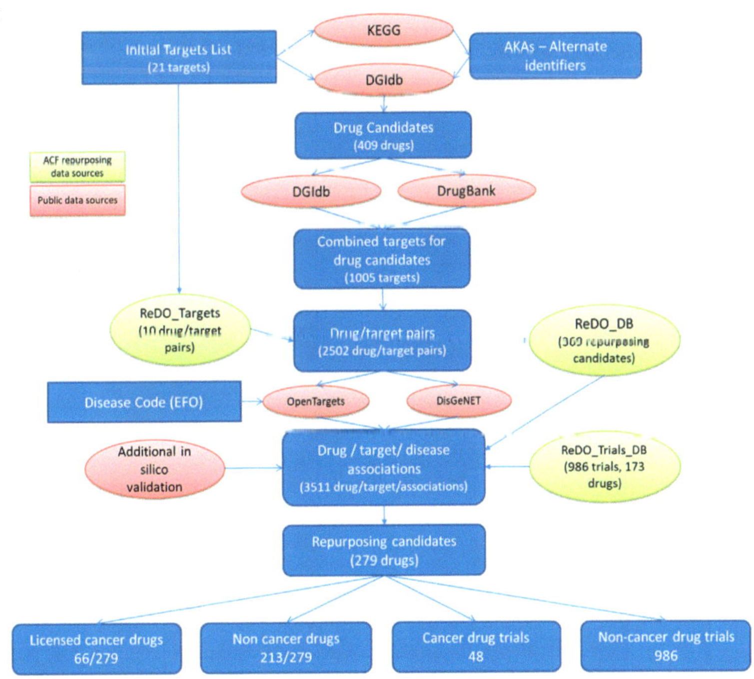

Figure 7. *Cont.*

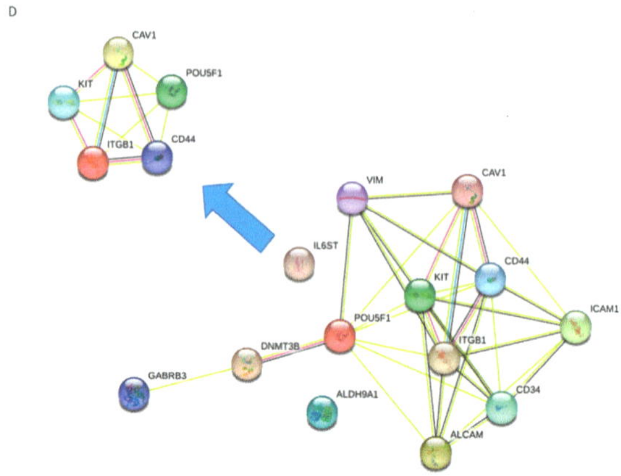

Figure 7. Identification of candidate shared stemness and chemoresistance therapeutic targets. (**A**) Summary of strategy to identify shared markers of stemness and MDR in ES preclinical models. (**B**) Stemness and MDR-associated genes increased in 3D spheroids and patient-derived ES-CSCs. The full and alternative gene names, the Ensembl gene abbreviation (Gene) and the data from which genes were identified (Source of the gene) are shown. The source of the gene is presented as 1 or 2, where 1 = target gene identified from transcriptome analysis of patient-derived ES-CSCs [17] and 2 = target gene identified by comparing transcriptome of SK-N-MC and TTC 466 cells grown in 3D and 2D (Figures 3 and 4). (**C**) Pipeline to identify candidate drugs based on 21 stemness and chemoresistance genes. Green ovals = drug-repurposing data sources, pink ovals = public data sources, blue squares = number of targets or drugs at each step. (**D**) Predicted interaction between the 13 shared stemness and chemoresistance genes (excluding p-glycoprotein and MRP1) that have corresponding candidate drugs using STRING [30]. Five of the genes are known to bind and regulate vital biological processes, and could be candidate prognostic biomarkers and/or therapeutic targets in ES. Pink lines = known interaction that has been experimentally determined, extracted from BIND, DIP, GRID, HPRD, IntAct, MINT and PID databases. Blue lines = known interaction based on data from curated databases Biocarta, BioCyc, GO, KEGG and Reactome. Grey lines = proteins reported to be co-expressed. Green lines = interactions identified by text mining. Filled nodes = 3D structure known or predicted. Coloured nodes = first shell of interacting proteins. White nodes = second shell of interacting proteins. Parental cells = patient-derived bulk population of ES cells from which the ES-CSCs were derived [14].

POU5F1/OCT-4 interacted with 10/13 of the identified stemness and chemoresistance genes (Figure 7D), suggesting shared functional protein–protein association networks. Five of these 10 gene products (POU5F1/OCT4, c-KIT, CAV1, ITGB1 and CD44) interact (Figure 7D), regulating vital cellular processes (Supplementary Data S9). All these proteins are highly expressed in ES-CSCs and so may represent candidate prognostic biomarkers and/or therapeutic targets in ES. We are currently investigating these possibilities. Using our customised pipeline, we identified two FDA approved small-molecule inhibitors, allopurinol and phenytoin, that target POU5F1/OCT4 and are used to treat gout and control seizures in epilepsy, respectively (Table 2, https://www.dgidb.org/genes/POU5F1#_interactions, accessed on 3 January 2023). These two licensed drugs have the highest DGIdb interaction scores of the three compounds that interact with POU5F1/OCT4, inhibit a range of other cancer-relevant molecular targets and are also currently in trials as cancer therapeutics, making them attractive candidate drugs for further in vitro analysis for ES. Several inhibitors targeting multiple receptor tyrosine kinases (TKIs) implicated in the pathogenesis of sarcomas including ES were also identified [52,53], including regorafenib, which is being evaluated in combination with chemotherapy (NCT02085148/2013-003579-36; NCT04055220/REGOSTA; 2021-005061-41/INTER-EWING-1; Table 2).

Table 2. Drugs directed against the shared stemness and chemoresistance therapeutic targets. Drugs directed against the shared ES stemness and chemoresistance targets that have been used in human clinical trials, identified using our in-house pipeline. Drug names, direct molecular targets, whether the drug has been previously used in the treatment of cancer patients and is on- or off-patent are shown. The number, trial phase and status, cancer type and detailed trial information of clinical trials are listed. NA = no phase 3 or phase 4 trials recorded for the drug. BNFC = British National Formulary- Children, a directory or all drugs approved for use in children. Y = yes, N = no. * = direct ES targets of fostamatinib (KIT, STK10, SRC, SIK1, RPS6KA1, ROCK2, RET, PTK2B, PTK2, PRKG2, PRKCD, PLK4, PLK3, PLK1, PIM3, PIK3CG, PIK3CD, PDGFRB, PDGFRA, PAK3, PAK1, NTRK3, NTRK2, NTRK1, NEK2, MST1R, MAPK14, MAP3K3, MAP2K2, LYN, LIMK1, KDR, ITK, INSRR, INSR, GSK3A, MTOR, FLT4, FLT3, FLT1, FGFR1, ERN1, ERBB4, ERBB2, EPHB4, EPHB2, EPHA3, EPHA2, EIF2AK4, EGFR, DYRK1B, DCLK3, DCLK1, CSK, CSF1R, CHEK2, CHEK1, CDK4, BTK, BRAF, AXL, AURKB, AURKA, ALK, ZAP70, WEE1, TNK2, TNK1, TIE1, TGFBR2, TEK, TAOK3, STK36, STK33, MET, JAK1, ABL1, SYK).

Drug	Ewings Sarcoma Targets	Cancer Drug	BNFC approval	Off-Patent	Trial Count	Max Trial Phase	Trial Phase (Number of Trials in Phase)	Trial Status (Number of Trials in Status)	Cancer Type (Number of Trials in Cancer Type)	Trial Information for Phase 3 and Phase 4 trials (Trial Number, Phase, Status, Recruitment Status, Cancer Type)
Cabozantinib	KIT, RET, KDR, MET	Y	N	N	6	Phase 2	Phase 1 (1), Phase 2 (5)	Recruiting (1), Active, not recruiting (3), Not yet recruiting (1), Other (1)	Multiple sarcomas, including Ewing's (1), Mixed paediatric, including Ewing's (3), Ewing's or osteosarcoma (2)	NA
Dasatinib	KIT, MAPK14, LYN, HSPA8, EPHB4, CSK, BCR, BTK, PDGFRB, EPHA2, SRC, ABL1	Y	Y	Y	2	Phase 2	Phase 1/2 (1), Phase 2 (1)	Active, not recruiting (1), Completed (1)	Mixed paediatric, including Ewing's (1), Multiple sarcomas, including Ewing's (1)	NA
Decitabine	DNMT3B, DNMT1	Y	N	Y	1	Phase 1	Phase 1 (1)	Completed (1)	Mixed, including sarcomas (1)	NA
Erdafitinib	KIT, KDR, PDGFRB, PDGFRA, CSF1R, RET, FGFR1	Y	N	N	1	Phase 2	Phase 2 (1)	Recruiting (1)	Mixed paediatric, including Ewing's (1)	NA
Imatinib	KIT, ABCB1, PDGFRB, ABL1, PDGFRA, CSF1R, NTRK1, RET, BCR	Y	Y	Y	4	Phase 2	Phase 2 (4)	Completed (4)	Mixed, including sarcomas (1), Multiple sarcomas, including Ewing's (1), Mixed paediatric, including Ewing's (1), Ewing's or DSRCT (1)	NI

Table 2. *Cont.*

Drug	Ewings Sarcoma Targets	Cancer Drug	BNFC approval	Off-Patent	Trial Count	Max Trial Phase	Trial Phase (Number of Trials in Phase)	Trial Status (Number of Trials in Status)	Cancer Type (Number of Trials in Cancer Type)	Trial Information for Phase 3 and Phase 4 trials (Trial Number, Phase, Status, Recruitment Status, Cancer Type)
Lenvatirib	KIT, RET, PDGFRA, FGFR1, FLT4, KDR, FLT1	Y	N	N	1	Phase 1/2	Phase 1/2 (1)	Active, not recruiting (1)	Mixed paediatric, including Ewing's (1)	NA
Pazopanib	KIT, ITK, PDGFRB, PDGFRA, FLT4, KDR, FLT1	Y	N	Y	1	Phase 2	Phase 2 (1)	Completed (1)	Mixed paediatric, including Ewing's (1)	NA
Regorafenib	KIT, RET, ABL1, BRAF, EPHA2, NTRK1, TEK, FGFR1, PDGFRB, PDGFRA, FLT4, KDR, FLT1	Y	N	N	5	Phase 2	Phase 1 (1), Phase 1/2 (1), Phase 2 (2), Other (1)	Recruiting (4), Active, not recruiting (1)	Multiple sarcomas, including Ewing's (3), Ewing's or osteosarcoma (1), Mixed paediatric, including Ewing's (1)	NA
Sorafenib	KIT, ABCB1, FLT1, RET, FGFR1, PDGFRB, FLT3, KDR, FLT4, BRAF	Y	N	Unclear	2	Phase 2	Phase 1 (1), Phase 2 (1)	Active, not recruiting (1), Completed (1)	Ewing's or DSRCT (1), Mixed paediatric, including Ewing's (1)	NA
Sunitinib	KIT, ABCB1, PDGFRA, CSF1R, FLT3, FLT4, KDR, FLT1, PDGFRB	Y	N	Y	2	Phase 2	Phase 1/2 (1), Phase 2 (1)	Ongoing (1), Completed (1)	Multiple sarcomas, including Ewing's (1), Mixed, including sarcomas (1)	NA
Temozolomide	ABCB1	Y	Y	Y	23	Phase 3	Phase 1 (10), Phase 1/2 (4), Phase 2 (8), Phase 3 (1)	Recruiting (8), Active, not recruiting (4), Completed (6), Not yet recruiting (2), Other (2), Terminated (1)	Mixed paediatric, including Ewing's (6), Ewing's (10), Mixed, including sarcomas (4), Ewing's or DSRCT (1), Ewing's or RMS (2)	NCT03495921 (Phase 3; Active, not recruiting)—Ewing sarcoma

Table 2. *Cont.*

Drug	Ewings Sarcoma Targets	Cancer Drug	BNFC approval	Off-Patent	Trial Count	Max Trial Phase	Trial Phase (Number of Trials in Phase)	Trial Status (Number of Trials in Status)	Cancer Type (Number of Trials in Cancer Type)	Trial Information for Phase 3 and Phase 4 trials (Trial Number, Phase, Status, Recruitment Status, Cancer Type)
Aliskiren	ABCB1	N	N	Y	1	Phase 2	Phase 2 (1)	Recruiting (1)	Multiple cancer types (1)	NA
Allopurinol	POU5F1	N	Y	Y	1	Phase 1/2	Phase 1/2 (1)	Not yet recruiting (1)	Lung (1)	NA
Atorvastatin	ABCB1, NR1I3, HDAC2, AHR, DPP4, HMGCR	N	Y	Y	16	Phase 3	Phase 1 (4), Phase 2 (6), Phase 2/3 (1), Phase 3 (5)	Recruiting (13), Ongoing (1), Not yet recruiting (2)	Multiple cancer types, Leukaemia (1), GI (4), Breast (8), Urological (3)	NCT03024684 (Phase 3; Recruiting)—Hepatocellular carcinoma, NCT03819101 (Phase 3; Recruiting)—Prostate, NCT03971019 (Phase 3; Recruiting)—Breast, NCT04601116 (Phase 3; Recruiting)—Breast, NCT04026230 (Phase 3; Recruiting)—Prostate
Azithromycin	ABCC1	N	Y	Y	3	Phase 2	Phase 2 (3)	Ongoing (3)	Lymphoma (2), Breast (1)	NA
Carvedilol	ABCB1, HIF1A, GJA1, VEGFA	N	Y	Y	2	Phase 2	Phase 1 (1), Phase 2 (1)	Active, not recruiting (1), Not yet recruiting (1)	Urological (1), CNS (1)	NA
Chlorpromazine	ABCB1	N	N	Y	3	Phase 2	Phase 1 (1), Phase 1/2 (1), Phase 2 (1)	Recruiting (2), Not yet recruiting (1)	GI (1), CNS (2)	NA
Citalopram	ABCB1	N	Y	Y	1	Phase 3	Phase 3 (1)	Ongoing (1)	CNS (1)	2013-004705-59 (Phase 3; Ongoing)—Glioblastoma

Table 2. *Cont.*

Drug	Ewings Sarcoma Targets	Cancer Drug	BNFC approval	Off-Patent	Trial Count	Max Trial Phase	Trial Phase (Number of Trials in Phase)	Trial Status (Number of Trials in Status)	Cancer Type (Number of Trials in Cancer Type)	Trial Information for Phase 3 and Phase 4 trials (Trial Number, Phase, Status, Recruitment Status, Cancer Type)
Clarithromycin	ABCB1	N	Y	Y	18	Phase 4	Phase 1/2 (2), Phase 2 (10), Phase 3 (3), Phase 4 (1), Other (2)	Recruiting (8), Active, not recruiting (8), Not yet recruiting (1), Other (1)	Lymphoma (1), Other Haem-onc (15), Multiple cancer types (1), GI (1)	ChiCTR2100047608 (Phase 4; Recruiting)—Multiple Myeloma, NCT02575144 (Phase 3; Active, not recruiting)—Multiple Myeloma, NCT02516696 (Phase 3; Active, not recruiting)—Multiple Myeloma, NCT04287660 (Phase 3; Recruiting)—Multiple Myeloma
Clopidogrel	ABCB1	N	N	Y	1	Phase 1	Phase 1 (1)	Recruiting (1)	Head and Neck (1)	NA
Colchicine	ABCB1, TUBB	N	N	Y	2	Phase 2	Phase 1 (1), Phase 2 (1)	Recruiting (2)	Urological, Multiple cancer types (1), GI (1)	NA
Cyclosporine	ABCC1, ABCB1	N	N	Y	2	Phase 2	Phase 1 (1), Phase 2 (1)	Recruiting (1), Active, not recruiting (1)	Leukaemia (1), Other Haem-onc (1)	NA
Deferoxamine	ABCB1	N	N	Y	4	Phase 3	Phase 1 (2), Phase 2 (1), Phase 3 (1)	Recruiting (3), Not yet recruiting (1)	Multiple cancer types (1), Breast (1), Leukaemia (1), GI (1)	IRCT20200313046756N2 (Phase 3; Pending)—Acute Myeloid Leukaemia
Digoxin	ABCB1	N	Y	Y	2	Phase 2	Phase 1 (1), Phase 2 (1)	Recruiting (2)	GI (1), Multiple cancer types (1)	NA

Table 2. *Cont.*

Drug	Ewings Sarcoma Targets	Cancer Drug	BNFC approval	Off-Patent	Trial Count	Max Trial Phase	Trial Phase (Number of Trials in Phase)	Trial Status (Number of Trials in Status)	Cancer Type (Number of Trials in Cancer Type)	Trial Information for Phase 3 and Phase 4 trials (Trial Number, Phase, Status, Recruitment Status, Cancer Type)
Disulfiram	ABCB1	N	N	Y	13	Phase 2/3	Phase 1 (4), Phase 1/2 (2), Phase 2 (6), Phase 2/3 (1)	Recruiting (7), Active, not recruiting (2), Ongoing (3), Not yet recruiting (1)	Multiple cancer types, GI (1), Soft-Tissue Sarcoma, Bone Sarcoma (1), CNS (3), Other Haem-onc (1), Urological (2), Multiple cancer types, Breast (1), GI (2), Breast (2)	NA
Doxycycline	ABCB1	N	Y	Y	10	Phase 2	Phase 1 (1), Phase 2 (9)	Recruiting (4), Active, not recruiting (4), Ongoing (1), Not yet recruiting (1)	Urological (1), Lymphoma (3), Multiple cancer types (1), Breast, Gynaecological (1), Breast (1), GI (2), Head and Neck (1)	NA
Fenofibrate	ABCB1, NR1I2	N	Y	Y	1	Phase 2	Phase 2 (1)	Recruiting (1)	CNS (1)	NA
Fostamatinib	*	N	N	N	2	Phase 1	Phase 1 (2)	Recruiting (2)	Gynaecological (1), Other Haem-onc, Leukaemia (1)	NA
Indomethacin	ABCC1, PTGS2	N	Y	Y	5	Phase 4	Phase 1 (1), Phase 1/2 (1), Phase 2 (1), Phase 3 (1), Phase 4 (1)	Recruiting (2), Active, not recruiting (3)	Urological (2), Breast (1), Head and Neck (2)	ChiCTR2000038968 (Phase 4; Recruiting)—Prostate, NCT01265849 (Phase 3; Active, not recruiting)—Cancer of the oral cavity
Itraconazole	ABCB1, ERBB2	N	Y	Y	15	Phase 3	Phase 1 (4), Phase 1/2 (2), Phase 2 (7), Phase 3 (1), Other (1)	Recruiting (9), Active, not recruiting (2), Completed (1), Not yet recruiting (1), Other (1), Suspended (1)	Leukaemia, Other Haem-onc (1), Lung (2), Multiple cancer types, Lymphoma, Leukaemia (1), follow-up continuing (1), Multiple cancer types (2), GI (4), Urological (1), Gynaecological (2), Skin (1)	NCT03458221 (Phase 3; Not yet recruiting)—Ovarian

Table 2. *Cont.*

Drug	Ewings Sarcoma Targets	Cancer Drug	BNFC approval	Off-Patent	Trial Count	Max Trial Phase	Trial Phase (Number of Trials in Phase)	Trial Status (Number of Trials in Status)	Cancer Type (Number of Trials in Cancer Type)	Trial Information for Phase 3 and Phase 4 trials (Trial Number, Phase, Status, Recruitment Status, Cancer Type)
Ivermectin	ABCB1	N	Y	Y	2	Phase 2	Phase 2 (2)	Recruiting (1), Not yet recruiting (1)	Breast (1), Multiple cancer types (1)	NA
Ketoconazole	ABCB1, NR1I3, NR1I2, AR	N	Y	Y	4	Phase 2	Phase 1 (2), Phase 2 (2)	Recruiting (1), Active, not recruiting (2), Ongoing (1)	Urological (2), Breast, CNS (1), CNS (1)	NA
Lansoprazole	ABCB1, MAPT	N	Y	Y	3	Phase 3	Phase 2 (1), Phase 3 (2)	Recruiting (2), Active, not recruiting (1)	Breast (1), Lymphoma (1), GI (1)	NCT04874935 (Phase 3; Recruiting)—Breast, NCT03647072 (Phase 3; Recruiting)—Non-Hodgkin Lymphoma
Levetiracetam	ABCB1	N	Y	Y	2	Phase 2	Phase 2 (1), Other (1)	Not yet recruiting (2)	CNS (2)	NA
Losartan	ABCB1	N	N	Y	12	Phase 3	Phase 1 (4), Phase 2 (7), Phase 3 (1)	Recruiting (9), Active, not recruiting (1), Not yet recruiting (2)	Multiple cancer types (3), Bone Sarcoma (1), GI (7), Breast (1)	CTRI/2021/05/033482 (Phase 3; Not Yet Recruiting)—Pancreatic
Lovastatin	ABCB1, HDAC2, HMGCR	N	N	Y	1	Other	Other (1)	Not yet recruiting (1)	GI (1)	NA
Maprotiline	ABCB1	N	N	Y	1	Phase 1	Phase 1 (1)	Not yet recruiting (1)	CNS (1)	NA
Mefloquine	ABCB1	N	Y	Y	1	Phase 1	Phase 1 (1)	Active, not recruiting (1)	CNS (1)	NA
Miconazole	ABCB1, NR1I2	N	Y	Y	2	Phase 2	Phase 1 (1), Phase 2 (1)	Recruiting (2)	Multiple cancer types (2)	NA
Midazolam	GABRB3, ABCB1	N	Y	Y	2	Phase 2	Phase 2 (2)	Recruiting (1), Not yet recruiting (1)	Urological (2)	NA

Table 2. *Cont.*

Drug	Ewings Sarcoma Targets	Cancer Drug	BNFC approval	Off-Patent	Trial Count	Max Trial Phase	Trial Phase (Number of Trials in Phase)	Trial Status (Number of Trials in Status)	Cancer Type (Number of Trials in Cancer Type)	Trial Information for Phase 3 and Phase 4 trials (Trial Number, Phase, Status, Recruitment Status, Cancer Type)
Mifepristone	ABCB1, NR1I2, KLK3, NR3C1, PGR	N	N	Y	2	Phase 3	Phase 2 (1), Phase 3 (1)	Active, not recruiting (1), Not yet recruiting (1)	Breast (2)	NCT05016349 (Phase 3; Not yet recruiting)—Breast
Miltefosine	ABCB1	N	N	Y	2	Phase 2	Phase 2 (2)	Recruiting (1), Active, not recruiting (1)	Breast (1), Multiple cancer types (1)	NA
Nelfinavir	ABCB1	N	N	Y	9	Phase 3	Phase 1 (2), Phase 1/2 (2), Phase 2 (3), Phase 3 (2)	Recruiting (5), Active, not recruiting (2), Ongoing (2)	Soft Tissue Sarcoma (1), Urological, Multiple cancer types (1), Gynaecological (3), Head and Neck (1), Other Haem-onc (1), GI (1), CNS (1)	NCT03256916 (Phase 3; Recruiting)—Advanced carcinoma of the cervix, CTRI/2017/08/009265 (Phase 3; Open to Recruitment)—Advanced carcinoma of the cervix
Nicardipine	ABCB1	N	N	Y	1	Other	Other (1)	Not yet recruiting (1)	CNS (1)	NA
Omeprazole	ABCB1, AHR	N	Y	Y	3	Phase 2	Phase 1 (2), Phase 2 (1)	Recruiting (2), Active, not recruiting (1)	Urological (1), Breast (1), GI (1)	NA
Pantoprazole	ABCB1	N	N	Y	2	Phase 2	Phase 1/2 (1), Phase 2 (1)	Active, not recruiting (1), Not yet recruiting (1)	Head and Neck (1), Urological (1)	NA
Phenytoin	POU5F1, SCN8A, SCN2A, SCN3A, NR1I2, SCN1A, SCN5A	N	Y	Y	1	Phase 2	Phase 2 (1)	Recruiting (1)	Bone Sarcoma, Soft-Tissue Sarcoma (1)	NA

Table 2. *Cont.*

Drug	Ewings Sarcoma Targets	Cancer Drug	BNFC approval	Off-Patent	Trial Count	Max Trial Phase	Trial Phase (Number of Trials in Phase)	Trial Status (Number of Trials in Status)	Cancer Type (Number of Trials in Cancer Type)	Trial Information for Phase 3 and Phase 4 trials (Trial Number, Phase, Status, Recruitment Status, Cancer Type)
Pravastatin	ABCB1, HDAC2, HMGCR	N	N	Y	2	Phase 4	Phase 2 (1), Phase 4 (1)	Active, not recruiting (1), Not yet recruiting (1)	Leukaemia (1), Breast (1)	ChiCTR2000034035 (Phase 4; Pending)—Breast
Propofol	GABRB3, ABCB1, SCN2A, SCN4A	N	Y	Y	29	Phase 4	Phase 2 (1), Phase 3 (2), Phase 4 (7), Other (19)	Recruiting (17), Active, not recruiting (2), Ongoing (1), Not yet recruiting (9)	GI (6), Breast, GI (1), Multiple cancer types (7), Breast (5), Urological (3), Lung (6), CNS (1)	NCT01975064 (Phase 4; Recruiting)—Breast, colon, rectal, NCT05331911 (Phase 4; Not yet recruiting)—Liver, NCT04475705 (Phase 4; Recruiting)—Paediatric solid tumours, NCT05141877 (Phase 4; Not yet recruiting)—Brain tumour, NCT03034096 (Phase 4; Recruiting)—Adult cancer, NCT04513808 (Phase 3; Recruiting)—Oesophageal caner, ChiCTR2000040604 (Phase 4; Pending)—Non-small cell lung cancer, AC-TRN12611000301965 (Phase 4; Not yet recruiting)—Breast, 2009-009114-40 (Phase 3; Ongoing)—Prostate

Table 2. *Cont.*

Drug	Ewings Sarcoma Targets	Cancer Drug	BNFC approval	Off-Patent	Trial Count	Max Trial Phase	Trial Phase (Number of Trials in Phase)	Trial Status (Number of Trials in Status)	Cancer Type (Number of Trials in Cancer Type)	Trial Information for Phase 3 and Phase 4 trials (Trial Number, Phase, Status, Recruitment Status, Cancer Type)
Propranolol	ABCB1, EGFR, ADRB3	N	N	Y	19	Phase 3	Phase 1 (3), Phase 1/2 (1), Phase 2 (13), Phase 2/3 (1), Phase 3 (1)	Recruiting (10), Active, not recruiting (1), Not yet recruiting (7), Other (1)	GI (8), Urological (2), Soft Tissue Sarcoma (2), Gynaecological (1), Skin (3), Other (1), Multiple cancer types (2)	CTRI/2019/11/021924 (Phase 3; Not Yet Recruiting)—Ovarian
Rifampicin	ABCB1, NR1I2	N	Y	Y	1	Phase 1	Phase 1 (1)	Recruiting (1)	Multiple cancer types, Lymphoma, Leukaemia (1)	NA
Ritonavir	ABCC1, ABCB1, NR1I2	N	Y	Y	1	Phase 1	Phase 1 (1)	Recruiting (1)	Leukaemia, Lymphoma (1)	NA
Sertraline	ABCB1, SLC29A4, SLC6A3	N	Y	Y	1	Phase 1	Phase 1 (1)	Recruiting (1)	Leukaemia (1)	NA
Simvastatin	ABCB1, HDAC2, HMGCR	N	Y	Y	18	Phase 4	Phase 1 (4), Phase 2 (11), Phase 3 (1), Phase 4 (2)	Recruiting (7), Active, not recruiting (3), Ongoing (3), Not yet recruiting (4), Suspended (1)	Breast (5), Other Haem-onc (1), Lung (3), Gynaecological (1), GI (6), Multiple cancer types (2)	ChiCTR2000034035 (Phase 4; Pending)—Breast, 2010-018491-24 (Phase 4; Ongoing)—Adults with bone metastasis, NCT03971019 (Phase 3; Recruiting)—Breast
Sirolimus	ABCB1, MTOR	N	Y	Y	28	Phase 4	Phase 1 (7), Phase 1/2 (4), Phase 2 (13), Phase 3 (2), Phase 4 (1), Other (1)	Recruiting (17), Active, not recruiting (7), Ongoing (3), Not yet recruiting (1)	Lung (2), Multiple cancer types, Lung (1), Multiple cancer types (6), Other (2), Soft-Tissue Sarcoma (3), Urological, Multiple cancer types (1), Bone Sarcoma, Soft Tissue Sarcoma (1), Gynaecological (1), Leukaemia (3), Endocrine (1), GI (3), CNS (1), Bone Sarcoma (2), Breast (1)	NCT04775173 (Phase 3; Recruiting)—Haemangioendothelioma, ChiCTR1900021896 (Phase 4; Recruiting)—Liver, NCT04736589 (Phase 3; Not yet recruiting)—Breast

Table 2. *Cont.*

Drug	Ewings Sarcoma Targets	Cancer Drug	BNFC approval	Off-Patent	Trial Count	Max Trial Phase	Trial Phase (Number of Trials in Phase)	Trial Status (Number of Trials in Status)	Cancer Type (Number of Trials in Cancer Type)	Trial Information for Phase 3 and Phase 4 trials (Trial Number, Phase, Status, Recruitment Status, Cancer Type)
Valganciclovir	ABCB1	N	Y	Y	5	Phase 2	Phase 1/2 (2), Phase 2 (2), Other (1)	Recruiting (4), Not yet recruiting (1)	GI (1), Lymphoma (1), Head and Neck (1), CNS (1), Multiple cancer types (1)	NA
Verapamil	ABCC1, ABCB1	N	N	Y	1	Phase 1	Phase 1 (1)	Active, not recruiting (1)	Lymphoma (1)	NA
Warfarin	ABCB1, AXL, NR1I2	N	N	Y	1	Phase 1	Phase 1 (1)	Recruiting (1)	GI (1)	NA
Zidovudine	ABCC1, ABCB1, TERT	N	Y	Y	1	Phase 2	Phase 2 (1)	Recruiting (1)	Leukaemia (1)	NA

4. Discussion

For the first time, we have identified expression of the membrane-associated ABC transporter protein ABCG1 on the surface of human ES cells. Levels of ABCG1 were increased in SK-N-MC cells forming 3D spheroids compared to cells in 2D culture. Expression of ABCG1 was particularly high in the outer rim of spheroids, suggesting a structural or transport function between cells within and outside the spheroid. Consistent with this hypothesis, knockdown of ABCG1 reduced the ability of ES cells to bind to each other and produce 3D spheroids, reminiscent of the ABCG1-dependent regulation across cellular and intracellular membranes [54]. ABCG1 protein was detected at lower levels in hypoxic cells surrounding the necrotic centre of spheroids, consistent with observations in mouse colon adenocarcinoma spheroids where expression of ABCG1 and hypoxia-inducible factor 1α are inversely correlated [55]. In contrast to studies in lung cancer [56] and normal haemopoietic stem cells [57], decreased ABCG1 had no direct effect on the proliferation or apoptosis of ES cells. Decreased ABCG1 is reported to promote apoptosis through increased expression of the endoplasmic reticulum (ER) stress proteins GRP78 and CHOP [58]. The ability to tolerate high levels of ER stress conferred by the *EWSR1-ETS* oncogene [59] may explain why knockdown of ABCG1 does not induce apoptosis in these ES cells. ABCG1 expression is higher in several cancer types compared to normal tissue [29,56], expression being associated with higher-grade tumours [60], metastasis [61] and poor response to chemotherapy [62]. In agreement with these observations, ABCG1 expression is increased in drug-resistant, self-renewing osteosarcoma cells [18], a second bone cancer characterised by recurrent disease and acquired MDR. Further studies are required to establish what effect this lipid ABC transporter protein has on plasma membrane organisation and recruitment of signalling processes that may regulate cell fate and contribute to tumour development, evasion and metastasis. Single-nucleotide polymorphisms (SNPs) of ABCG1 in the first cytoplasmic domain (within intron 2) are associated with survival of patients with non-small-cell lung cancer [63]. Mutagenesis studies have highlighted the importance of this region for effective trafficking of ABCG1 to the plasma membrane and regulation of cholesterol efflux [64,65]. However, the clinical relevance of the 17 SNPs identified within the cytoplasmic region of ABCG1 (amino acids 178–195, exon 5, www.nvbi.nlm.nih.gov/snp, accessed on 13 July 2022) has yet to be established. In ES 3D spheroids, we identified two novel ABCG1 transcripts, the most abundant transcript (transcript 1) missing exon 5 of the cytoplasmic region. The cellular functional and clinical significance of this deletion and these transcripts requires further investigation. These hypotheses and data are consistent with the premise that ABCG1 identifies ES cells that are capable of surviving chemotherapy and may be responsible for progression and relapse in some patients. We are currently investigating the expression and functional role ABCG1 mRNAs and proteins in ES.

In agreement with the premise that CD133 can identify some cells with a CSC phenotype [14,66], the number of CD133-positive cells and level of CD133 expression was higher in ES cells from 3D spheroids compared to cells grown in 2D. CD133-positive cells shared common characteristics of cancer stem-like cells including increased colony formation from a single cell. However, in patient-derived cells there was no correlation between CD133 expression and self-renewing ability from a single cell. This highlights the difference between established ES cell lines and cultures more recently derived from patient tumours and supports the premise that CD133 does not identify all ES cell populations with a self-renewing phenotype. We are currently investigating the difference in the genotype and phenotype of ES cell lines and patient-derived cells. Although CD133 is reported to be a biomarker of stem cells, this is controversial, which may arise from heterogeneity and uncertainty about its physiological role(s) [67]. In contrast to studies in metastatic melanoma [68], we found no enrichment of the canonical Wnt pathway in CD133-positive ES cells compared to CD133-negative cells [69] and the level of CD133 did not predict outcomes. The increased drug resistance of TC-32 CD133-positive cells compared to CD133-negative cells might instead be effected through greater levels of the drug efflux protein MRP1, overexpression of which induces resistance to chemotherapy [23] and high membrane expression predicts a worse clinical outcome [27]. This suggests the need for further studies on the intrinsic and acquired resistance mechanisms in ES.

Increasing evidence suggests that ES can arise in neurally derived mesenchymal stem cells (MSCs; [70]) or in cells of the neural crest [71]. The interaction between the permissive cellular environment and the EWSR1-ETS tumour-specific chimeric transcription factors leading to cellular transformation and ES is poorly understood, although is no doubt important illustrated by the high incidence of ES in Europeans compared to Africans [72–74]. Rewiring of the transcriptome and epigenome, in addition to rare oncogenic driver events including STAG2, TP53 and CDKN2A are also known contributing factors [75]. Interestingly, five of six cell lines produced 3D progeny with projections similar to those displayed by tissue organoids [76], not previously reported in bone cancer. In contrast, the SKES-1 cells generated a cellular core with surrounding dissociating cells reminiscent of ES spheroids from CD133-positive STA-ET 8.2 cells [15]. Heterogeneity of spheroid compactness, growth and stability over time may reflect the amount of extracellular matrix components produced by the different cell lines [77], suggesting these models may be useful tools to evaluate the efficacy of therapeutic candidates. In this study, we found that the ES-CSC phenotype was associated with the expression of ABC transporter proteins and stemness features. However, because of the bystander stemness pathways associated with the cell of origin, unravelling the functional molecular mechanisms leading to Ewing sarcomagenesis requires further investigation.

We identified POU5F1 as a promising candidate therapeutic target, interacting with 10 of the prioritised targets. The protein product of this gene (octamer-binding transcription factor 4, OCT4) regulates several functional characteristics of CSCs, including self-renewal, survival, drug resistance, epithelial–mesenchymal transition and metastasis [78]. This is consistent with emerging roles for this protein in the tumorigenesis of adult cancers and stimulation of expression by the EWSR1 proto-oncogene in several human cancers, including ES [79]. Using our bespoke pipeline, we identified two FDA-approved drugs that are off patent and target POU5F1/OCT4, allopurinol and phenytoin. The efficacy and safety of phenytoin is currently being evaluated in a phase 2 clinical trial as part of a combination maintenance therapy in patients with clinically advanced sarcoma including ES [80] and in patients with metastatic pancreatic cancer [81]. A trial combining allopurinol and mycophenolate mofetil with chemotherapy in patients with relapsed small-cell lung cancer is expected to start recruitment later this year (the CLAMP trial, NCT05049863). We are currently investigating the effect of allopurinol and phenytoin alone and in combination with additional novel drugs and standard of care chemotherapy [82] in ES.

The majority of drugs we identified target the ABC transporter proteins p-glycoprotein and/or MRP1, which contribute to MDR by reducing the level of drug within cells. Several ABC transporter proteins have been described in ES, including MRP1, p-glycoprotein, ABCG1, ABCF1, ABCA6 and ABCA7 [83]. P-glycoprotein has been most frequently studied, although it does not predict outcome [27,84–86]. Despite the development of third-generation inhibitors of p-glycoprotein, to date, these drugs are of limited or no clinical value, and the majority of trials have been stopped due to unacceptable toxicity. In contrast, high membrane expression of MRP1 in tumours at diagnosis predicts reduced event-free and overall survival for patients [27]. Interestingly, low levels of ABCF1 in combination with high levels of IGF2BP3 also predicts poor outcome for patients [87]. Despite the clinical failure of inhibitors to ABC transporter proteins, several targeted small molecules, including some TKIs, interact with one or more ABC transporters, suggesting inhibitors may in some cases be beneficial [88]. Increased understanding of the functional role of ABC transporter proteins, including ABCG1, in the cells responsible for progression and relapse is needed to establish which, if any, are worthwhile candidate therapeutic targets to overcome MDR.

5. Conclusions

In summary, we have evidence that ABCG1 may be a candidate biomarker that could be used to select ES patients for treatment. This hypothesis requires validation. We have identified proteins expressed by ES-CSCs that might be therapeutic targets and used our bespoke pipeline to identify repurposing drug candidates that have the potential to inhibit these targets and eradicate ES-CSCs. These drugs include FDA-approved small molecules to ABC transporter proteins and two inhibitors of POU5F1/OCT-4.

Supplementary Materials: The following supporting information can be downloaded at https://www.mdpi.com/article/10.3390/cancers15030769/s1. Supplementary Methods. Methods relating to data described in Supplementary Data S1 are described. Supplementary Data S1. Results of experiments analysing CD133-positive and CD133-negative TC-32 and A673 populations. Supplementary Data S2 (A) Colony formation, (B) proliferation, (C) cell growth, (D) cell cycle, (E) telomere length analyses and (F) migration in A673 CD133-positive and CD133-negative cells. (G) Protein expression of MRP1 in A673 CD133-positive and -negative cells. Equal protein loading was confirmed by expression of α-tubulin. (H). Drug efflux activity of MRP1 was quantified by efflux of calcein F compared with initial fluorescence following loading with calcein AM. Results are shown as means ± SEM of three independent experiments. Calcein F efflux was compared using a non-parametric Mann–Whitney two-tailed *t*-test. (I) The effect of doxorubicin (3.5–224 nm) on A673 CD133-positive and -negative viable cell numbers at 48 h was quantified using the trypan blue exclusion assay. Viable cell number is presented as the percentage of cells after treatment with doxorubicin relative to the vehicle control (DMSO); results are shown as the mean ± SEM of three independent experiments. EC_{50} values were calculated using linear regression and compared using the extra sum of squares F test. Supplementary Data S3. (A) Colony formation, (B) proliferation, (C) cell growth, (D) cell cycle, (E) telomere length analyses and (F) migration in TC-32 CD133-positive and CD133-negative cells. (G) Protein expression of MRP1 in TC-32 CD133-positive and -negative cells. (H) Drug efflux activity of MRP1 was quantified by efflux of calcein F. (I) The effect of doxorubicin (3.5-224 nm) on TC-32 CD133-positive and -negative viable cell numbers at 48 h was quantified using the trypan blue exclusion assay. Supplementary Data S4. Expression of genes associated with differentiation, pluripotency and stemness, ABC transporter protein mRNAs and those linked with the Wnt signalling pathway in A673 and TC-32 CD133-positive and -negative cells. Supplementary Data S5. (A) Expression of genes at the RNA level associated with differentiation, pluripotency and stemness in SK-N-MC and TTC 466 cells in 2D and 3D cultures. (B) Protein expression of ABCG1 in ES cell lines (n = 6) was evaluated by Western blot. Equal protein loading was confirmed by probing the blots for GRP75. Supplementary Data S6. Prognostic significance of CD133 and ABCG1 RNA in diagnosis ES tissue. Supplementary Data S7. Complete list of drugs (n = 279) directed against the shared stemness and chemoresistance therapeutic targets. Supplementary Data S8. Drugs directed against the shared stemness and chemoresistance therapeutic targets and corresponding clinical trial identifiers. Supplementary Data S9. Statistically significant (by false discovery rate) GO terms and biological processes driven by POU5F1/OCT4, c-KIT, CAV1, ITGB1 and CD44. Refs. [18,23,31,34,89–91] are cited in the supplementary materials.

Author Contributions: Conceptualization, S.A.B.; methodology, E.A.R. and S.A.B.; software, E.A.R.; validation, E.A.R., P.P., D.E.L., L.A.S., G.A. and S.A.B.; formal analysis, E.A.R., P.P., D.E.L., L.A.S., G.A. and S.A.B.; investigation, E.A.R., D.E.L., L.A.S. and G.A.; resources, P.P. and S.A.B.; data curation, E.A.R., P.P., D.E.L., L.A.S., G.A. and S.A.B.; writing—original draft preparation, E.A.R. and S.A.B.; writing—review and editing, E.A.R., P.P. and S.A.B.; visualization, E.A.R. and S.A.B.; supervision, S.A.B.; project administration, S.A.B.; funding acquisition, S.A.B. All authors have read and agreed to the published version of the manuscript.

Funding: This research was funded by Yorkshire Cancer Research award L352, PPO12; Bone Cancer Research Trust award BCRT 5918; Ewing Sarcoma Research Trust award BCRT/36/13, Little Princess Trust award CCLGA 2021 09 Burchill, the Candlelighters Trust, and the University of Leeds. L.A.S. was the recipient of a University of Leeds scholarship.

Institutional Review Board Statement: The study was conducted in accordance with the Declaration of Helsinki. The study was approved and performance monitored by the Integrated Research Application System (IRAS167880) and CRN Yorkshire & Humber NIHR Clinical Research Network (EDGE 79301), respectively.

Informed Consent Statement: Informed consent was obtained from all subjects involved in the study.

Data Availability Statement: The FASTQ files of sequenced ES cell lines, 2D and 3D SK-N-MC cells are available in the Research Data Leeds Repository (University of Leeds), Burchill Susan and Roundhill Elizabeth (2022): Total RNA sequencing of ES cell lines, TTC 466 and SK-N-MC 2D and 3D cultures. [Dataset]. https://doi.org/10.5518/1210.

Acknowledgments: The authors thank Kim Cass, Tom Maisey, Samantha Brownhill and Andrea Berry for technical assistance.

Conflicts of Interest: The authors declare no conflict of interest.

References

1. Balamuth, N.J.; Womer, R.B. Ewing's sarcoma. *Lancet Oncol.* **2010**, *11*, 184–192. [CrossRef] [PubMed]
2. Cotterill, S.J.; Ahrens, S.; Paulussen, M.; Jurgens, H.F.; Voute, P.A.; Gadner, H.; Craft, A.W. Prognostic factors in Ewing's tumor of bone: Analysis of 975 patients from the European Intergroup Cooperative Ewing's Sarcoma Study Group. *J. Clin. Oncol.* **2000**, *18*, 3108–3114. [CrossRef] [PubMed]
3. Ahrens, S.; Hoffmann, C.; Jabar, S.; Braun-Munzinger, G.; Paulussen, M.; Dunst, J.; Rube, C.; Winkelmann, W.; Heinecke, A.; Gobel, U.; et al. Evaluation of prognostic factors in a tumor volume-adapted treatment strategy for localized Ewing sarcoma of bone: The CESS 86 experience. Cooperative Ewing Sarcoma Study. *Med. Pediatr. Oncol.* **1999**, *32*, 186–195. [CrossRef]
4. Rodriguez-Galindo, C.; Navid, F.; Liu, T.; Billups, C.A.; Rao, B.N.; Krasin, M.J. Prognostic factors for local and distant control in Ewing sarcoma family of tumors. *Ann. Oncol.* **2007**, *19*, 814–820. [CrossRef] [PubMed]
5. Stahl, M.; Ranft, A.; Paulussen, M.; Bolling, T.; Vieth, V.; Bielack, S.; Gortitz, I.; Braun-Munzinger, G.; Hardes, J.; Jurgens, H.; et al. Risk of recurrence and survival after relapse in patients with Ewing sarcoma. *Pediatr. Blood Cancer* **2011**, *57*, 549–553. [CrossRef]
6. Barker, L.M.; Pendergrass, T.W.; Sanders, J.E.; Hawkins, D.S. Survival after recurrence of Ewing's sarcoma family of tumors. *J. Clin. Oncol. Off. J. Am. Soc. Clin. Oncol.* **2005**, *23*, 4354–4362. [CrossRef]
7. Huang, M.; Lucas, K. Current therapeutic approaches in metastatic and recurrent ewing sarcoma. *Sarcoma* **2011**, *2011*, 863210. [CrossRef]
8. Joo, K.M.; Kim, S.Y.; Jin, X.; Song, S.Y.; Kong, D.S.; Lee, J.I.; Jeon, J.W.; Kim, M.H.; Kang, B.G.; Jung, Y.; et al. Clinical and biological implications of CD133-positive and CD133-negative cells in glioblastomas. *Lab. Investig.* **2008**, *88*, 808–815. [CrossRef]
9. Ogden, A.T.; Waziri, A.E.; Lochhead, R.A.; Fusco, D.; Lopez, K.; Ellis, J.A.; Kang, J.; Assanah, M.; McKhann, G.M.; Sisti, M.B.; et al. Identification of A2B5+CD133- tumor-initiating cells in adult human gliomas. *Neurosurgery* **2008**, *62*, 505–514. [CrossRef]
10. Tirino, V.; Desiderio, V.; d'Aquino, R.; De Francesco, F.; Pirozzi, G.; Graziano, A.; Galderisi, U.; Cavaliere, C.; De Rosa, A.; Papaccio, G.; et al. Detection and characterization of CD133+ cancer stem cells in human solid tumours. *PLoS ONE* **2008**, *3*, e3469. [CrossRef]
11. Walter, D.; Satheesha, S.; Albrecht, P.; Bornhauser, B.C.; D'Alessandro, V.; Oesch, S.M.; Rehrauer, H.; Leuschner, I.; Koscielniak, E.; Gengler, C.; et al. CD133 positive embryonal rhabdomyosarcoma stem-like cell population is enriched in rhabdospheres. *PLoS ONE* **2011**, *6*, e19506. [CrossRef]
12. Wang, J.; Sakariassen, P.O.; Tsinkalovsky, O.; Immervoll, H.; Boe, S.O.; Svendsen, A.; Prestegarden, L.; Rosland, G.; Thorsen, F.; Stuhr, L.; et al. CD133 negative glioma cells form tumors in nude rats and give rise to CD133 positive cells. *Int. J. Cancer* **2008**, *122*, 761–768. [CrossRef]
13. Cornaz-Buros, S.; Riggi, N.; DeVito, C.; Sarre, A.; Letovanec, I.; Provero, P.; Stamenkovic, I. Targeting cancer stem-like cells as an approach to defeating cellular heterogeneity in Ewing sarcoma. *Cancer Res.* **2014**, *74*, 6610–6622. [CrossRef]
14. Suva, M.L.; Riggi, N.; Stehle, J.C.; Baumer, K.; Tercier, S.; Joseph, J.M.; Suva, D.; Clement, V.; Provero, P.; Cironi, L.; et al. Identification of cancer stem cells in Ewing's sarcoma. *Cancer Res.* **2009**, *69*, 1776–1781. [CrossRef]
15. Jiang, X.; Gwye, Y.; Russell, D.; Cao, C.; Douglas, D.; Hung, L.; Kovar, H.; Triche, T.J.; Lawlor, E.R. CD133 expression in chemo-resistant Ewing sarcoma cells. *BMC Cancer* **2010**, *10*, 116. [CrossRef]
16. Fujii, H.; Honoki, K.; Tsujiuchi, T.; Kido, A.; Yoshitani, K.; Takakura, Y. Sphere-forming stem-like cell populations with drug resistance in human sarcoma cell lines. *Int. J. Oncol.* **2009**, *34*, 1381–1386.
17. Roundhill, E.A.; Chicon-Bosch, M.; Jeys, L.; Parry, M.; Rankin, K.S.; Droop, A.; Burchill, S.A. RNA sequencing and functional studies of patient-derived cells reveal that neurexin-1 and regulators of this pathway are associated with poor outcomes in Ewing sarcoma. *Cell Oncol. (Dordr)* **2021**, *44*, 1065–1085. [CrossRef]
18. Roundhill, E.A.; Jabri, S.; Burchill, S.A. ABCG1 and Pgp identify drug resistant, self-renewing osteosarcoma cells. *Cancer Lett.* **2019**, *453*, 142–157. [CrossRef]
19. Charafe-Jauffret, E.; Ginestier, C.; Iovino, F.; Wicinski, J.; Cervera, N.; Finetti, P.; Hur, M.H.; Diebel, M.E.; Monville, F.; Dutcher, J.; et al. Breast cancer cell lines contain functional cancer stem cells with metastatic capacity and a distinct molecular signature. *Cancer Res.* **2009**, *69*, 1302–1313. [CrossRef]
20. Suzuki, E.; Chiba, T.; Zen, Y.; Miyagi, S.; Tada, M.; Kanai, F.; Imazeki, F.; Miyazaki, M.; Iwama, A.; Yokosuka, O. Aldehyde dehydrogenase 1 is associated with recurrence-free survival but not stem cell-like properties in hepatocellular carcinoma. *Hepatol. Res.* **2012**, *42*, 1100–1111. [CrossRef]

21. Bertolini, F.; Sukhatme, V.P.; Bouche, G. Drug repurposing in oncology–patient and health systems opportunities. *Nat. Rev. Clin. Oncol.* **2015**, *12*, 732–742. [CrossRef] [PubMed]
22. Pantziarka, P.; Verbaanderd, C.; Huys, I.; Bouche, G.; Meheus, L. Repurposing drugs in oncology: From candidate selection to clinical adoption. *Semin. Cancer Biol.* **2021**, *68*, 186–191. [CrossRef]
23. Roundhill, E.A.; Burchill, S.A. Detection and characterisation of multi-drug resistance protein 1 (MRP-1) in human mitochondria. *Br. J. Cancer* **2012**, *106*, 1224–1233. [CrossRef] [PubMed]
24. Aflatoonian, B.; Ruban, L.; Shamsuddin, S.; Baker, D.; Andrews, P.; Moore, H. Generation of Sheffield (Shef) human embryonic stem cell lines using a microdrop culture system. *In Vitro Cell Dev. Biol. Anim.* **2010**, *46*, 236–241. [CrossRef] [PubMed]
25. Fischer, M.; Skowron, M.; Berthold, F. Reliable transcript quantification by real-time reverse transcriptase-polymerase chain reaction in primary neuroblastoma using normalization to averaged expression levels of the control genes HPRT1 and SDHA. *J. Mol. Diagn.* **2005**, *7*, 89–96. [CrossRef]
26. Lastowska, M.; Viprey, V.; Santibanez-Koref, M.; Wappler, I.; Peters, H.; Cullinane, C.; Roberts, P.; Hall, A.G.; Tweddle, D.A.; Pearson, A.D.; et al. Identification of candidate genes involved in neuroblastoma progression by combining genomic and expression microarrays with survival data. *Oncogene* **2007**, *26*, 7432–7444. [CrossRef]
27. Roundhill, E.; Burchill, S. Membrane expression of MRP-1, but not MRP-1 splicing or Pgp expression, predicts survival in patients with ESFT. *Br. J. Cancer* **2013**, *109*, 195–206. [CrossRef]
28. Kodani, M.; Yang, G.; Conklin, L.M.; Travis, T.C.; Whitney, C.G.; Anderson, L.J.; Schrag, S.J.; Taylor, T.H., Jr.; Beall, B.W.; Breiman, R.F.; et al. Application of TaqMan low-density arrays for simultaneous detection of multiple respiratory pathogens. *J. Clin. Microbiol.* **2011**, *49*, 2175–2182. [CrossRef]
29. Demidenko, R.; Razanauskas, D.; Daniunaite, K.; Lazutka, J.R.; Jankevicius, F.; Jarmalaite, S. Frequent down-regulation of ABC transporter genes in prostate cancer. *BMC Cancer* **2015**, *15*, 683. [CrossRef]
30. Szklarczyk, D.; Gable, A.L.; Lyon, D.; Junge, A.; Wyder, S.; Huerta-Cepas, J.; Simonovic, M.; Doncheva, N.T.; Morris, J.H.; Bork, P.; et al. STRING v11: Protein-protein association networks with increased coverage, supporting functional discovery in genome-wide experimental datasets. *Nucleic Acids Res.* **2019**, *47*, D607–D613. [CrossRef]
31. Brownhill, S.C.; Taylor, C.; Burchill, S.A. Chromosome 9p21 gene copy number and prognostic significance of p16 in ESFT. *Br. J. Cancer* **2007**, *96*, 1914–1923. [CrossRef]
32. Anders, S.; Huber, W. Differential expression analysis for sequence count data. *Genome Biol.* **2010**, *11*, R106. [CrossRef]
33. Brownhill, S.; Cohen, D.; Burchill, S. Proliferation index: A continuous model to predict prognosis in patients with tumours of the Ewing's sarcoma family. *PLoS ONE* **2014**, *9*, e104106. [CrossRef]
34. Myatt, S.S.; Redfern, C.P.; Burchill, S.A. p38MAPK-Dependent sensitivity of Ewing's sarcoma family of tumors to fenretinide-induced cell death. *Clin. Cancer Res.* **2005**, *11*, 3136–3148. [CrossRef]
35. Wishart, D.S.; Feunang, Y.D.; Guo, A.C.; Lo, E.J.; Marcu, A.; Grant, J.R.; Sajed, T.; Johnson, D.; Li, C.; Sayeeda, Z.; et al. DrugBank 5.0: A major update to the DrugBank database for 2018. *Nucleic Acids Res.* **2018**, *46*, D1074–D1082. [CrossRef]
36. Freshour, S.L.; Kiwala, S.; Cotto, K.C.; Coffman, A.C.; McMichael, J.F.; Song, J.J.; Griffith, M.; Griffith, O.L.; Wagner, A.H. Integration of the Drug-Gene Interaction Database (DGIdb 4.0) with open crowdsource efforts. *Nucleic Acids Res.* **2021**, *49*, D1144–D1151. [CrossRef]
37. Pantziarka, P.; Verbaanderd, C.; Sukhatme, V.; Rica Capistrano, I.; Crispino, S.; Gyawali, B.; Rooman, I.; Van Nuffel, A.M.; Meheus, L.; Sukhatme, V.P.; et al. ReDO_DB: The repurposing drugs in oncology database. *Ecancermedicalscience* **2018**, *12*, 886. [CrossRef]
38. Ochoa, D.; Hercules, A.; Carmona, M.; Suveges, D.; Gonzalez-Uriarte, A.; Malangone, C.; Miranda, A.; Fumis, L.; Carvalho-Silva, D.; Spitzer, M.; et al. Open Targets Platform: Supporting systematic drug-target identification and prioritisation. *Nucleic Acids Res.* **2021**, *49*, D1302–D1310. [CrossRef]
39. Pinero, J.; Ramirez-Anguita, J.M.; Sauch-Pitarch, J.; Ronzano, F.; Centeno, E.; Sanz, F.; Furlong, L.I. The DisGeNET knowledge platform for disease genomics: 2019 update. *Nucleic Acids Res.* **2020**, *48*, D845–D855. [CrossRef]
40. Pantziarka, P.; Capistrano, I.R.; De Potter, A.; Vandeborne, L.; Bouche, G. An Open Access Database of Licensed Cancer Drugs. *Front. Pharmacol.* **2021**, *12*, 627574. [CrossRef]
41. Pantziarka, P.; Vandeborne, L.; Bouche, G. A Database of Drug Repurposing Clinical Trials in Oncology. *Front. Pharmacol.* **2021**, *12*, 790952. [CrossRef] [PubMed]
42. Shi, W.; Kwon, J.; Huang, Y.; Tan, J.; Uhl, C.G.; He, R.; Zhou, C.; Liu, Y. Facile Tumor Spheroids Formation in Large Quantity with Controllable Size and High Uniformity. *Sci. Rep.* **2018**, *8*, 6837. [CrossRef] [PubMed]
43. Zhang, Q.; Yu, S.; Lam, M.M.T.; Poon, T.C.W.; Sun, L.; Jiao, Y.; Wong, A.S.T.; Lee, L.T.O. Angiotensin II promotes ovarian cancer spheroid formation and metastasis by upregulation of lipid desaturation and suppression of endoplasmic reticulum stress. *J. Exp. Clin. Cancer Res.* **2019**, *38*, 116. [CrossRef] [PubMed]
44. Lee, J.W.; Sung, J.S.; Park, Y.S.; Chung, S.; Kim, Y.H. Isolation of spheroid-forming single cells from gastric cancer cell lines: Enrichment of cancer stem-like cells. *Biotechniques* **2018**, *65*, 197–203. [CrossRef] [PubMed]
45. Ivascu, A.; Kubbies, M. Rapid generation of single-tumor spheroids for high-throughput cell function and toxicity analysis. *J. Biomol. Screen* **2006**, *11*, 922–932. [CrossRef]
46. Lee, C.; Lee, C.; Atakilit, A.; Siu, A.; Ramos, D.M. Differential spheroid formation by oral cancer cells. *Anticancer Res.* **2014**, *34*, 6945–6949.
47. Ayob, A.Z.; Ramasamy, T.S. Cancer stem cells as key drivers of tumour progression. *J. Biomed. Sci.* **2018**, *25*, 20. [CrossRef]

48. Ribeiro-Dantas, M.D.C.; Oliveira Imparato, D.; Dalmolin, M.G.S.; de Farias, C.B.; Brunetto, A.T.; da Cunha Jaeger, M.; Roesler, R.; Sinigaglia, M.; Siqueira Dalmolin, R.J. Reverse Engineering of Ewing Sarcoma Regulatory Network Uncovers PAX7 and RUNX3 as Master Regulators Associated with Good Prognosis. *Cancers* **2021**, *13*, 1860. [CrossRef]
49. Cidre-Aranaz, F.; Li, J.; Holting, T.L.B.; Orth, M.F.; Imle, R.; Kutschmann, S.; Ammirati, G.; Ceranski, K.; Carreno-Gonzalez, M.J.; Kasan, M.; et al. Integrative gene network and functional analyses identify a prognostically relevant key regulator of metastasis in Ewing sarcoma. *Mol. Cancer* **2022**, *21*, 1. [CrossRef]
50. Miller, H.E.; Gorthi, A.; Bassani, N.; Lawrence, L.A.; Iskra, B.S.; Bishop, A.J.R. Reconstruction of Ewing Sarcoma Developmental Context from Mass-Scale Transcriptomics Reveals Characteristics of EWSR1-FLI1 Permissibility. *Cancers* **2020**, *12*, 948. [CrossRef]
51. Khoogar, R.; Li, F.; Chen, Y.; Ignatius, M.; Lawlor, E.R.; Kitagawa, K.; Huang, T.H.; Phelps, D.A.; Houghton, P.J. Single-cell RNA profiling identifies diverse cellular responses to EWSR1/FLI1 downregulation in Ewing sarcoma cells. *Cell Oncol.* **2022**, *45*, 19–40. [CrossRef] [PubMed]
52. Grunewald, T.G.; Alonso, M.; Avnet, S.; Banito, A.; Burdach, S.; Cidre-Aranaz, F.; Di Pompo, G.; Distel, M.; Dorado-Garcia, H.; Garcia-Castro, J.; et al. Sarcoma treatment in the era of molecular medicine. *EMBO Mol. Med.* **2020**, *12*, e11131. [CrossRef] [PubMed]
53. Wilding, C.P.; Elms, M.L.; Judson, I.; Tan, A.C.; Jones, R.L.; Huang, P.H. The landscape of tyrosine kinase inhibitors in sarcomas: Looking beyond pazopanib. *Expert Rev. Anticancer Ther.* **2019**, *19*, 971–991. [CrossRef] [PubMed]
54. Wu, A.; Wojtowicz, K.; Savary, S.; Hamon, Y.; Trombik, T. Do ABC transporters regulate plasma membrane organization? *Cell Mol. Biol. Lett.* **2020**, *25*, 37. [CrossRef]
55. Namba, Y.; Sogawa, C.; Okusha, Y.; Kawai, H.; Itagaki, M.; Ono, K.; Murakami, J.; Aoyama, E.; Ohyama, K.; Asaumi, J.I.; et al. Depletion of Lipid Efflux Pump ABCG1 Triggers the Intracellular Accumulation of Extracellular Vesicles and Reduces Aggregation and Tumorigenesis of Metastatic Cancer Cells. *Front. Oncol.* **2018**, *8*, 376. [CrossRef]
56. Tian, C.; Huang, D.; Yu, Y.; Zhang, J.; Fang, Q.; Xie, C. ABCG1 as a potential oncogene in lung cancer. *Exp. Ther. Med.* **2017**, *13*, 3189–3194. [CrossRef]
57. Yvan-Charvet, L.; Pagler, T.A.; Seimon, T.A.; Thorp, E.; Welch, C.L.; Witztum, J.L.; Tabas, I.; Tall, A.R. ABCA1 and ABCG1 protect against oxidative stress-induced macrophage apoptosis during efferocytosis. *Circ. Res.* **2010**, *106*, 1861–1869. [CrossRef]
58. Xue, J.; Wei, J.; Dong, X.; Zhu, C.; Li, Y.; Song, A.; Liu, Z. ABCG1 deficiency promotes endothelial apoptosis by endoplasmic reticulum stress-dependent pathway. *J. Physiol. Sci.* **2013**, *63*, 435–444. [CrossRef]
59. Apfelbaum, A.A.; Wrenn, E.D.; Lawlor, E.R. The importance of fusion protein activity in Ewing sarcoma and the cell intrinsic and extrinsic factors that regulate it: A review. *Front. Oncol.* **2022**, *12*, 1044707. [CrossRef]
60. Elsnerova, K.; Mohelnikova-Duchonova, B.; Cerovska, E.; Ehrlichova, M.; Gut, I.; Rob, L.; Skapa, P.; Hruda, M.; Bartakova, A.; Bouda, J.; et al. Gene expression of membrane transporters: Importance for prognosis and progression of ovarian carcinoma. *Oncol. Rep.* **2016**, *35*, 2159–2170. [CrossRef]
61. Litviakov, N.V.; Garbukov, E.; Slonimskaia, E.M.; Tsyganov, M.M.; Denisov, E.V.; Vtorushin, S.V.; Khristenko, K.; Zav'ialova, M.V.; Cherdyntseva, N.V. Correlation of metastasis-free survival in patients with breast cancer and changes in the direction of expression of multidrug resistance genes during neoadjuvant chemotherapy. *Vopr. Onkol.* **2013**, *59*, 334–340.
62. Litviakov, N.V.; Cherdyntseva, N.V.; Tsyganov, M.M.; Slonimskaya, E.M.; Ibragimova, M.K.; Kazantseva, P.V.; Kzhyshkowska, J.; Choinzonov, E.L. Deletions of multidrug resistance gene loci in breast cancer leads to the down-regulation of its expression and predict tumor response to neoadjuvant chemotherapy. *Oncotarget* **2016**, *7*, 7829–7841. [CrossRef]
63. Wang, Y.; Liu, H.; Ready, N.E.; Su, L.; Wei, Y.; Christiani, D.C.; Wei, Q. Genetic variants in ABCG1 are associated with survival of nonsmall-cell lung cancer patients. *Int. J. Cancer* **2016**, *138*, 2592–2601. [CrossRef]
64. Gu, H.M.; Li, G.; Gao, X.; Berthiaume, L.G.; Zhang, D.W. Characterization of palmitoylation of ATP binding cassette transporter G1: Effect on protein trafficking and function. *Biochim. Biophys. Acta* **2013**, *1831*, 1067–1078. [CrossRef]
65. Gu, H.M.; Wang, F.Q.; Zhang, D.W. Caveolin-1 interacts with ATP binding cassette transporter G1 (ABCG1) and regulates ABCG1-mediated cholesterol efflux. *Biochim. Biophys. Acta* **2014**, *1841*, 847–858. [CrossRef]
66. Riggi, N.; Suva, M.L.; Stamenkovic, I. The cancer stem cell paradigm in Ewing's sarcoma: What can we learn about these rare cells from a rare tumor? *Expert Rev. Anticancer Ther.* **2011**, *11*, 143–145. [CrossRef]
67. Glumac, P.M.; LeBeau, A.M. The role of CD133 in cancer: A concise review. *Clin. Transl. Med.* **2018**, *7*, 18. [CrossRef]
68. Rappa, G.; Fodstad, O.; Lorico, A. The stem cell-associated antigen CD133 (Prominin-1) is a molecular therapeutic target for metastatic melanoma. *Stem Cells* **2008**, *26*, 3008–3017. [CrossRef]
69. Danieau, G.; Morice, S.; Redini, F.; Verrecchia, F.; Royer, B.B. New Insights about the Wnt/beta-Catenin Signaling Pathway in Primary Bone Tumors and Their Microenvironment: A Promising Target to Develop Therapeutic Strategies? *Int. J. Mol. Sci.* **2019**, *20*, 3751. [CrossRef]
70. Tirode, F.; Laud-Duval, K.; Prieur, A.; Delorme, B.; Charbord, P.; Delattre, O. Mesenchymal stem cell features of Ewing tumors. *Cancer Cell* **2007**, *11*, 421–429. [CrossRef]
71. von Levetzow, C.; Jiang, X.; Gwye, Y.; von Levetzow, G.; Hung, L.; Cooper, A.; Hsu, J.H.; Lawlor, E.R. Modeling initiation of Ewing sarcoma in human neural crest cells. *PLoS ONE* **2011**, *6*, e19305. [CrossRef] [PubMed]
72. Jawad, M.U.; Cheung, M.C.; Min, E.S.; Schneiderbauer, M.M.; Koniaris, L.G.; Scully, S.P. Ewing sarcoma demonstrates racial disparities in incidence-related and sex-related differences in outcome: An analysis of 1631 cases from the SEER database, 1973-2005. *Cancer* **2009**, *115*, 3526–3536. [CrossRef] [PubMed]

73. Machiela, M.J.; Grunewald, T.G.P.; Surdez, D.; Reynaud, S.; Mirabeau, O.; Karlins, E.; Rubio, R.A.; Zaidi, S.; Grossetete-Lalami, S.; Ballet, S.; et al. Genome-wide association study identifies multiple new loci associated with Ewing sarcoma susceptibility. *Nat. Commun.* **2018**, *9*, 3184. [CrossRef] [PubMed]
74. Randall, R.L.; Lessnick, S.L.; Jones, K.B.; Gouw, L.G.; Cummings, J.E.; Cannon-Albright, L.; Schiffman, J.D. Is There a Predisposition Gene for Ewing's Sarcoma? *J. Oncol.* **2010**, *2010*, 397632. [CrossRef]
75. Grunewald, T.G.P.; Cidre-Aranaz, F.; Surdez, D.; Tomazou, E.M.; de Alava, E.; Kovar, H.; Sorensen, P.H.; Delattre, O.; Dirksen, U. Ewing sarcoma. *Nat. Rev. Dis. Prim.* **2018**, *4*, 5. [CrossRef]
76. Kumar, S.V.; Er, P.X.; Lawlor, K.T.; Motazedian, A.; Scurr, M.; Ghobrial, I.; Combes, A.N.; Zappia, L.; Oshlack, A.; Stanley, E.G.; et al. Kidney micro-organoids in suspension culture as a scalable source of human pluripotent stem cell-derived kidney cells. *Development* **2019**, *146*, dev172361. [CrossRef]
77. Keller, F.; Rudiger, R.; MAthias, H. Towards optimized breast cancer 3D spheroid mono- and co-culture models for pharmacological research and screening. *J. Cell. Biotechnol.* **2019**, *5*, 89–101. [CrossRef]
78. Wang, Y.J.; Herlyn, M. The emerging roles of Oct4 in tumor-initiating cells. *Am. J. Physiol. Cell Physiol.* **2015**, *309*, C709–C718. [CrossRef]
79. Kim, J.; Kim, H.S.; Shim, J.J.; Lee, J.; Kim, A.Y.; Kim, J. Critical role of the fibroblast growth factor signalling pathway in Ewing's sarcoma octamer-binding transcription factor 4-mediated cell proliferation and tumorigenesis. *FEBS J.* **2019**, *286*, 4443–4472. [CrossRef]
80. Chua, V.; Chawla, S.P.; Zheng, K.; Kim, T.; Del Priore, G.; Kim, S. Phase II study of SM-88 in Ewing's and other sarcomas. In Proceedings of the 2021 ASCO Annual Meeting I, Chicago, IL, USA, 28 May 2021.
81. Chung, V.; Chawla, S.P.; Dong, H.; Kim, S.; Korn, R.L.; Lim, K.-H.; Noel, M.S.; Noonan, A.M.; Oberstein, P.E.; Ocean, A.J.; et al. Phase II/III study of SM-88 in patients with metastatic pancreatic cancer. In Proceedings of the 2021 Gastrointestinal Cancers Symposium, Online, 20 January 2021.
82. Brennan, B.; Kirton, L.; Marec-Berard, P.; Martin-Broto, J.; Gelderblom, H.; Gaspar, N.; Strauss, S.; Sastre Urgelles, A.; Anderton, J.; Laurence, V.; et al. Comparison of two chemotherapy regimens in Ewing sarcoma (ES): Overall and subgroup results of the Euro Ewing 2012 randomized trial (EE2012). In Proceedings of the 2020 ASCO Annual Meeting I, Virtual, 25 May 2020; p. 11500.
83. Serra, M.; Hattinger, C.M.; Pasello, M.; Casotti, C.; Fantoni, L.; Riganti, C.; Manara, M.C. Impact of ABC Transporters in Osteosarcoma and Ewing's Sarcoma: Which Are Involved in Chemoresistance and Which Are Not? *Cells* **2021**, *10*, 2461. [CrossRef]
84. Perri, T.; Fogel, M.; Mor, S.; Horev, G.; Meller, I.; Loven, D.; Issakov, J.; Kollender, Y.; Smirnov, A.; Zaizov, R.; et al. Effect of P-glycoprotein expression on outcome in the Ewing family of tumors. *Pediatr. Hematol. Oncol.* **2001**, *18*, 325–334. [CrossRef]
85. Scotlandi, K.; Manara, M.C.; Hattinger, C.M.; Benini, S.; Perdichizzi, S.; Pasello, M.; Bacci, G.; Zanella, L.; Bertoni, F.; Picci, P.; et al. Prognostic and therapeutic relevance of HER2 expression in osteosarcoma and Ewing's sarcoma. *Eur. J. Cancer* **2005**, *41*, 1349–1361. [CrossRef]
86. Roessner, A.; Ueda, Y.; Bockhorn-Dworniczak, B.; Blasius, S.; Peters, A.; Wuisman, P.; Ritter, J.; Paulussen, M.; Jurgens, H.; Bocker, W. Prognostic implication of immunodetection of P glycoprotein in Ewing's sarcoma. *J. Cancer Res. Clin. Oncol.* **1993**, *119*, 185–189. [CrossRef]
87. Mancarella, C.; Pasello, M.; Ventura, S.; Grilli, A.; Calzolari, L.; Toracchio, L.; Lollini, P.L.; Donati, D.M.; Picci, P.; Ferrari, S.; et al. Insulin-Like Growth Factor 2 mRNA-Binding Protein 3 is a Novel Post-Transcriptional Regulator of Ewing Sarcoma Malignancy. *Clin. Cancer Res.* **2018**, *24*, 3704–3716. [CrossRef]
88. Eisenmann, E.D.; Talebi, Z.; Sparreboom, A.; Baker, S.D. Boosting the oral bioavailability of anticancer drugs through intentional drug-drug interactions. *Basic Clin. Pharmacol. Toxicol.* **2022**, *130* (Suppl. S1), 23–35. [CrossRef]
89. Kang, Y. Analysis of cancer stem cell metastasis in xenograft animal models. *Methods Mol. Biol.* **2009**, *568*, 7–19. [CrossRef]
90. Sampieri, K.; Fodde, R. Cancer stem cells and metastasis. *Semin. Cancer Biol.* **2012**, *22*, 187–193. [CrossRef]
91. O'Callaghan, N.; Dhillon, V.; Thomas, P.; Fenech, M. A quantitative real-time PCR method for absolute telomere length. *Biotechniques* **2008**, *44*, 807–809. [CrossRef]

cancers

MDPI

Communication

Routine EWS Fusion Analysis in the Oncology Clinic to Identify Cancer-Specific Peptide Sequence Patterns That Span Breakpoints in Ewing Sarcoma and DSRCT

Peter M. Anderson [1,*], Zheng Jin Tu [2], Scott E. Kilpatrick [3], Matteo Trucco [1], Rabi Hanna [1] and Timothy Chan [4]

1 Pediatric Hematology and Bone Marrow Transplant, Pediatric and Taussig Cancer Institutes, Cleveland Clinic, Cleveland, OH 44195, USA
2 Bioinformatics, Molecular Pathology and Cytogenomics, Department of Laboratory Medicine, Pathology and Laboratory Medicine Institute, Cleveland Clinic, Cleveland, OH 44195, USA
3 Orthopedic Pathology and Center for ePathology, Pathology and Laboratory Medicine Institute, Cleveland Clinic, Cleveland, OH 44195, USA
4 Center for Immuno-Oncology, Department of Radiation Oncology, Taussig Cancer Institute and Lerner Research Institute, Cleveland Clinic, Cleveland, OH 44195, USA
* Correspondence: andersp@ccf.org; Tel.: +1-216-308-2706

Simple Summary: EWS-based fusions are aberrantly fused genes that drive Ewing sarcoma and desmoplastic small round cell tumor (DSRCT). Fusion data may be useful for personalized mRNA vaccines but are not yet routinely obtained in clinical practice. We present our workflow for the characterization of EWS driver fusions in a real-world pediatric oncology setting. We use rapid targeted sequencing of the breakpoint and genetic analysis to determine fusion sequences. We report amino acid fusion sequences from the EWS gene and the fusion partner gene (FLI1, ERG, FEV, WT1). Our workflow allows easy discernment of clinically relevant similarities and differences of EWS fusions. This simple analysis allows an understanding of molecular features of driver fusions underlying Ewing sarcoma or DSRCT in a real-world setting. This workflow is being utilized to obtain fusion neoantigen data used in "personalized" cancer vaccine trials under designs that seek to stimulate an immune response against Ewing sarcoma or DSRCT.

Abstract: (1) Background: EWS fusion genes are associated with Ewing sarcoma and other Ewing family tumors including desmoplastic small round tumor, DSRCT. We utilize a clinical genomics workflow to reveal real-world frequencies of EWS fusion events, cataloging events that are similar, or divergent at the EWS breakpoint. (2) Methods: EWS fusion events from our next-generation sequencing panel (NGS) samples were first sorted by breakpoint or fusion junctions to map out the frequency of breakpoints. Fusion results were illustrated as in-frame fusion peptides involving EWS and a partner gene. (3) Results: From 2471 patient pool samples for fusion analysis at the Cleveland Clinic Molecular Pathology Laboratory, we identified 182 fusion samples evolved with the EWS gene. They are clustered in several breakpoints: chr22:29683123 (65.9%), and chr22:29688595 (2.7%). About 3/4 of Ewing sarcoma and DSRCT tumors have an identical EWS breakpoint motif at Exon 7 (SQQSSSYGQQ-) fused to a specific part of FLI1 (NPSYDSVRRG or-SSLLAYNTSS), ERG (NLPYEPPRRS), FEV (NPVGDGLFKD) or WT1 (SEKPYQCDFK). Our method also worked with Caris transcriptome data, too. Our primary clinical utility is to use this information to identify neoantigens for therapeutic purposes. (4) Conclusions and future perspectives: our method allows interpretation of what peptides result from the in-frame translation of EWS fusion junctions. These sequences, coupled with HLA-peptide binding data, are used to identify potential sequences of cancer-specific immunogenic peptides for Ewing sarcoma or DSRCT patients. This information may also be useful for immune monitoring (e.g., circulating T-cells with fusion-peptide specificity) to detect vaccine candidates, responses, or residual disease.

Citation: Anderson, P.M.; Tu, Z.J.; Kilpatrick, S.E.; Trucco, M.; Hanna, R.; Chan, T. Routine EWS Fusion Analysis in the Oncology Clinic to Identify Cancer-Specific Peptide Sequence Patterns That Span Breakpoints in Ewing Sarcoma and DSRCT. *Cancers* **2023**, *15*, 1623. https://doi.org/10.3390/cancers15051623

Academic Editors: Maria Paola Paronetto and Armita Bahrami

Received: 18 January 2023
Revised: 14 February 2023
Accepted: 28 February 2023
Published: 6 March 2023

Keywords: fusion gene analysis; cancer-specific sequence; EWS-FLI1; EWS-ERG; EWS-WT1; mRNA; polypeptide; breakpoint; next generation sequencing (NGS); information and education virtual visit

1. Introduction

Ewing sarcoma treatment has become relatively standardized with the vincristine, dexrazoxane + doxorubicin, and cyclophosphamide + mesna cycles alternating with ifosfamide + mesna and etoposide cycles [1–3]. Dose intensification using higher doses of chemotherapy or consolidation with high-dose chemotherapy and autologous transplant has been performed with no or modest improvement, respectively [4–8]. Despite apparent immunogenicity in a study of EWS-FLI1 consolidative immunotherapy [9], Ewing sarcoma vaccines against EWS peptides or mRNA coding for EWS peptides have not become part of standard-of-care or are not in clinical trials yet. Patients with excellent (100%) necrosis have a much better prognosis; Ewing sarcoma patients with pelvic tumors or poor chemotherapy response have a poor prognosis [10–13]. Similarly, poor necrosis, axial location, relapsed, and/or metastases continue to have poor survival [11–16]. How the EWS-fusion genes translate into biologic behavior is an active area of research (this issue of *Cancers,* Ewing Sarcoma: Basic Biology, Clinical Challenges and Future Perspectives, and references [17–30]).

In our pediatric and medical oncology practice, we see many high-risk and metastatic Ewing sarcoma and DSRCT patients, both in person and also using informational/educational virtual visits. Our pathologists also have extensive experience in diagnosing bone and soft tissue sarcomas, including the use of an in-house sarcoma NGS panel (refs. [31–34], Figure 1) and the use of whole exome RNA sequencing reports from Caris since January 2022. An unmet need exists to not just diagnose Ewing sarcoma and DSRCT using molecular tools but also to learn more about the nature of the EWS fusion genes and potential best targets in the context of HLA in each Ewing sarcoma or DSRCT patient [9,28–30]. Collecting real-world information about EWS gene fusions and resulting peptides may be leveraged to help design future immunologic approaches including identifying in a more precise manner (e.g., with HLA binding assays, T-call and B-cell recognition assays) what constructs to include in personalized cancer vaccines. The first step is to understand which cancer-specific peptides (neoantigens) are present in a real-world clinical practice setting and then seek to routinely obtain more information (e.g., HLA type). The HLA type is now routinely available on Caris reports. Then, we can better understand similarities and differences between how patients –may possibly stimulate the most effective T cell and B-cell responses to EWS neoantigens.

Different breakpoints, or "types", in the fusion genes have been identified for some time now, and their prognostic significance has been speculated upon previously [35] With the routine availability in our clinical practice of NGS and transcriptome sequencing panels (e.g., Caris) as part of sarcoma pathology and molecular reports detailing the presence of fusion genes, we have developed practical tools to investigate whether identical, similar, or very different peptides are made from the EWS-partner fusion genes in Ewing sarcoma and DSRCT in our clinical population. Our team is also interested in immunogenicity analysis of EWS fusion peptides and other cancer-specific peptides (neoantigens) to use as neoantigen targets for new therapies [36,37]. From this perspective, we share current efforts and data on EWS-FLI1, EWS-ERG, EWS-FEV, and EWS-WT1 gene fusions from our Cleveland Clinic NGS sarcoma panel as well as Caris transcriptome data. Thus, we are at an early stage of using this information to facilitate ongoing analysis and prediction of the immunogenicity of EWS fusion peptides in the context of HLA. This information may help to overcome barriers to effective new immunologic approaches against cancer-specific neoantigens in Ewing sarcoma and DSRCT [36–39].

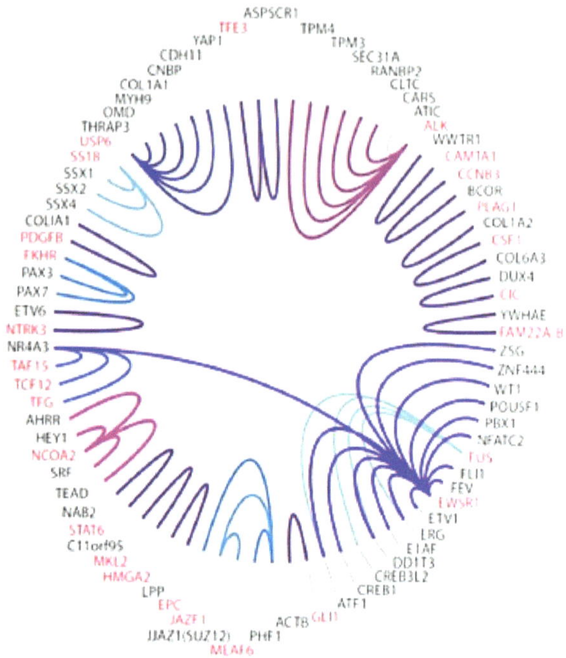

Figure 1. Current sarcoma NGS panel at Cleveland Clinic detects many gene fusions (same color) including 14 involving the EWS gene (Purple at 4 o'clock) [32]. Clinical NGS reports detail which exons are involved in a fusion gene such as EWS-FI1, EWS-ERG, and EWS-WT1. Since January 2022 Caris transcriptome reports are also obtained for additional information.

2. Materials and Methods

Pathology reports and gene fusions were analyzed using the Cleveland Clinic NGS sarcoma panel from January 2019. Since January 2022, we also have available data from Caris reports for transcriptome analysis. The Cleveland Clinic sarcoma NGS panel is based on anchored multiplex polymerase chain reaction (PCR) enriched for 34 gene targets [32]. The amplicons were subjected to massively parallel sequencing with 151x2 cycle pair-end reads. An informatics pipeline was used for read alignment (GRCh37 as reference genome), fusion identification, and annotation. For in-house NGS or Caris analysis, the sequenced short reads were aligned to the reference genome (hg19 or GRCh37) by either Archer Analysis or Caris transcriptome analysis pipeline. To obtain DNA and peptide sequences around the fusion junction, result alignment bam file (s) were loaded into Integrative Genome Viewer (IGV). Breakpoints were manually examined, and DNA sequences were extracted. This sequence was then translated to a peptide sequence. For ease of comparison, both fusion partner DNA and peptide sequences were represented with a different color to identify the exact point of fusion. Figure 2 shows EWS genomic (hg19) breakpoint distribution.

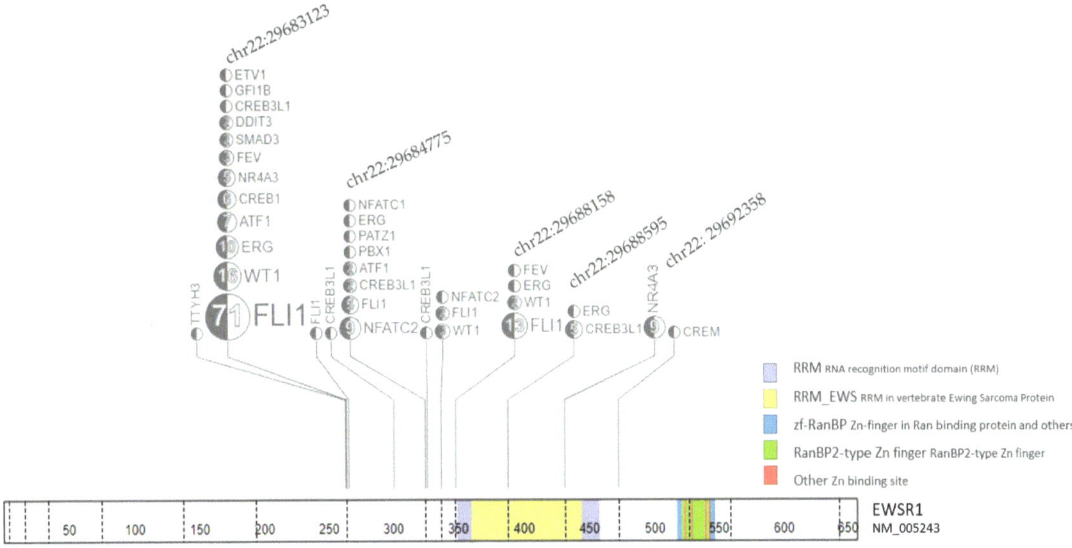

Figure 2. EWS fusion genomic (hg19) breakpoint distribution from 190 Fusion samples. Each breakpoints on the genomic location with samples/total are labeled on the EWS gene. The top five locations are further annotated with fusion gene in the box linked with dash line.

This in-house reporting system was used to display data as the DNA sequence spanning the fusion gene breakpoint and the amino acid sequence of an in-frame peptide spanning the gene fusion breakpoint. For ease of interpretation, the data was displayed to allow "at-a-glance" pattern recognition of identical, similar, or unique EWS-FLI1, EWS-ERG, EWS-FEV, and EWS-WT1 gene fusions (Figure 3) through visual examine fusion data via Integrative Genome Viewer [40]. DNA sequences across the gene fusion breakpoint and translational amino acid sequences were compared in Ewing sarcoma and DSRCT patients seen in the clinic and during virtual visits [33,34] who had provided paraffin blocks of tumor samples for analysis using our in-house NGS and/or Caris analysis. We also had both In-house NGS using Archer and Caris transcriptome using STARfusion data compared in 18 patients [EWS-FLI1 ($n = 14$), EWS-WT1 ($n = 3$), and EWS FEV ($n = 1$)] This allowed us to demonstrate that the generalization of our methods pertain not only out NGS but other platforms that can help oncologists and others know "at-a-glance" similarities and difference in individual EWS gene fusions.

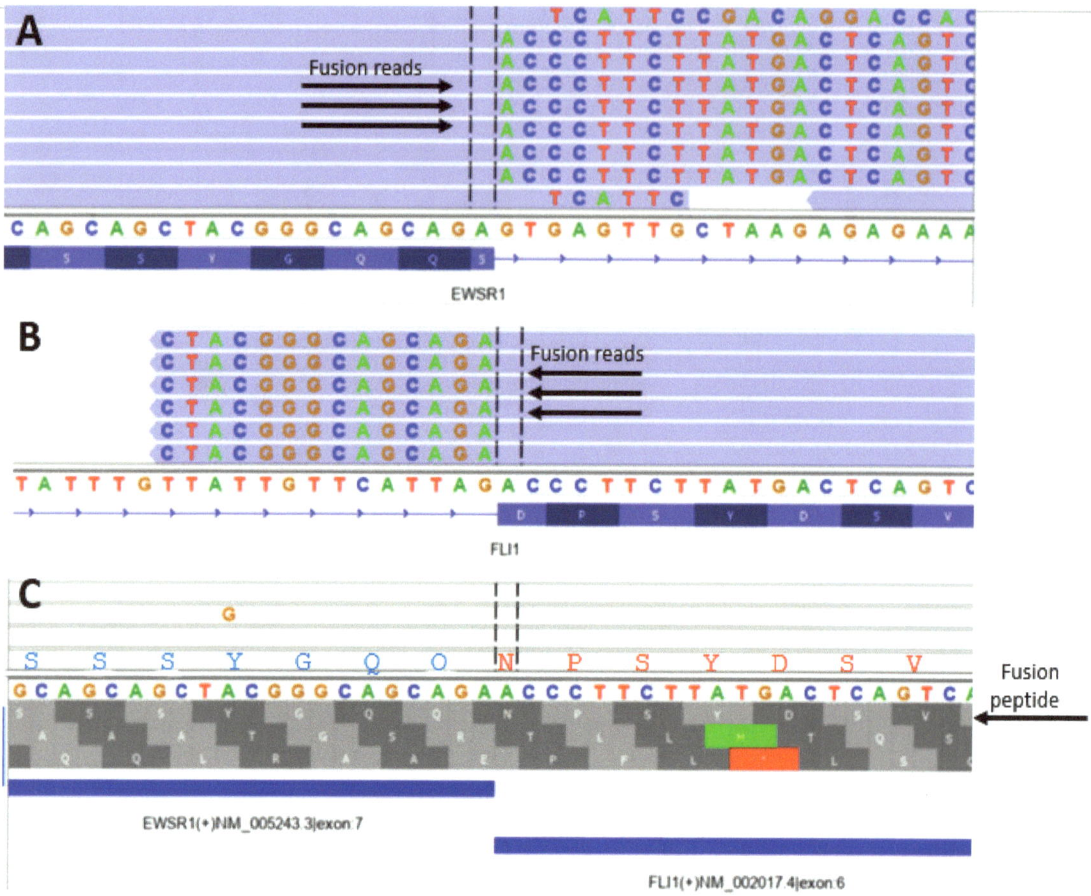

Figure 3. Determination of breakpoint and in-frame analysis to determine polypeptide sequences corresponding to fusion gene products in a single patient with Ewing sarcoma. (**A**) EWS gene IGV view against hg19 reference genome; (**B**) FLI1 gene IGV view against hg19 reference genome; (**C**) fusion reads IGV view against fusion contig reference. The first translational frame is the in-frame fusion with peptide sequence of SSSYGQQNPSYDSV between EWS and FLI1 gene.

3. Results

From 190 identified fusion samples with the EWS gene, the breakpoints in a genomic location are plotted in Figure 2. Most of the fusions are clustered in several breakpoints including chr22:29683123 (65.9%), chr22:29684775 (9.3%), chr22:29688158 (8.2%), chr22: 29692358 (4.9%), and chr22:29688595 (2.7%), (Figure 2). Additional comparison data using Caris transcriptome was available for EWS-FLI1 (n = 14), EWS-WT1 (n = 3), and EWS-FEV (n = 1).

A typical analysis is depicted in Figure 3. The presence of start (green) or stop (red) codons at the bottom of Figure 3 shows out-of-frame sequences compared to read-through regions for the in-frame sequences. Table 1 shows the frequencies and sequence details of the most common 24 amino acid polypeptides that span the EWS breakpoints in EWS-FLI1 fusion. Table 2 details the in-frame sequence of EWS-ERG, and EWS-WT1 fusions. In-frame analysis showed the EWS motif from exon 7 (SQQSSSYGQQ) accounted for over 80% of gene fusions involving EWS-FLI1, EWS-ERG, EWS-FEV, and EWS WT1, while the remaining gene fusions involved other parts of the EWS gene.

Table 1. Frequencies of Common EWS-FLI1 Fusion genes and corresponding polypeptides that span the EWS breakpoint.

EWS-FLI1 Breakpoint Fusion Peptide Analysis	Number (Per Cent)
SQQSSSYGQQNPSYDSVRRG	19
SQQSSSYGQQSSLLAYNTTS	12
SQQSSSYGQQNPYQILGPTS	1
SQQSSSYGQQRSGQIQLWQF	1
	33/40 (83%)
Other EWS-FLI1 Fusions Peptides	
PMDEGPDLDLGSLLAYNTTS	5
GERGGFNKPGGPPLGGAQTI	1
CVEFSSLIDQPVYPDVLASG	1
	7/40 (17%)

Table 2. EWS-ERG, EWS-FEV, and EWS-WT1 sequences and corresponding fusion polypeptides that span the breakpoints in Ewing sarcoma and DSRCT.

Ewing Sarcomas with EWS-ERG Fusions
EWS-ERG exon 7-9 (N = 3/6, 50%) SQQSSSYGQQNLPYEPPRRS
Ewing sarcomas with EWS-FEV fusions
EWS-FEV1 exon 7-2 (2/3; 67%) SQQSSSYGQQNPVGDGLFKD
DSRCT with EWS-WT1 fusions
EWS-WT1 exon 7-7 (N = 9/12, 75%) SQQSSSYGQQSEKPYQCDFK
EWS-WT1 exon 9-7 (N = 3/12 25%) GERGGFNKPGGEKPYQCDFK

In a real-world clinical practice (PA), using in-frame analysis of EWS-FLI1 peptides that span the breakpoint showed the EWS breakpoint sequence SQQSSSYGQQ to be the most common in our analysis (Table 1). The most common FLI1 sequence was NPSYDSVRRG. Some samples yielded an aspartate (D) instead of asparagine (N) at the FLI1 fusion point, but on manual inspection, this is related to breakpoint yielding asparagine. Although many EWS-FLI1 gene fusion events are possible, it seems that a limited number occur. Furthermore, the majority have identical breakpoints as identified by in-frame peptide analysis. Using Caris transcriptome data, we confirmed that our method is a general one as depicted in Figure 4.

Similarly, for EWS-ERG, EWS-FEV, and EWS-WT1, gene fusion analysis using in-frame peptide sequences, a common EWS sequence in EWS-ERG, EWS-FEV, and EWS-WT1 fusions was also SQQSSSYGQQ (Table 2). Thus, although many EWS-ERG, EWS-FEV, and EWS-WT1 gene fusion events are possible, it appears a few are more frequently associated with the development of Ewing sarcoma and DSRCT, respectively.

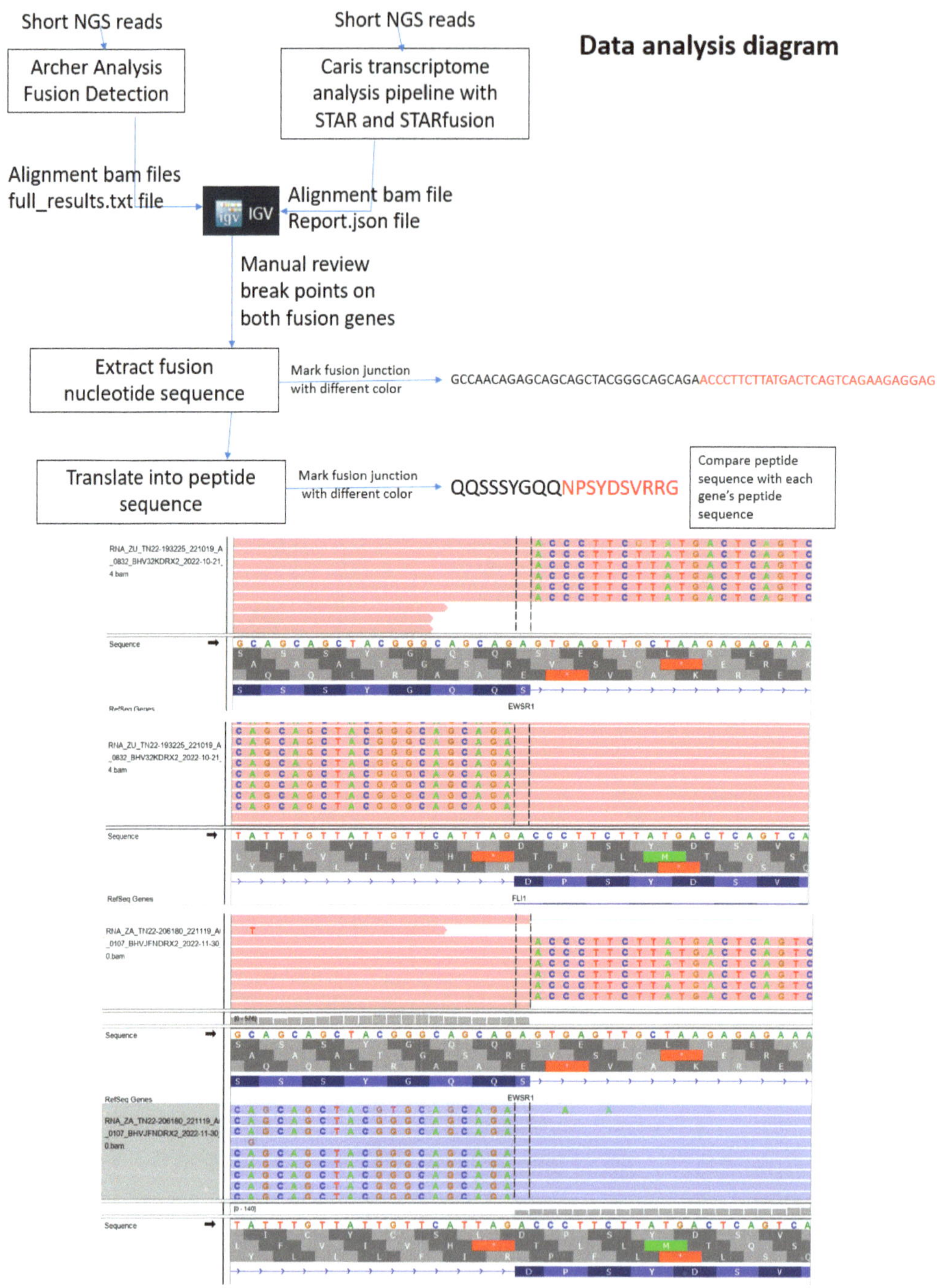

Figure 4. Data analysis diagram of how to depict EWS fusion junction breakpoints and corresponding amino acids. Top: flow diagram of how genomic data is depicted for in-house NGS using Archer Analysis or Caris transcriptome with STAR fusion detection. Middle and bottom: two examples of in-frame analysis of Caris transcriptome data to yield fusion peptide data with manual review of breakpoints of both fusion genes.

4. Discussion

EWS fusion genes are critical mutations associated with Ewing sarcoma and DSRCT. How these events translate into the malignant phenotype has and continues to be an area of very active investigation with many downstream effects resulting from the action of the EWS fusion genes [9,18,19,22–25,28–30,41–45]. We sought to try to learn whether EWS-FLI1, EWS-ERG, EWS-FEV, and EWS-WT1 gene fusions would result in identical, similar, or very different polypeptides using in-frame analysis. It seems that our method identified identical polypeptides for a particular fusion event (e.g., EWS–FLI1 exon 7-7). As expected, some types of fusions were more common than others (Tables 1 and 2). In particular, the EWS gene breaks and fuses to a partner in a way to often yields SQQSSSYGQQ fused to a partner sequence for both Ewing sarcoma and DSRCT in 83% of cases. Nevertheless, both Ewing sarcoma and DSRCT can result from a variety of different fusions involving EWS and a partner gene. Thus, a unifying pathway of fusion gene action resulting in protein action (s) may be one of the numerous actions of fusion genes and/or polypeptides leading to the Ewing sarcoma and DSRCT cancer phenotypes (many paths lead to Rome analogy). Since effective epitopes for B cells may be different from T cells as reported by Liu et al. [28], precise knowledge of sequences that span EWS breakpoints to augment the additional study of best neoantigens will be important for both humoral as well as T-cell responses to neoantigens.

Our in-frame analysis could have several important future clinical applications including (1) identification of different sequences and lengths of peptides that span the specific breakpoint to test for patient-specific HLA binding to the cancer-specific peptide (e.g., selection of best neoantigens for personalized cancer vaccines); (2) analysis of T-cell and B-cell immune responses prior to immunization, during therapy, after therapy, and in response to chemotherapy, radiation, cryoablation, or immunologic interventions. These could include immune-stimulatory drugs, reduction in the tumor inhibitory microenvironment (e.g., inflammation reduction, stereotactic body radiotherapy (SBRT), freezing of tumors, and depletion of regulatory t-cells [25,46–48]). This information could also increase the success of priming and boosting the immune system with vaccines composed of fusion peptides mRNA for the unique fusion gene sequence in the context of a particular HLA type and neoantigen binding (personalized mRNA vaccines using precision data). Furthermore, using Caris transcriptome data, we showed that our method has general applicability for others interested in helping physicians, genetic counselors, patients, and caregivers to increase their understanding about cancer-specific sequences resulting from somatic genetic events in Ewing sarcoma and DSRCT in the oncology clinic. Thus, we are in at an early stage in the process of starting to obtain more personal and precise data for better future interventions, the next generation of evidenced based medicine in oncology as recently described by Subbiah [49].

Not only gene fusions, but also frameshift mutations create neoantigens that start with a normal sequence and then transition to an unexpected cancer-specific sequence as part of early or late mutational events in cancer. Thus, our method may have applications not only for gene fusions such as EWS-FLII1, EWS-ERG, EWS-FEV, and EWS- WT1 but also for other gene fusions which are commonly found in sarcomas and also somatic frameshift mutations. Since the best choice of neoantigens in the context of HLA is vital, the collection of both in-frame fusion peptide data as well as HLA type may someday become a more routine Ewing sarcoma and DSRCT practice as part of tumor and host testing to f additional immunologic interventions in higher-risk, metastatic, and relapsed Ewing sarcoma, DSRCT, and other solid tumor patents (such as fusion-related sarcomas). Since acid decalcification does not allow for DNA analysis, this is one barrier to general applications for bone sarcomas such as Ewing sarcoma. Thus, we are currently in the process of implementing improved methods of tissue fixation (e.g., EDTA) that will facilitate both pathologic and genetic analysis of bone biopsies and resected samples. This should facilitate future studies in not only Ewing sarcoma but also osteosarcoma and analysis of samples from bone metastases, too.

5. Conclusions

Our current sarcoma NGS panel as well as Caris transcriptome data was suitable f learning additional basic biology about EWS fusion genes including the exact breakpoi and intra- and intergroup similarities and differences involving EWS-FLI1, EWS-ER(EWS-FEV, and EWS-WT1 in-frame polypeptides predicted by the fusion genes. Furthe more, our method of analysis also has possible future applications including analysis o immunogenicity of fusion peptides and cancer-specific genes such as frameshift mutation point mutations, insertions, and deletions to yield cancer-specific neoantigens in the context of a patient's HLA type. Although at an early stage, our paradigm of obtaining in-frame information may then facilitate the better design of more personal, precise, and effective data-driven immunologic interventions against Ewing sarcoma and DSRCT and other solid tumors including designing mRNA vaccines using the sequences and peptide size and HLA binding affinities with the highest chances of success [38,49].

6. Patents

The TC and Cleveland Clinic Foundation has a patent pending for the prediction of cancer-specific neoantigens such as EWS fusion peptides to HLA.

Author Contributions: Conceptualization, P.M.A., Z.J.T., R.H. and T.C. methodology, P.M.A., S.E.K. and Z.J.T.; software, Z.J.T.; validation, P.M.A., M.T. and Z.J.T.; analysis P.M.A. and Z.J.T.; investigation, P.M.A. and S.E.K.; resources, R.H. and T.C.; data curation, Z.J.T. and P.M.A.; writing—original draft preparation, P.M.A.; writing—review and editing, M.T.; visualization, R.H. and T.C.; supervision, R.H.; project administration, P.M.A. and R.H.; funding acquisition, P.M.A. and T.C. All authors have read and agreed to the published version of the manuscript.

Funding: This research was funded by the Cleveland Clinic Anderson Sarcoma Research Fund T56428 with contributions by the Morden Foundation, Spin for Gin, Little Warriors Foundation, and a Velosano Kids Impact award.

Informed Consent Statement: Patient consent was waived due to no interventions and analysis of existing data.

Data Availability Statement: Data supporting reported results can be obtained by contacting P.M.A. and Z.J.T. Zheng Jin Tu can provide additional details about in-frame analysis upon request (tuz@ccf.org).

Acknowledgments: We acknowledge the support of the Cleveland Clinic Sarcoma program and Laboratory Medicine and Pathology Institute, Learner Research Institute, and the Pediatric Institute to learn more about Ewing family tumors.

Conflicts of Interest: The authors declare no conflict of interest.

References

1. Womer, R.B.; West, D.C.; Krailo, M.D.; Dickman, P.S.; Pawel, B.R.; Grier, H.E.; Marcus, K.; Sailer, S.; Healey, J.H.; Dormans, J.P.; et al. Randomized controlled trial of interval-compressed chemotherapy for the treatment of localized Ewing sarcoma: A report from the Children's Oncology Group. *J. Clin. Oncol.* **2012**, *30*, 4148–4154. [CrossRef]
2. Gaspar, N.; Hawkins, D.S.; Dirksen, U.; Lewis, I.J.; Ferrari, S.; Le Deley, M.-C.; Kovar, H.; Grimer, R.; Whelan, J.; Claude, L.; et al. Ewing Sarcoma: Current Management and Future Approaches Through Collaboration. *J. Clin. Oncol.* **2015**, *33*, 3036–3046. [CrossRef] [PubMed]
3. Riggi, N.; Suva, M.; Stamenkovic, I. Ewing's Sarcoma. *N. Engl. J. Med.* **2021**, *384*, 154–164. [CrossRef] [PubMed]
4. Granowetter, L.; Womer, R.; Devidas, M.; Krailo, M.; Wang, C.; Bernstein, M.; Marina, N.; Leavey, P.; Gebhardt, M.; Healey, J.; et al. Dose-intensified compared with standard chemotherapy for nonmetastatic Ewing sarcoma family of tumors: A Children's Oncology Group Study. *J. Clin. Oncol.* **2009**, *27*, 2536–2541. [CrossRef] [PubMed]
5. Dirksen, U.; Brennan, B.; Le Deley, M.-C.; Cozic, N.; Van Den Berg, H.; Bhadri, V.; Brichard, B.; Claude, L.; Craft, A.; Amler, S.; et al. High-Dose Chemotherapy Compared With Standard Chemotherapy and Lung Radiation in Ewing Sarcoma With Pulmonary Metastases: Results of the European Ewing Tumour Working Initiative of National Groups, 99 Trial and EWING 2008. *J. Clin. Oncol.* **2019**, *37*, 3192–3202. [CrossRef]

Whelan, J.; Le Deley, M.-C.; Dirksen, U.; Le Teuff, G.; Brennan, B.; Gaspar, N.; Hawkins, D.S.; Amler, S.; Bauer, S.; Bielack, S.; et al. High-Dose Chemotherapy and Blood Autologous Stem-Cell Rescue Compared With Standard Chemotherapy in Localized High-Risk Ewing Sarcoma: Results of Euro-E.W.I.N.G.99 and Ewing-2008. *J. Clin. Oncol.* **2018**, *36*, 3110–3119. [CrossRef] [PubMed]
7. Rasper, M.; Jabar, S.; Ranft, A.; Jürgens, H.; Amler, S.; Dirksen, U. The value of high-dose chemotherapy in patients with first relapsed Ewing sarcoma. *Pediatr. Blood Cancer* **2014**, *61*, 1382–1386. [CrossRef] [PubMed]
8. Windsor, R.; Hamilton, A.; McTiernan, A.; Dileo, P.; Michelagnoli, M.; Seddon, B.; Strauss, S.J.; Whelan, J. Survival after high-dose chemotherapy for refractory and recurrent Ewing sarcoma. *Eur. J. Cancer* **2022**, *170*, 131–139. [CrossRef]
9. Mackall, C.L.; Rhee, E.H.; Read, E.J.; Khuu, H.M.; Leitman, S.F.; Bernstein, D.; Tesso, M.; Long, L.M.; Grindler, D.; Merino, M.; et al. A pilot study of consolidative immunotherapy in patients with high-risk pediatric sarcomas. *Clin. Cancer Res.* **2008**, *14*, 4850–4858. [CrossRef]
10. Ahmed, S.K.; Witten, B.G.; Harmsen, W.S.; Rose, P.S.; Krailo, M.; Marcus, K.J.; Randall, R.L.; DuBois, S.G.; Janeway, K.A.; Womer, R.B.; et al. Ewing's sarcoma: Only patients with 100% of necrosis after chemotherapy should be classified as having a good response. *Bone Joint J.* **2016**, *98*, 1138–1144.
11. Bosma, S.E.; Lancia, C.; Rueten-Budde, A.J.; Ranft, A.; Gelderblom, H.; Fiocco, M.; van de Sande, M.A.J.; Dijkstra, P.D.S.; Dirksen, U. Easy-to-use clinical tool for survival estimation in Ewing sarcoma at diagnosis and after surgery. *Sci. Rep.* **2019**, *9*, 11000. [CrossRef]
12. Ahmed, S.K.; Witten, B.G.; Harmsen, W.S.; Rose, P.S.; Krailo, M.; Marcus, K.J.; Randall, R.L.; DuBois, S.G.; A Janeway, K.; Womer, R.B.; et al. Analysis of local control outcomes and clinical prognostic factors in localized pelvic Ewing sarcoma patients treated with radiation therapy: A Report from the Children's Oncology Group. *Int. J. Radiat. Oncol. Biol. Phys.* **2022**, *115*, 337–346. [CrossRef]
13. Lozano-Calderón, S.A.; Albergo, J.I.; Groot, O.Q.; Merchan, N.A.; El Abiad, J.M.; Salinas, V.; Mier, L.C.G.; Montoya, C.S.; Ferrone, M.L.; Ready, J.E.; et al. Complete tumor necrosis after neoadjuvant chemotherapy defines good responders in patients with Ewing sarcoma. *Cancer* **2022**, *129*, 60–70. [CrossRef]
14. Stahl, M.; Ranft, A.; Paulussen, M.; Bölling, T.; Vieth, V.; Bielack, S.; Görtitz, I.; Braun-Munzinger, G.; Hardes, J.; Jürgens, H.; et al. Risk of recurrence and survival after relapse in patients with Ewing sarcoma. *Pediatr. Blood Cancer* **2011**, *57*, 549–553. [CrossRef] [PubMed]
15. Bosma, S.E.; Ayu, O.; Fiocco, M.; Gelderblom, H.; Dijkstra, P.D.S. Prognostic factors for survival in Ewing sarcoma: A systematic review. *Surg. Oncol.* **2018**, *27*, 603–610. [CrossRef] [PubMed]
16. Grunewald, T.G.P.; Cidre-Aranaz, F.; Surdez, D.; Tomazou, E.M.; de Alava, E.; Kovar, H.; Sorensen, P.H.; Delattre, O.; Dirksen, U. Ewing sarcoma. *Nat. Rev. Dis. Primers.* **2018**, *4*, 5. [CrossRef]
17. A Showpnil, I.; Selich-Anderson, J.; Taslim, C.; A Boone, M.; Crow, J.C.; Theisen, E.R.; Lessnick, S.L. EWS/FLI mediated reprogramming of 3D chromatin promotes an altered transcriptional state in Ewing sarcoma. *Nucleic Acids Res.* **2022**, *50*, 9814–9837. [CrossRef] [PubMed]
18. Shulman, D.S.; Whittle, S.B.; Surdez, D.; Bailey, K.M.; de Álava, E.; Yustein, J.T.; Shlien, A.; Hayashi, M.; Bishop, A.J.R.; Crompton, B.D.; et al. An international working group consensus report for the prioritization of molecular biomarkers for Ewing sarcoma. *NPJ Precis. Oncol.* **2022**, *6*, 65. [CrossRef]
19. Apfelbaum, A.A.; Wu, F.; Hawkins, A.G.; Magnuson, B.; Jiménez, J.A.; Taylor, S.D.; Wrenn, E.D.; Waltner, O.; Pfaltzgraff, E.R.; Song, J.Y.; et al. EWS: FLI1 and HOXD13 Control Tumor Cell Plasticity in Ewing Sarcoma. *Clin. Cancer Res.* **2022**, *28*, 4466–4478. [CrossRef]
20. Weaver, D.T.; Pishas, K.I.; Williamson, D.; Scarborough, J.; Lessnick, S.L.; Dhawan, A.; Scott, J.G. Network potential identifies therapeutic miRNA cocktails in Ewing sarcoma. *PLoS Comput. Biol.* **2021**, *17*, e1008755. [CrossRef]
21. Boone, M.A.; Taslim, C.; Crow, J.C.; Selich-Anderson, J.; Byrum, A.K.; Showpnil, I.A.; Sunkel, B.D.; Wang, M.; Stanton, B.Z.; Theisen, E.R.; et al. The FLI portion of EWS/FLI contributes a transcriptional regulatory function that is distinct and separable from its DNA-binding function in Ewing sarcoma. *Oncogene* **2021**, *40*, 4759–4769. [CrossRef]
22. Deng, Q.; Natesan, R.; Cidre-Aranaz, F.; Arif, S.; Liu, Y.; Rasool, R.U.; Wang, P.; Mitchell-Velasquez, E.; Das, C.K.; Vinca, E.; et al. Oncofusion-driven de novo enhancer assembly promotes malignancy in Ewing sarcoma via aberrant expression of the stereociliary protein LOXHD1. *Cell Rep.* **2022**, *39*, 110971. [CrossRef] [PubMed]
23. Jiménez, J.A.; Apfelbaum, A.A.; Hawkins, A.G.; Svoboda, L.K.; Kumar, A.; Ruiz, R.O.; Garcia, A.X.; Haarer, E.; Nwosu, Z.C.; Bradin, J.; et al. EWS-FLI1 and Menin Converge to Regulate ATF4 Activity in Ewing Sarcoma. *Mol. Cancer Res.* **2021**, *19*, 1182–1195. [CrossRef]
24. Theisen, E.R.; Miller, K.R.; Showpnil, I.A.; Taslim, C.; Pishas, K.I.; Lessnick, S.L. Transcriptomic analysis functionally maps the intrinsically disordered domain of EWS/FLI and reveals novel transcriptional dependencies for oncogenesis. *Genes Cancer* **2019**, *10*, 21–38. [CrossRef] [PubMed]
25. Hawkins, A.G.; Basrur, V.; Leprevost, F.D.V.; Pedersen, E.; Sperring, C.; Nesvizhskii, A.I.; Lawlor, E.R. The Ewing Sarcoma Secretome and Its Response to Activation of Wnt/beta-catenin Signaling. *Mol. Cell. Proteom.* **2018**, *17*, 901–912. [CrossRef] [PubMed]

26. Kovar, H.; Amatruda, J.; Brunet, E.; Burdach, S.; Cidre-Aranaz, F.; de Alava, E.; Dirksen, U.; van der Ent, W.; Grohar, P.; Grünewald, T.G.P.; et al. The second European interdisciplinary Ewing sarcoma research summit–A joint effort to deconstructing the multiple layers of a complex disease. *Oncotarget* **2016**, *7*, 8613–8624. [CrossRef]
27. Scopim-Ribeiro, R.; Lizardo, M.M.; Zhang, H.-F.; Dhez, A.-C.; Hughes, C.S.; Sorensen, P.H. NSG Mice Facilitate ex vivo Characterization of Ewing Sarcoma Lung Metastasis Using the PuMA Model. *Front.Oncol.* **2021**, *11*, 645757. [CrossRef]
28. Liu, H.; Huang, L.; Luo, J.; Chen, W.; Zhang, Z.; Liao, X.; Dai, M.; Shu, Y.; Cao, K. Prediction and identification of B cell epitopes derived from EWS/FLI-l fusion protein of Ewing's sarcoma. *Med. Oncol.* **2012**, *29*, 3421–3430. [CrossRef]
29. Mackall, C.L. In search of targeted therapies for childhood cancer. *Front. Oncol.* **2011**, *1*, 18. [CrossRef]
30. Orentas, R.J.; Lee, D.; Mackall, C. Immunotherapy targets in pediatric cancer. *Front. Oncol.* **2012**, *2*, 3. [CrossRef]
31. Kilpatrick, S.E.; Reith, J.D.; Rubin, B. Ewing Sarcoma and the History of Similar and Possibly Related Small Round Cell Tumors: From Whence Have We Come and Where are We Going? *Adv. Anat. Pathol.* **2018**, *25*, 314–326. [CrossRef]
32. Cheng, Y.-W.; Meyer, A.; A Jakubowski, M.; O Keenan, S.; E Brock, J.; Azzato, E.M.; Weindel, M.; Farkas, D.H.; Rubin, B.P. Gene Fusion Identification Using Anchor-Based Multiplex PCR and Next-Generation Sequencing. *J. Appl. Lab. Med.* **2021**, *6*, 917–930. [CrossRef] [PubMed]
33. Anderson, P.M.; Trucco, M.M.; Garzone, S.; Sartoski, S.; Nystrom, L.; Zahler, S.; Thomas, S.; Murphy, E.S. Virtual visits for sarcomas and other rare cancers including desmoids: The evolution of informaton and educational opportunities to improve health: 2017–2022. In Proceedings of the Connective Tissue Oncology Society (CTOS), Vancouver, BC, Canada, 16–19 November 2022.
34. Anderson, P.M.; Hanna, R. Defining Moments: Making Time for Virtual Visits and Catalyzing Better Cancer Care. *Health Commun.* **2020**, *35*, 787–791. [CrossRef]
35. Le Deley, M.-C.; Delattre, O.; Schaefer, K.-L.; Burchill, S.A.; Koehler, G.; Hogendoorn, P.C.; Lion, T.; Poremba, C.; Marandet, J.; Ballet, S.; et al. Impact of EWS-ETS fusion type on disease progression in Ewing's sarcoma/peripheral primitive neuroectodermal tumor: Prospective results from the cooperative Euro-E.W.I.N.G. 99 trial. *J. Clin. Oncol.* **2010**, *28*, 1982–1988. [CrossRef] [PubMed]
36. Yang, W.; Lee, K.-W.; Srivastava, R.M.; Kuo, F.; Krishna, C.; Chowell, D.; Makarov, V.; Hoen, D.; Dalin, M.G.; Wexler, L.; et al. Immunogenic neoantigens derived from gene fusions stimulate T cell responses. *Nat. Med.* **2019**, *25*, 767–775. [CrossRef]
37. Palmer, C.D.; Rappaport, A.R.; Davis, M.J.; Hart, M.G.; Scallan, C.D.; Hong, S.-J.; Gitlin, L.; Kraemer, L.D.; Kounlavouth, S.; Yang, A.; et al. Individualized, heterologous chimpanzee adenovirus and self-amplifying mRNA neoantigen vaccine for advanced metastatic solid tumors: Phase 1 trial interim results. *Nat. Med.* **2022**, *28*, 1619–1629. [CrossRef]
38. Sellars, M.C.; Wu, C.J.; Fritsch, E.F. Cancer vaccines: Building a bridge over troubled waters. *Cell* **2022**, *185*, 2770–2788. [CrossRef]
39. Schoenmaker, L.; Witzigmann, D.; Kulkarni, J.A.; Verbeke, R.; Kersten, G.; Jiskoot, W.; Crommelin, D.J.A. mRNA-lipid nanoparticle COVID-19 vaccines: Structure and stability. *Int. J. Pharm.* **2021**, *601*, 120586. [CrossRef] [PubMed]
40. Robinson, J.T.; Thorvaldsdóttir, H.; Winckler, W.; Guttman, M.; Lander, E.S.; Getz, G.; Mesirov, J.P. Integrative genomics viewer. *Nat. Biotechnol.* **2011**, *29*, 24–26. [CrossRef]
41. Graham, G.T.; Selvanathan, S.P.; Zöllner, S.K.; Stahl, E.; Shlien, A.; Caplen, N.J.; Üren, A.; A Toretsky, J. Comprehensive profiling of mRNA splicing indicates that GC content signals altered cassette exon inclusion in Ewing sarcoma. *NAR Cancer* **2022**, *4*, zcab052. [CrossRef]
42. Flores, G.; Grohar, P.J. One oncogene, several vulnerabilities: EWS/FLI targeted therapies for Ewing sarcoma. *J. Bone Oncol.* **2021**, *31*, 100404. [CrossRef] [PubMed]
43. Pishas, K.I.; Drenberg, C.D.; Taslim, C.; Theisen, E.R.; Johnson, K.M.; Saund, R.S.; Pop, I.L.; Crompton, B.D.; Lawlor, E.R.; Tirode, F.; et al. Therapeutic Targeting of KDM1A/LSD1 in Ewing Sarcoma with SP-2509 Engages the Endoplasmic Reticulum Stress Response. *Mol. Cancer Ther.* **2018**, *17*, 1902–1916. [CrossRef] [PubMed]
44. Harlow, M.L.; Maloney, N.; Roland, J.; Navarro, M.J.G.; Easton, M.K.; Kitchen-Goosen, S.M.; Boguslawski, E.A.; Madaj, Z.B.; Johnson, B.K.; Bowman, M.J.; et al. Lurbinectedin Inactivates the Ewing Sarcoma Oncoprotein EWS-FLI1 by Redistributing It within the Nucleus. *Cancer Res.* **2016**, *76*, 6657–6668. [CrossRef]
45. Volchenboum, S.L.; Andrade, J.; Huang, L.; Barkauskas, D.A.; Krailo, M.; Womer, R.B.; Ranft, A.; Potratz, J.; Dirksen, U.; Triche, T.J.; et al. Gene Expression Profiling of Ewing Sarcoma Tumors Reveals the Prognostic Importance of Tumor-Stromal Interactions: A Report from the Children's Oncology Group. *J. Pathol. Clin. Res.* **2015**, *1*, 83–94. [CrossRef] [PubMed]
46. Hong, M.M.Y.; Maleki Vareki, S. Addressing the Elephant in the Immunotherapy Room: Effector T-Cell Priming versus Depletion of Regulatory T-Cells by Anti-CTLA-4 Therapy. *Cancers* **2022**, *14*, 1580. [CrossRef]
47. Li, Y.-J.; Yang, X.; Zhang, W.-B.; Yi, C.; Wang, F.; Li, P. Clinical implications of six inflammatory biomarkers as prognostic indicators in Ewing sarcoma. *Cancer Manag. Res.* **2017**, *9*, 443–451. [CrossRef]
48. Gassmann, H.; Schneider, K.; Evdokimova, V.; Ruzanov, P.; Schober, S.J.; Xue, B.; von Heyking, K.; Thiede, M.; Richter, G.H.S.; Pfaffl, M.W.; et al. Ewing Sarcoma-Derived Extracellular Vesicles Impair Dendritic Cell Maturation and Function. *Cells* **2021**, *10*, 2081. [CrossRef]
49. Subbiah, V. The next generation of evidence-based medicine. *Nat. Med.* **2023**, *29*, 49–58. [CrossRef]

MDPI AG
Grosspeteranlage 5
4052 Basel
Switzerland
Tel.: +41 61 683 77 34

Cancers Editorial Office
E-mail: cancers@mdpi.com
www.mdpi.com/journal/cancers

www.ingramcontent.com/pod-product-compliance
Lightning Source LLC
Chambersburg PA
CBHW041356210926
51850CB00013B/3099
* 9 7 8 3 7 2 5 8 3 4 4 7 1 *